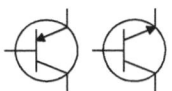

Solving Electronic Circuits

in MATLAB and SIMULINK

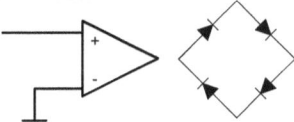

Mohammad Nuruzzaman

Electrical Engineering Department
King Fahd University of Petroleum & Minerals
Dhahran, Saudi Arabia

CreateSpace
4900 LaCross Road, North Charleston
SC 29406, USA
www.createspace.com

Dr. Mohammad Nuruzzaman
Electrical Engineering Department
King Fahd University of Petroleum and Minerals
KFUPM BOX 1286
Dhahran 31261, Saudi Arabia
Email: nzaman@kfupm.edu.sa, nzaman@ymail.com, mzamandr@gmail.com
Skype: nzaman1769   Twitter: @nzaman00
Web Link: http://faculty.kfupm.edu.sa/EE/NZAMAN/

ISBN-13: 978-1516918959
ISBN-10: 1516918959

Printed in the United States of America

This book is printed on acid-free paper.

To my parents
Mohammad Shamsul Haque & Nurbanu Begum

Other titles by the author:

1. M. Nuruzzaman, *"Control System Analysis & Design in MATLAB and SIMULINK"*, June, 2014, Lulu Press, Inc., North Carolina.
2. M. Nuruzzaman, *"Finite Difference Fundamentals in MATLAB"*, July, 2013, CreateSpace, South Carolina.
3. M. Nuruzzaman, *"Digital Image: Theories, Algorithms, and Applications"*, June, 2012, CreateSpace, Washington.
4. M. Nuruzzaman, *"Digital Audio Fundamentals in MATLAB"*, July, 2010, CreateSpace, California.
5. M. Nuruzzaman, *"Modern Approach to Solving Electromagnetics in MATLAB"*, January, 2009, BookSurge Publishing, Charleston, South Carolina.
6. M. Nuruzzaman, *"Signal and System Fundamentals in MATLAB and SIMULINK"*, July, 2008, BookSurge Publishing, Charleston, South Carolina.
7. M. Nuruzzaman, *"Electric Circuit Fundamentals in MATLAB and SIMULINK"*, October, 2007, BookSurge Publishing, Charleston, South Carolina.
8. M. Nuruzzaman, *"Technical Computation and Visualization in MATLAB for Engineers and Scientists"*, February, 2007, AuthorHouse, Bloomington, Indiana.
9. M. Nuruzzaman, *"Digital Image Fundamentals in MATLAB"*, September, 2005, AuthorHouse, Bloomington, Indiana.
10. M. Nuruzzaman, *"Modeling and Simulation in SIMULINK for Engineers and Scientists"*, January, 2005, AuthorHouse, Bloomington, Indiana.
11. M. Nuruzzaman, *"Tutorials on Mathematics to MATLAB"*, March, 2003, AuthorHouse, Bloomington, Indiana.

Preface

Studying microelectronics requires simulation tool which is mostly conducted in $\underline{S}$imulated $\underline{P}$rogram $\underline{I}$ntegrated $\underline{C}$ircuit $\underline{E}$mphasis or SPICE. *"Solving Electronic Circuits in MATLAB and SIMULINK"* introduces a startling alternate computer aided tool to SPICE. In today's hi-tech systems perhaps microelectronics is the most important engineering element. Everyday we use so many gadgets or devices that we are about to forget that those are originated from micro or nano electronics. Without electronics following could not have been achieved:

- Desktops and laptops make use of high speed chips which are incredibly compact electronic circuitry,
- Any sort of controller semiautomatic or automatic must employ microelectronics,
- Cell phones and mobile devices like iPhone again contain high speed electronic chipset,
- Home entertainment devices for instance CD player, LCD TV, DVD, etc can not be imagined without electronic circuits,
- Low/high power systems also employ electronics for smooth conversion and protection,
- Electronics is at the heart of digital measuring instruments,
- Electronics are part and parcel of communication and navigation means,
- In most toys sensory elements and electronics are hidden,
- Internet linking devices are basically electronics mostly in nano and pico scale,
- Identification or financial transaction by dint of Smart card, Visa Card, Master Cad, etc are composed of electronic chipsets, and needless to mention numerous applications are all around.

Since the inception of MATLAB (whose elaboration is matrix laboratory) during 1990s, it has been incorporating many sciences and engineering branches. Embedded functions in MATLAB are plenty in number and do not call for reprogramming, nor does require lumbering compiling often encountered in base language such as in FORTRAN or C. Hi tech industry and electronics are almost inseparable. The more is the number of computer aided tools, the better is the analyzing convenience of the discipline which is why the text is devised. In its intrinsic form the processing is matrix based in MATLAB while in SIMULINK (short for SIMULATION and LINK), an additional part of MATLAB, the maneuvering is box or block oriented. As a tutoring tool of electronics MATLAB in conjunction with SIMULINK forms an excellent combination. We aim at presenting introductory electronic circuit analysis and design training through academic approach.

It is no doubt that electronics is a matured as well as continually evolving discipline, numerical analysis and design of which are an indispensable part. Most electronic circuit physics are not obliged with mono variation for example diode characteristic is neither linear nor exponential thoroughly. This sort of input-output relationship is not lenient but MATLAB and SIMULINK at least ease the computing or graphing whether symbolic or numeric. Quick solving and graphing is a reality due to continually increasing computer processor speed which was time intensive in earlier days of the engineering. One notorious fact about electronic circuitry is most voltage-current characteristics inherit some form of constraint. Despite assumptions, limitations, or constraints of electronics, attempts have been made to illustrate the verification of established theories or solution of user-defined academic/design problems. Engaging computer tools engage users too for satisfactory results, our approach and effort have been exercised in this facilitation. Independent solving skill certainly needs a supplemental tool on whose account the text is a perfect read. Nevertheless some students may shun asking classroom professors to lay a hand on all sorts of instructional materials or teaching tools due to different barriers. The text renders enough illustrations and clues to break the ice and aids to carry out electronic problem solving or circuit design.

Chapter 1 presents a brief introduction to MATLAB's and SIMULINK's getting-started features. An electronic circuit is an electrical circuit too that is why chapter 2 addresses basic electrical circuit related topic implementation in MATLAB and SIMULINK. Chapter 3 merely demonstrates diode focused electronic circuit modeling. Implementation on bipolar junction transistor (BJT) is addressed in Chapter 4. Operational amplifier (OP AMP) is nothing but arithmetic centric electronic circuitry which is considered in chapter 5. Despite the similarity of metal oxide semiconductor (MOS) devices to BJT chapter 6 explicates MOS transistor in order to point out relevant intricacy. Frequency response is widely exercised in electronic system design which is exclusively taken care of in chapter 7. An important outcome of the course is the reader should be able to handle design elements on electronic circuits on whose account chapter 8 is devised. Pertinent tools are exercised to get the reader first hand experience on designs. Nevertheless appendices A through G explain electronic system related coding, function, or embedded graphing tool to the context of MATLAB/SIMULINK.

My words of acknowledgement are due to the King Fahd University of Petroleum and Minerals (KFUPM). I am especially appreciative of library facilities, electronic system reading materials, and MATLAB software that I received from the university.

Mohammad Nuruzzaman

Acknowledgements

I sincerely thank the following for their inputs, encouragements, comments, and suggestions about my MATLAB/SIMULINK text's development:

Mohsin M. Jamali (Department of Electrical Engineering and Computer Science, The University of Toledo, Ohio, USA), Rama VenKat (Electrical and Computer Engineering, University of Nevada, Las Vegas, Nevada, USA), Gilberto E. Urroz (Department of Civil & Environmental Engineering, Utah Water Research Laboratory, Utah State University, Utah, USA), G. P. Rangaiah (Department of Chemical and Environmental Engineering, National University of Singapore, Singapore), Flaminio Squazzoni (Department of Social Sciences, University of Brescia, Italy), G. N. Reddy (Department of Electrical Engineering, Lamar University, Texas, USA), Kashif Javaid (Audio Applications, National Semiconductor), Bahram Shahian (Electrical Engineering, California State University, Long Beach, USA), Lihong Li (Department of Engineering Science and Physics, Staten Island College, New York, USA), Rollins Turner (Department of Computer Science and Engineering, University of South Florida, USA), Sally Hawkins (Computer Science and Engineering, University of Nebraska-Lincoln, Nebraska, USA), Clinton Fookes (School of Engineering Systems, Faculty of Built Environment and Engineering, Queensland University of Technology, Brisbane, Australia), Swamy Jagannatham (University of California: Extension School, Los Angeles, USA), Ahmed Mabrouk (Computer Science Department, Islamic University of Malaysia, Malaysia), Jimmy Thomas (School of Biological Sciences, University of Canterbury, New Zealand), Irineu Antunes Junior (CECS, UFABC, Brazil), Kamel Alboaouh (Alumni, King Fahd University of Petroleum and Minerals, Saudi Arabia), John Bofarull Guix (London Metropolitan, UK), Ronald Tangelder (www.freenet.com, Germany), Nopparat Seemuang (Department of Mechanical Engineering, The University of Sheffield, UK), Andy Blanco (Southern California Institute of Technology, USA), Jonathan Ibera (Southern California Institute of Technology, Anaheim, USA), Young Sup Lee (Department of Embedded Systems Engineering, Incheon National University, South Korea), and Ismat Al-Dmour (Computer Science and Information Technology, Al-Baha University, Saudi Arabia).

Table of Contents

Chapter 4
Modeling Transistor Circuits

Chapter 5
Modeling Operational Amplifiers

Chapter 6
Modeling Field Effect Transistors

Chapter 1

Introduction to MATLAB and SIMULINK

MATLAB is a computing software, which provides the quickest and easiest way to compute scientific and technical problems and visualize the solutions. As worldly standard for simulation and analysis, engineers, scientists, and researchers are becoming more and more affiliated with MATLAB and SIMULINK. The general questionnaires about MATLAB or SIMULINK before one gets started with are contents of this chapter. SIMULINK is designed to function over MATLAB. Much of MATLAB computing approach presupposes that the element to be handled is a vector or matrix. Whereas SIMULINK maps a technical problem into computer model through elementary blocks. Our highlight covers the following:

- ✤ MATLAB and SIMULINK features at the command window
- ✤ Getting started in MATLAB/SIMULINK from scratch
- ✤ Frequently encountered questions on MATLAB/SIMULINK
- ✤ Relevant introductory topics and forms of assistance about MATLAB/SIMULINK

1.1 What is MATLAB?

MATLAB is mainly a scientific and technical computing software whose elaboration is matrix laboratory. Command prompt of MATLAB (>>) provides an interactive system. In the workspace of MATLAB, most data element is dealt as a matrix without dimensioning. The package is incredibly advantageous for matrix-oriented computing. MATLAB's easy-to-use platform enables us to compute and manipulate matrices, perform numerical analyses, and visualize different variety of one/two/three dimensional graphics in a matter of second or seconds without conventional programming as conducted in FORTRAN, PASCAL, or C.

1.1.1 MATLAB's opening window features

If you do not have MATLAB installed in your personal computer, contact MathWorks (owner and developer, www.mathworks.com) for the installation CD. If you know how to get in MATLAB and its basics, you may skip the chapter. Assuming the package is installed in your system, run MATLAB from the Start of the Microsoft Windows. Let us get familiarized with MATLAB's opening window features. Figure 1.1(a) shows a typical firstly opened MATLAB window. Depending on desktop setting or MATLAB version, your MATLAB window may not look like figure 1.1(a) but descriptions of the features by and large are appropriate.

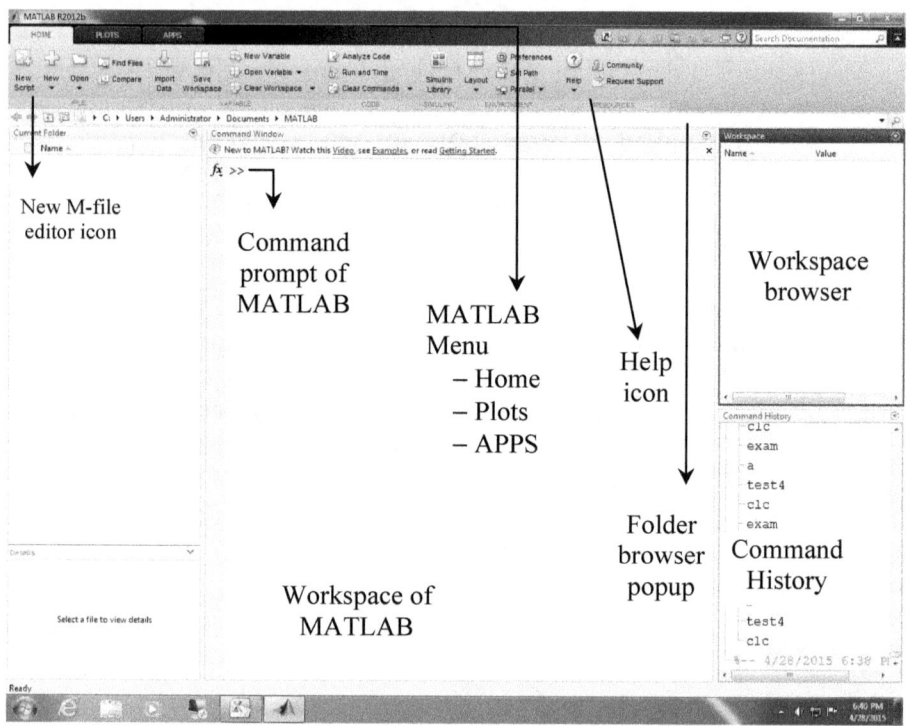

Figure 1.1(a)   Typical features of MATLAB's firstly opened window

♦ Command prompt of MATLAB

Command prompt means that you tell MATLAB to do something from here (>> in figure 1.1(a)). As an interactive system, MATLAB responds to user through this prompt. MATLAB cursor will be blinking after >> prompt once you open MATLAB i.e. MATLAB is ready to take your commands. To enter any command, type executable MATLAB statements from keyboard and to execute that, press Enter key (symbol ↵ for 'Hit the Enter Key' operation).

✦ MATLAB Menu

MATLAB is accompanied with three submenus namely HOME, PLOTS, and APPS. Each submenu has its own features. Use the mouse to click different submenus and their brief descriptions are as follows:

Submenu HOME: This is basically the firstly opened default window i.e. figure 1.1(a). It allows us to open a new script or M-file, model, or Graphical User Inter-face (GUI) layout maker, open a file which was saved before, load a saved workspace, import data from a file, save the workspace variables, set the required path to execute a file, print the workspace, and keep provision for changing the command window property.

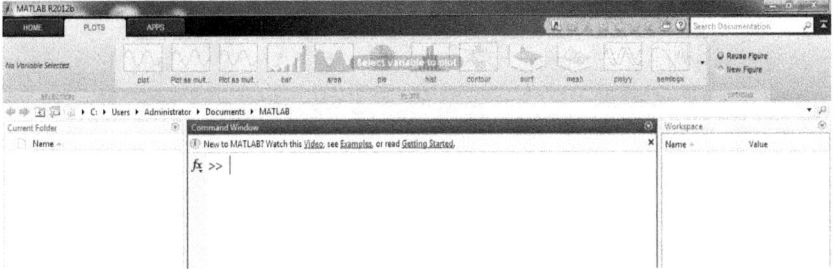

Figure 1.1(b) Submenu PLOTS

Submenu PLOTS: The second submenu PLOTS (figure 1.1(b)) includes some of available graphical tools embedded in MATLAB. MATLAB has vast graphics supports, above is just sample on that.

Submenu APPS: MATLAB has numerous built-in libraries which conduct specific discipline oriented simulations, these libraries are called Toolbox or APPS. The submenu APPS (figure 1.1(c)) just displays some of many APPS available. For example in its own menu bar you find Curve Fitting, Optimization, etc.

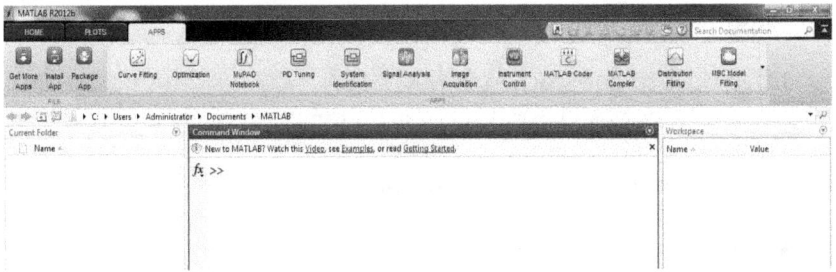

Figure 1.1(c) Submenu APPS

♦ MATLAB workspace

Workspace (figure 1.1(a)) is the platform of MATLAB where one executes MATLAB commands. During execution of commands, one may have to deal with some input and output variables. These variables can be one-dimensional array, multi-dimensional array, characters, symbolic objects, etc. Again to deal with graphics window, we have texts, graphics, or object handles. Workspace holds all those variables or handles for you. As a subwindow of figure 1.1(a), its browser exhibits the types or properties of those variables or handles. If the browser is not found in the opening window of MATLAB, click the Layout down Workspace in the HOME bar to bring the subwindow (figure 1.1(d)).

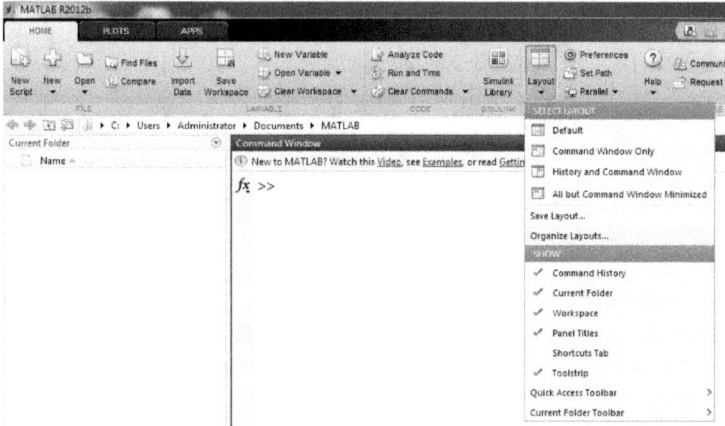

Figure 1.1(d) Layout pulldown menu under HOME

♦ MATLAB command history

There is a subwindow in figure 1.1(a) called Command History which holds all previously used commands at command prompt. Depending on desktop setting, it may or may not appear during the opening of MATLAB. If it does not, click the Command History from figure 1.1(d) under the Layout. Used commands of this window you may copy-paste in an editor for later use.

1.1.2 How to get started in MATLAB?

New MATLAB users face a common question how to get started in MATLAB? This tutorial is for beginners in MATLAB. Here we address the terms under the following bold headings.

♦ How can I enter a vector/matrix?

The first step is the user has to be in the command window of MATLAB. Look for the command prompt >> in the command window. Row

or column matrices are termed as vectors. We intend to enter the row matrix
R=[2   3   4   −2   0] into the workspace of MATLAB. Type the
following from the keyboard at the command prompt:

>>R=[2 3 4 -2 0]   ← Arial font set for executable commands i.e. R⇔R

There is one space gap between two elements of the matrix R but no space
gap at the edge elements. All elements are placed under the [ ]. Press Enter
key after the third brace] from the keyboard and we see

R =

       2   3   4   -2   0
>>   ← command prompt is ready again

It means we assigned the row matrix to the workspace variable R. Whenever
we call R, MATLAB understands the whole row matrix. Matrix R is having
five elements. Even if R had 100 elements, it would understand the whole
matrix that is one of many appreciative features of MATLAB. Next we wish
to enter the column matrix C=$\begin{bmatrix} 7 \\ 8 \\ 10 \\ -11 \end{bmatrix}$. Again type the following from the
keyboard at the blinking cursor:

>>C=[7;8;10;-11] ↵   you will see (↵ means 'Press the Enter Key'),

C =

       7
       8
      10
     −11
>>   ← command prompt is ready again

This time we also assigned the column matrix to the workspace variable C.
For the column matrix, there is one semicolon ; between two consecutive
elements of the matrix C but no space gap is necessary. As another option,
the matrix C could have been entered by writing C=[7 8 10 -11]'. The
operator ' of keyboard is matrix transposition operator in MATLAB. As if
you entered a row matrix but at the end just the transposition operator ' is
attached. After that the rectangular matrix A=$\begin{bmatrix} 20 & 6 & 7 \\ 5 & 12 & -3 \\ 1 & -1 & 0 \\ 19 & 3 & 2 \end{bmatrix}$ is to be entered:

>>A=[20 6 7;5 12 -3;1 -1 0;19 3 2] ↵   you will see,

A =

      20   6   7
       5 12   −3
       1 −1   0
      19   3   2

Two consecutive rows of A are separated by semicolon ; and consecutive elements in a row are separated by one space gap. Instead of typing all elements in a row, one can type the first row, press Enter key, the cursor blinks in the next line, type the second row, and so on.

◆ How can I use colon and semicolon operators?

The operators semicolon ; and colon : have special significance in MATLAB. Most MATLAB statements and M-file programming use these two operators almost in every line. Generation of vectors can easily be performed by the colon operator no matter how many elements we need. Let us carry out the following at the command prompt to see the importance of the colon operator:

>>A=1:4 ↵   you will see,

A =
   1 2 3 4   ← We created a vector A or row matrix
                     where A=[1 2 3 4]

Let us interact with MATLAB by the following commands:

>>R=1:3:10 ↵   you will see,

R =
   1 4 7 10   ← We created a vector or row matrix R whose elements form an
                    arithmetic progression with first element 1, last element 10,
                    and common difference or increment 3

Vector with decrement can also be generated:

>>C=[0:-2:-10]' ↵   you will see,

C =

   0
  -2
  -4   ← We created a vector or column matrix C whose
  -6   consecutive elements have the decrement 2 with the
  -8   first element 0 and the last element −10
 -10

MATLAB is also capable of producing vectors whose elements are decimal numbers. Let us form a row matrix R whose first element is 3, last element is 6, and increment is 0.5 which we accomplish as follows:

>>R=3:0.5:6 ↵   you will see,
R =
   3.0000   3.5000   4.0000   4.5000   5.0000   5.5000   6.0000

Then, what is the use of semicolon operator? Append a semicolon at the end in the last command and execute that:

>>R=3:0.5:6; ↵   you will see,
>>   ← Assignment is not shown

Type **R** at the command prompt and press Enter:
```
>>R ↵
```

R =

    3.0000 3.5000 4.0000 4.5000 5.0000 5.5000 6.0000

It indicates that the semicolon operator prevents MATLAB from displaying the contents of the workspace variable **R**.

✦ How can I call a built-in MATLAB function?

    In MATLAB, thousands of M-files or built-in function files are embedded. Knowing descriptions of the function, numbers of input and output arguments, and nature of the arguments is mandatory in order to execute a built-in function. Let us start with a simplest example. We intend to find $\sin x$ for $x = \dfrac{3\pi}{2}$ which should be -1. The MATLAB counterpart (appendix A) of $\sin x$ is **sin(x)** where **x** can be any real or complex number in radians and can be a matrix too. The angle $\dfrac{3\pi}{2}$ is written as **3*pi/2**(π is coded by **pi**) and let us perform it as follows:
```
>>sin(3*pi/2) ↵
```

ans =
    −1

By default the return from any function is assigned to workspace **ans**. If you wanted to assign the return to **S**, you would write **S=sin(3*pi/2);**.

    As another example, let us factorize the integer 84 ($84=2\times2\times3\times7$). The MATLAB built-in function **factor** finds the factors of an integer and the implementation is as follows:
```
>>f=factor(84) ↵
```

f =
    2   2   3   7

Output of the **factor** is a row matrix which we assigned to workspace **f** in fact the **f** can be any user-given name. Thus you can call any other built-in function from the command prompt provided that you have the knowledge about the calling of inputs to and outputs from the function.

✦ How can one open and execute an M-file or a script file?

    This is the most important start up for the beginners. An M-file can be regarded as a text or script file. A collection of executable MATLAB statements are the contents of an M-file. Ongoing discussion made you familiarize with entering a matrix, computing a sine value, and factorizing an integer. These three executions took place at the command prompt. They can be executed from an M-file as well. This necessitates opening the M-file editor. Referring to figure 1.1(a), you find the icon for **New Script** in the

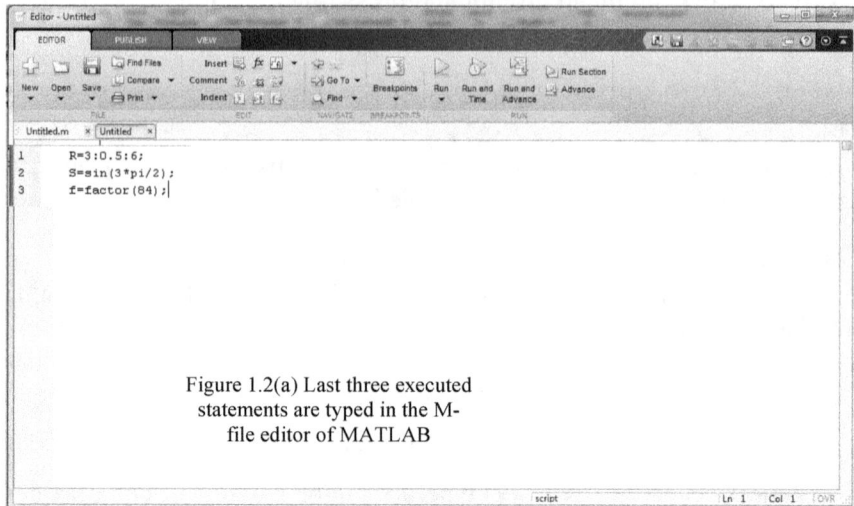

Figure 1.2(a) Last three executed
statements are typed in the M-
file editor of MATLAB

upper left corner of the
window, click it to see
the new untitled script
editor. After opening the
new script editor, we
typed the last three
executable statements in
the untitled file as shown
in figure 1.2(a). The next
step is to save the untitled
file by clicking the Save
icon of figure 1.2(a).
Figure 1.2(b) presents the
File Save dialog window.
We typed the file name as

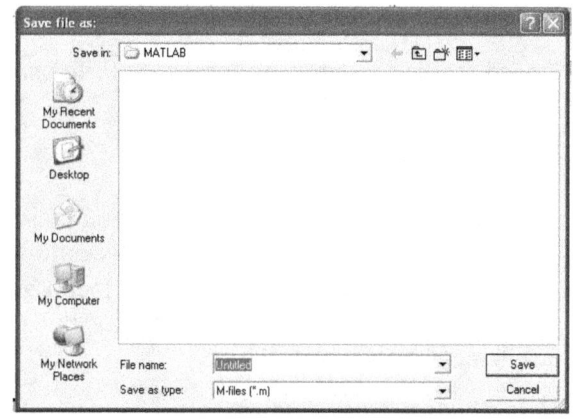

Figure 1.2(b)   Save dialog window for naming the M-file

test (can be any name of your choice) in the slot of File name in the window.
The script file has the extension .m but we do not type .m only the file name
is enough. After saving the file, let us move on to MATLAB command
prompt and conduct the following:

    >>test ↵
    >>   ← command prompt is ready again

It indicates that MATLAB executed the M-file by name test and is ready for
next command. We can check calling the assignees whether the previously
performed executions occurred exactly as follows:

    >>R ↵

    R =
          3.0000   3.5000   4.0000   4.5000   5.0000   5.5000   6.0000

>>S ⏎   >>f ⏎

S =   f =
   -1   2   2   3   7

This is what we found before. Thus one can run any executable statements in script file. The reader might ask in which folder or path the **test** was saved. Figure 1.1(a) shows the bar (down the menu bar in the upper portion) for Current Folder which is C:\Users\Administrator\Documents\MATLAB. That is the location of your file. If you want to save the script file in other folder or directory, change path by clicking the path browser icon before saving the file. When you call the **test** or any other file from the command prompt, the prompt must be in the same directory where the file is in or its path must be defined to MATLAB.

◆ **What are input and output arguments of a function file?**

MATLAB is a collection of thousands of script files. Some files are executed without any return and some return results which are called function files (appendix F). You have seen the use of function **sin(x)** before, which has one input argument **x**. The statement **test(x,y)** means that the **test** is a function file which has two input arguments – **x** and **y**. Again the **test(x,y,z)** means the **test** is a function file which needs three input arguments – **x**, **y**, and **z**. Similar style also follows for the return but under the third brace. The **[a,b]=test(x,y)** means there are two output arguments from the **test** which are **a** and **b** and the **[a,b,c]=test(x,y)** means three returns from the **test** which are **a**, **b**, and **c**.

◆**How can I plot a graph?**

MATLAB is very convenient for plotting different sorts of graphs. The graphs are plotted either from mathematical expression or from data. Let us plot the function $y = -2\sin 2x$. MATLAB function **ezplot** plots y versus x type graph taking the expression as its input argument. MATLAB code (appendix A) for the $-2\sin 2x$ is -

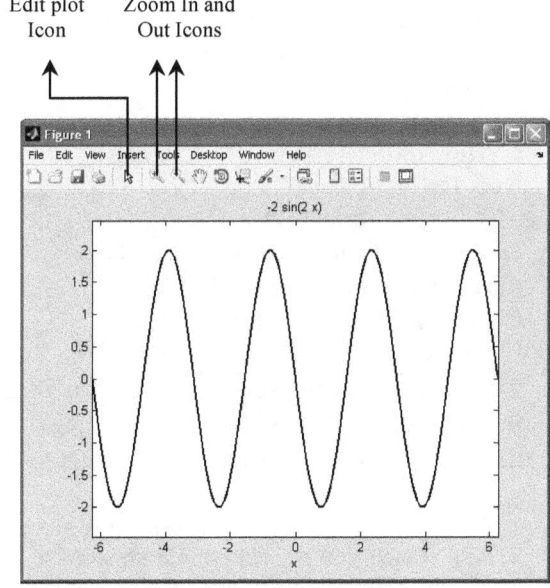

Edit plot Icon   Zoom In and Out Icons

Figure 1.2(c) Graph of $-2\sin 2x$ versus x

2*sin(2*x). The functional code is input argumented by using single inverted comma hence we conduct the following at command prompt:

>>ezplot('-2*sin(2*x)') ↵

Figure 1.2(c) presents the outcome from above execution. The window in which the graph is plotted is called MATLAB figure window. Any graphics is plotted in the figure window, which has its own menu (such as File, Edit, etc) as shown in figure 1.2(c).

1.1.3 Some queries about MATLAB environment

Users need to know the answers to some questions when they start working in MATLAB. MATLAB environment related some queries are presented in the sequel.

⎅ **How to change the numeric format?**

When you perform any computation at the command prompt, the output is returned up to four decimal display due to short numeric format which is the default one. There are other numeric formats too. To reach the numeric format dialog box, the clicking operation sequence is HOME ⇒ Preferences ⇒ Command Window ⇒ Text Display ⇒ Numeric Format (select from the popup menu e.g. long).

⎅ **How to change the font or background color settings?**

One might be interested to change the background color or font color while working in the command window. The clicking sequence is HOME ⇒ Preferences ⇒ MATLAB. You find Desktop Tool Colors in the right half window, uncheck that and select any Text or Background Color from the popup.

⎅ **How to delete some/all variables from the workspace?**

In order to delete all variables present at the workspace, the clicking sequence is HOME ⇒ Clear Workspace (figure 1.1(a)). If you want to delete a particular

Figure 1.2(d) Workspace browser displays variable information

workspace variable, select the concern variable by using the mouse pointer in the workspace browser (assuming that it is open like the figure 1.2(d)) and then rightclick ⇒ delete.

⎅ **How to clear workspace but not the variables?**

Once you conduct some sessions at the command prompt, monitor screen keeps all interactive sessions. You can clear the screen contents without removing the variables by command clc or performing the clicking operation HOME ⇒ Clear Commands (figure 1.1(a)).

⊟ **How to know the current path?**

In the upper portion of figure 1.1(a), the Current Folder bar is located that indicates in which path the command prompt is or execute cd (abbreviation for the current directory) at the command prompt.

⊟ **How to see different variables at the workspace?**

There are two ways of viewing – either use the command who or look at the workspace browser (like figure 1.2(d)) which exhibits information about workspace variables for example R is the name of variable which holds some values. One can view, change, or edit the contents of a variable by doubleclicking the concern variable situated at workspace browser.

⊟ **How to enter a long command line?**

MATLAB statements can be too long to fit in one line. Giving a break in the middle of a statement is accomplished by the ellipsis (three dots are called ellipsis). We show that considering the entering of vector x=[1:3:10] as follows:

```
>>x=[1:3: . . . ┘
        10] ┘
x =
        1   4   7   10
```

Typing takes place in two lines and there is one space gap before the ellipsis.

⊟ **Editing at the command prompt**

This is advantageous specially for those who work frequently in the command window without opening a script file. Keyboard has different arrow keys marked by ← ↑ → ↓. One may type a misspelled command at the command prompt causing error message to appear. Instead of retyping the entire line, press uparrow (for previous line) or downarrow (for next line) to edit the MATLAB statement. Or you can reexecute any past statement this way. For example we generated a row vector 1 through 10 with increment 2 and assigned the vector to x. The necessary command is x=1:2:10. Mistakenly you typed x+1:2:10. The response is as follows:

```
>>x+1:2:10 ┘
??? Undefined function or variable 'x'.
```

You discovered the mistake and want to correct that. Press ↑ key to see,

```
>>x+1:2:10
```

Edit the command going to the + sign by using the left arrow key or mouse pointer. At the prompt, if you type x and press ↑ again and again, you see used commands that start with x.

⊟ **Saving and loading data**

User can save workspace variables or data in a binary file having the extension .mat. Suppose you have the matrix $A = \begin{bmatrix} 3 & 4 & 8 \\ 0 & 2 & 1 \end{bmatrix}$ and wish to save

A in a file by the name **data.mat**. Let us carry out the following:

>>A=[3 4 8;0 2 1]; ⏎   ← Assigning the A to A

Now move on to the workspace browser (figure 1.2(d)) and you see the variable A including its information located in the subwindow. Bring mouse pointer on A, rightclick the mouse, and click the **Save As**. The Save dialog window appears and type only **data** (not **data.mat**) in the slot of **File name**. If it is necessary, you can save all workspace variables by using the same action but clicking **HOME** ⇒ **Save Workspace** (figure 1.1(a)). One retrieves the data file by clicking **HOME** ⇒ **Import Data** (figure 1.1(a)). Another option is use the command **load data** at the command prompt.

⊟ **How to delete a file from the command prompt?**

Let us delete just mentioned **data.mat** by executing the command **delete data.mat** at the command prompt.

⊟ **How to see the data held in a variable?**

Figure 1.2(d) presents some variable information in which you find R. Doubleclick the R or your variable in the workspace browser and find the matrix contents of R in a data sheet.

1.2 What is SIMULINK?

SIMULINK is an additional part of MATLAB which provides an easeful way to model, simulate, and analyze many dynamic systems which are characterized by some inputs and outputs. Without opening MATLAB we can not turn SIMULINK operational.

Elaboration of SIMULINK is simulation and link. A particular input-output relationship can be assigned to some block. One can interpret that SIMULINK is a vast collection of this kind of blocks. Although

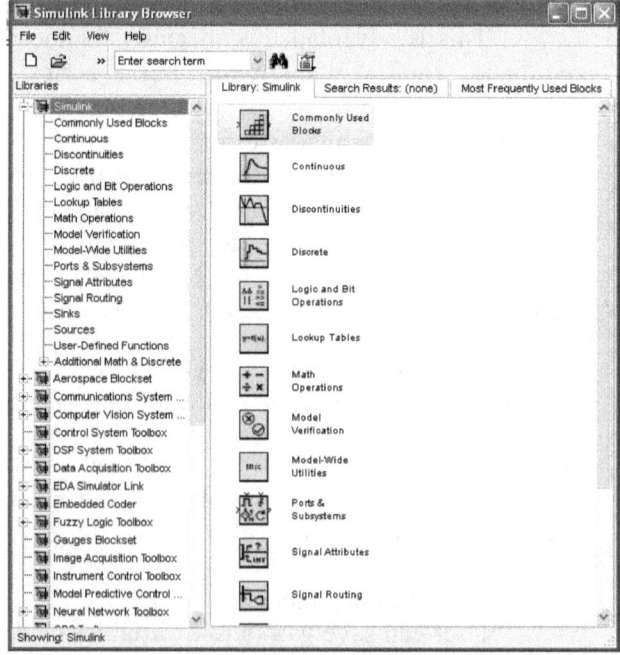

Figure 1.3(a) SIMULINK library contents

-12-

the blocks stand for simple mathematical relationship but being concatenated they build a much complicated system. Initially SIMULINK was intended particularly to handle the linear time invariant continuous systems. With the progress of time discrete time systems as well as hybrid ones surfaced in SIMULINK to adapt it to more pragmatic modeling of real world's dynamic systems. In a simplistic way if MATLAB is a world of matrices, then SIMULINK is a world of blocks. In most scientific and engineering systems three types of constituent elements are seen – source, system, and sink. For example in electrical engineering, applied DC voltage to a circuit, R - C filter, and voltmeter correspond to the source-system-sink terminology. SIMULINK is the best tool for technical analysis if we can characterize a scientific problem in terms of source, system, and sink.

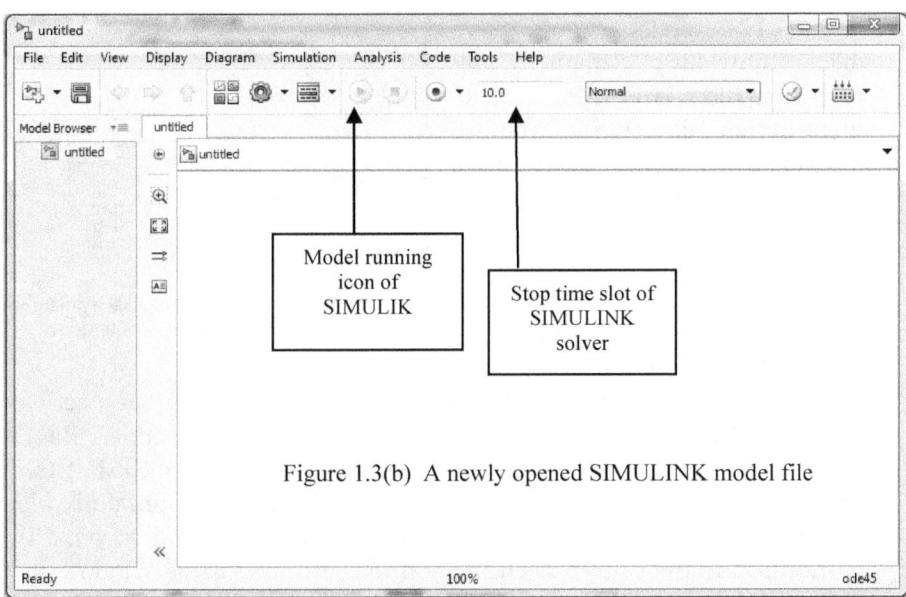

Figure 1.3(b) A newly opened SIMULINK model file

1.2.1 How can I get into SIMULINK?

Since SIMULINK is an extension of MATLAB, first we have to get into MATLAB. Both the MATLAB command prompt and its menu bar provide means of getting into SIMULINK. Figure 1.1(a) shows the indication of command prompt. Either you type simulink at the command prompt and press enter or click the Simulink Library icon shown in figure 1.1(a) then click new Model from the Menu. SIMULINK is an aggregation of functional blocks arranged in a tree structure which you see like the figure 1.3(a) by just mentioned either action.

In figure 1.3(a) you find different families of block set. For example Aerospace Blockset keeps source/system/sink analysis blocks for aerospace engineering. Again the Communications System keeps source/system/sink analysis blocks for communications engineering and so on. In figure 1.3(a) left half window you find many families e.g. **Computer Vision System, Control System Toolbox,** etc, each of which is one full-fledged SIMULINK package.

1.2.2 Where can I build a SIMULINK model?

Like every software there should be a file where we can build SIMULINK model. The model we intend to build is entirely problem dependent. Get the newly untitled SIMULINK model file opened like figure 1.3(b) as quoted earlier. This is the platform where we build any SIMULINK model.

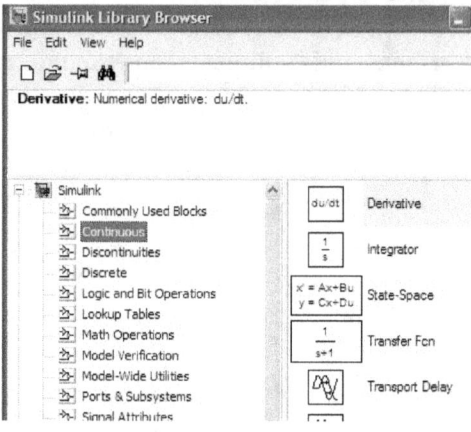

The reader should have the knowledge about any particular block's function and its input-output descriptions before bringing it from SIMULINK library to the untitled model file. A model is defined as the interconnected blocks found in SIMULINK library to work out a scientific/engineering problem.

Figure 1.3(c) Different blocks' availability under the subfamily **Continuous**

Let us bring a block from SIMULINK library in the untitled model file. To perform such action, click the SIMULINK library browser icon at the menu bar thus SIMULINK library of figure 1.3(a) appears and the cursor is residing in SIMULINK. The right half part of the window in figure 1.3(a) displays the subfamily blocksets for example **Commonly Used Blocks, Continuous,** and so on. Click the **Continuous** down the **SIMULINK** on the left part of the window in figure 1.3(a). We see various blocks available under the subclass **Continuous** like the figure 1.3(c) for example **Derivative, Integrator, State-Space,** etc. We intend to bring the **Derivative** in our untitled model file. Bring the mouse pointer on the **Derivative** block, move the mouse pointer keeping your finger pressed in the left button of mouse to any convenient area of the untitled model, and release the left button of the mouse. Now you see the **Derivative** block at the untitled model file as shown in figure 1.3(d). Another way of bringing the block is

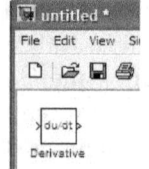

Figure 1.3(d) **Derivative** block in the untitled model file

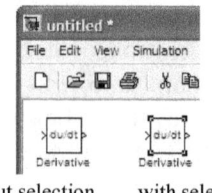

without selection   with selection

Figure 1.3(e) **Derivative** block with and without selection

click the **Derivative**, rightclick the mouse, see the **Add to untitled** in the popup, click the **Add to untitled**, and find the block in your model file. Therefore we say the link for the **Derivative** block is **SIMULINK** → **Continuous** → **Derivative**. We maintain this style of locating a block in the appendix B. However we have been successful in bringing the **Derivative** in the untitled model file. Now we can save the model by any convenient name in working directory. By the way MATLAB source code file has the extension .m but SIMULINK model does .mdl. Keep in mind that *one should know the link of a block to bring it from SIMULINK library to a model file.*

1.2.3 Block manipulation in SIMULINK

During a SIMULINK model building, we need some manipulations of blocks to construct a seemly, well-placed, and well-devised model, most frequently encountered ones of which are the following.

⊟ **How to select a block?**

Let us say you brought the **Derivative** block in an untitled model file following the link Simulink → Continuous → Derivative. Bring the mouse pointer on the **Derivative** block and click the left button of the mouse to see the selection as shown in figure 1.3(e).

Input Port → du/dt → Output Port

Derivative

Figure 1.3(f) **Derivative** block
mentioning the input and output ports

Real-Imag to
Complex

Complex to
Magnitude-Angle

Figure 1.3(g) Blocks with multiple
input and output ports

⊟ **How to detect the input and output ports of a block?**

Referring to **Derivative** of figure 1.3(d), we see that left and right sides of the block contain the symbol **>**. One identifies input and output ports as presented in figure 1.3(f). The number of input and output ports is not always 1, blocks **Real-Imag to Complex** and **Complex to Magnitude-Angle** of figure 1.3(g) have two input ports and two output ports respectively.

⊟ **How to delete a block?**

You brought the **Derivative** in an untitled model file but want to delete that. There are several options for this. First select the block, then press the **Delete** button from keyboard, click the **Cut** icon in model menu bar, or click the **Cut** followed by the rightclick of the mouse on the block.

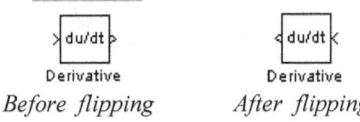

du/dt

Derivative

Before flipping

du/dt

Derivative

After flipping

Figure 1.4(a) Flipping the **Derivative** block

⬚ How to flip a block?

We have the ongoing **Derivative** block at our model file and wish to flip the block like figure 1.4(a). First we bring mouse pointer on the block, click right button of the mouse, and then click **Flip Block** via **Format**. The action is shown in figure 1.4(b).

⬚ How to rotate a block?

Suppose we wish to rotate just mentioned **Derivative** block. Bring mouse pointer on the block, click right button of the mouse, and click **Rotate Block** via **Format** (figure 1.4(b) shows the action too). You see the change as shown in figure 1.4(c). Figure 1.4(c) seen rotation indicates the operation that the block is rotated clockwise 90^0 at a time. If you intend to rotate the block 270^0 clockwise (indicated in figure 1.4(d)), you need the operation three times.

⬚ How to remove a block name?

The name of **Derivative** block can be removed as presented in figure 1.4(e) by clicking **Hide Name** of figure 1.4(b). A clumsy model might

Figure 1.4(b) Pulldown menu of the **Derivative** following the rightclick on the block

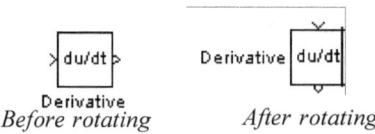

Before rotating   After rotating

Figure 1.4(c) Rotating the **Derivative** block by 90^0 clockwise

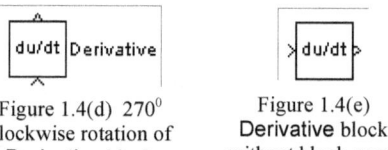

Figure 1.4(d) 270^0 clockwise rotation of **Derivative** block

Figure 1.4(e) **Derivative** block without block name

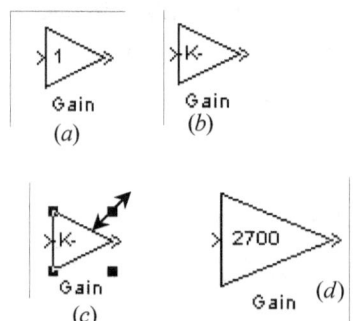

Figure 1.5 **Gain** block a) with default gain, b) with gain 2700, c) with selection, and d) after enlargement

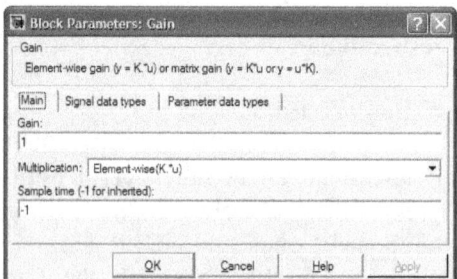

Figure 1.5(e) Block parameter window of **Gain** block

give better look on removing block names if the reader is well acquainted with the blocks' operation.

How to enlarge or contract a block?

Let us bring a **Gain** block as shown in figure 1.5(a) following the link **Simulink** → **Math Operations** → **Gain** in an untitled SIMULINK model file, **Gain** has the default gain 1. If we have five or six digits gain, the default size does not allow to display that. Doubleclick the block to see its parameter window like figure 1.5(e), let us enter the **Gain** slot value of figure 1.5(e) from default 1 to 2700 (as a four digit example) by using the keyboard, and click OK. The block displays the inside gain as shown in figure 1.5(b). Select the block, bring your mouse pointer on the upper right square target to see the figure 1.5(c), move the mouse pointer to the right keeping the left button of the mouse pressed, and release the left button of the mouse. You should see the enlarged **Gain** block of the figure 1.5(d). In a similar way for the oversize block, we can reduce the block size by moving towards the inside of the block after selection.

Figure 1.5(f) **Derivative** block by the name D

How to rename a block?

During SIMULINK model building, it may be necessary to use the same kind of block twice or more. Then it requires renaming the block for identification. Let us say we brought a **Derivative** block in an untitled model file as shown in figure 1.3(d). We wish to write just **D** as

Derivative block in a model

Figure 1.5(g) **Derivative** block with the annotation

the block name instead of **Derivative** (like figure 1.5(f)). Bring the mouse pointer on the word **Derivative**, click left button of the mouse (word is selected), and delete the other letters except **D** by using **Delete** button of keyboard, bring mouse pointer outside the block, and leftclick it. After completely deleting the name **Derivative**, you can even enter any word of your choice from the keyboard.

How to include the annotation to a block?

Suppose we have **Derivative** block at an untitled model file as shown in figure 1.3(d) and wish to write the

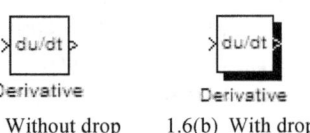

1.6(a) Without drop shadow   1.6(b) With drop shadow

Figures 1.6(a)-(b) **Derivative** block without and with the drop shadow

line **Derivative block in a model** down the block in the model file. Bring mouse pointer at the desired position in the model file, doubleclick the mouse to see the blinking cursor, type the **Derivative block in a model** from keyboard, bring mouse pointer out of the block, and click the left button of the mouse. Figure 1.5(g) shows the action we performed. Once typed, we can even drag the whole text to move anywhere in the model file.

⊟ **How to add drop shadow to a block?**

Some reader might be interested to see the drop shadow form of SIMULINK block rather than plain shape. Suppose we have the **Derivative** block at an untitled model file (figure 1.3(d)). Rightclick on the block to see pulldown menu of figure 1.4(b) and click the **Show Drop Shadow** via **Format**. The necessary change is depicted in figures 1.6(a)-(b). To remove the shadow, again rightclick on the block and

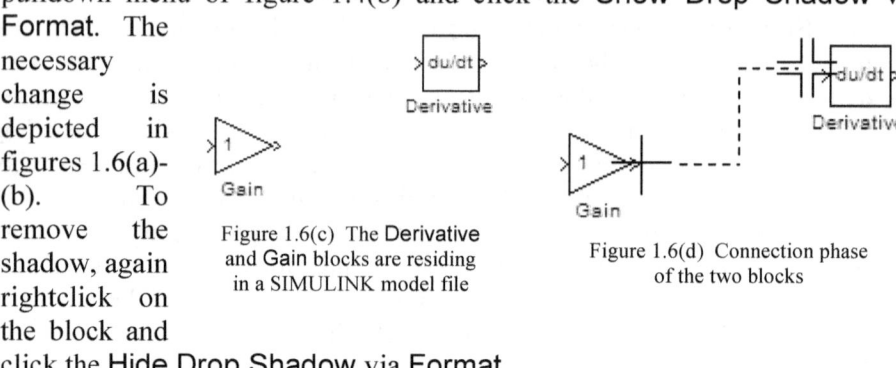

Figure 1.6(c) The **Derivative** and **Gain** blocks are residing in a SIMULINK model file

Figure 1.6(d) Connection phase of the two blocks

click the **Hide Drop Shadow** via **Format**.

⊟ **How to change SIMULINK model file background color?**

Rightclick anywhere in the model file and see the **Screen Color** in a prompt menu. From the popup of **Screen Color**, you can choose any background color for the model.

⊟ **How to copy a block within a SIMULINK model?**

Select the block, click the **Copy** icon at the model menu bar (this action is called copy in the clipboard), and paste it as many times as you want.

⊟ **How to connect two blocks?**

This manipulation is very important in the sense that we frequently need to connect blocks in a model while working in SIMULINK. The reader is familiar with **Gain**

Figure 1.6(e) The **Derivative** and **Gain** blocks are connected

and **Derivative** blocks from previous discussions. Suppose the two blocks are residing in a SIMULINK model file as shown in figure 1.6(c). We intend to connect the two blocks. Connection of the blocks must be correct syntactically. The output port of any block can only be connected to the input port of any other block but not to the output port of others. The same syntax is also true for the input port. However bring the mouse pointer on the output port of **Gain** block, find the single cross target as shown in figure 1.6(d), press left button of the mouse, move the single cross target anywhere in the model keeping your finger pressed, bring the single cross target close to input port of the **Derivative**, find the double cross target as shown in figure 1.6(d), and release the left button of the mouse. You should find the two blocks connected as shown in figure 1.6(e).

✂ **What is a parameter or block parameter window?**

In the following chapters we are going to mention the term parameter window many times. After bringing any block in a SIMULINK model file and then doubleclicking it, always do we see a prompt or dialog window in which we find one or more slots for value or parameter taking depending on the purpose of the block. That dialog window is termed as the parameter window for example window of the **Gain** block in figure 1.5(e).

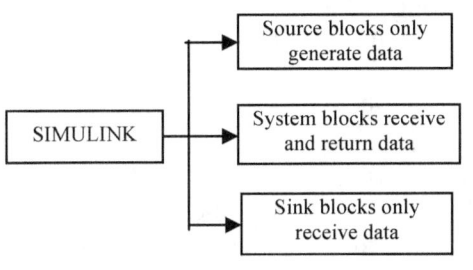

Figure 1.6(f) Basic block types in SIMULINK

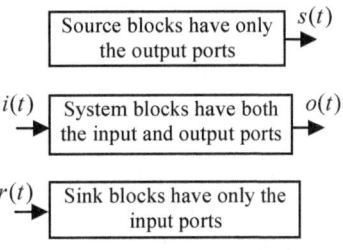

Figure 1.6(g) SIMULINK data flow in various blocks

1.2.4 Basic block categories of SIMULINK

By now we know that SIMULINK is a vast collection of blocks. These blocks follow three basic characteristics. Some blocks only generate data which are called source blocks, some blocks receive data, perform mathematical operation depending on the problem, and then return data which are called system blocks, and the third types only receive data which are called sink blocks. Figure 1.6(f) presents the basic block types found in SIMULINK. Figure

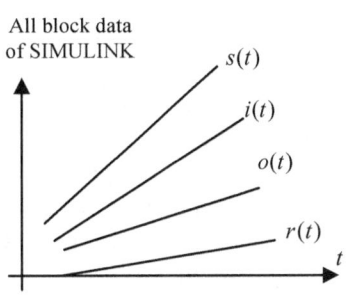

Figure 1.6(h) State or t dependency of SIMULINK data

1.6(g) shows the data flow as a function of t where $s(t)$ generated, $i(t)$ input and $o(t)$ output, and $r(t)$ received correspond to source, system, and sink blocks respectively.

Note: It is extremely important to mention that all block data whether source, system, or sink present in SIMULINK model shares the common t variation. This is called the state or t dependency of SIMULINK block data. Figure 1.6(h) illustrates this sort of dependency assuming all linear data.

1.2.5 Display and Scope blocks of SIMULINK

A practical model contains dozens of blocks which are interconnected by functional lines in SIMULINK. When a model is being run, blocks **Display** and **Scope** (appendix B for link and appearance) show how the functions are changing. The functional data flow or computation may or may not be seen during run time because it happens so rapidly – in a fraction of second. It also depends on the problem whether it is time consuming.

The **Display** block is convenient only for showing a single scalar or few matrix data output at the end of simulation. The block is designed to show the instantaneous value flowing through the functional line which it is connected to. Once SIMULINK has finished a simulation, the block shows the last value. Default size of the block is for a single

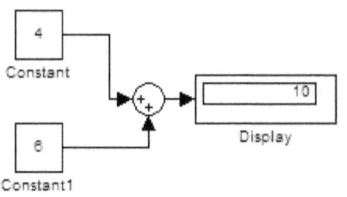

Figure 1.7(a) Adding two constants and displaying the result

scalar. For matrix data output one needs to enlarge the block to view all in it.

If the turnout of a SIMULINK block is in the form of a long row or column matrix which may hold hundreds of data elements, it is not feasible to see results through a **Display** block. The graphical plot is a better way to observe the output. The **Scope** in all sense mimics an oscilloscope that essentially displays the signal variation with time. The **Scope** has two axes – horizontal and vertical. The horizontal and vertical axes simulate the independent and dependent variables respectively. The horizontal axis does not have to be time even though it is originated in that name, any physical quantity such as displacement, frequency, speed or other can be assigned to the horizontal axis.

1.2.6 How to get started in SIMULINK?

In previous sections the reader has gone through bringing

Figure 1.7(b) Block parameter window for the Constant

and connecting blocks in a SIMULINK model file. Now we present simple modeling lessons for beginners in SIMULINK aiming to illustrate simulation style in this platform. Whatever operations such as manipulations, computations, assignments, or comparisons are carried out in conventional software can be conducted in SIMULINK through various blocks and functional lines. Since most algorithms are hidden in functional lines and blocks of SIMULINK, initially one might feel it complicated. Most of model building in SIMULINK happens through mouse operation rather than writing the source codes.

Let us go through the following three tutorials as a quick start in SIMULINK.

✦ Tutorial one

Two numbers are to be added – 4 and 6. The output should be 10 that is the problem statement.

The first question is where we should keep the numbers. In SIMULINK every programming aspect happens through blocks. You find a block called **Constant** through link **SIMULINK → Sources → Constant** (figure 1.3(a)). Open a new SIMULINK model file (subsection 1.2.2) and bring the **Constant** block in the untitled model file. The default value in the block is 1. Doubleclick the block to see its parameter window like figure 1.7(b) and enter 4 in the slot of **Constant value** from keyboard after deleting default 1 but leaving other parameters unchanged in the window. It means the first number 4 is going to be generated by the **Constant**. Similarly bring another **Constant** block from the same link and enter 6 as the **Constant value** after doubleclicking it. You see the latter block by the name **Constant1**. If you bring one more **Constant** block in the model, SIMULINK names that as **Constant2**. This style of naming is followed for all other blocks.

However we need a **Sum** block to add the two numbers that can be reached via **SIMULINK → Math Operations → Sum**. Bring the **Sum** in the model file. To see the computation, we need a **Display** block whose link is **SIMULINK → Sinks → Display** and also bring the block in the model file. Place the four blocks relatively and connect them (subsection 1.2.3 for connection) according to figure 1.7(a). Click the start simulation icon ▶ at the icon bar of model and SIMULINK

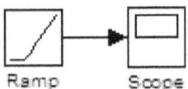

Figure 1.7(c) **Ramp** block connected with **Scope** block

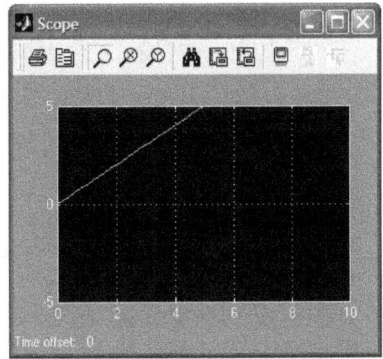

Figure 1.7(d) **Scope** block shows ramp function

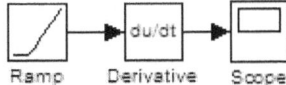

Figure 1.7(e) **Derivative** block differentiates output generated by **Ramp** and **Scope** shows **Derivative** output

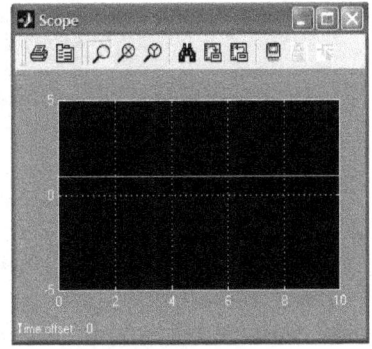

Figure 1.7(f) **Scope** output for model of figure 1.7(e)

responds showing the summation 10 in the **Display** block like figure 1.7(a). You can also run the model file from menu bar of SIMULINK by clicking first the **Simulation** and then the **Start** in the pulldown menu.

MATLAB command window provides alternative for running a SIMULINK model from its command prompt. Let us say we have the SIMULINK model saved by the name **test.mdl**. To run it from the command prompt, we carry out the following (**sim** is a built-in command for running a SIMLINK model file and the file name must be under quote):

>>sim('test') ↵

This action would show **10** in the **Display** block too.

✦ **Tutorial two**

We intend to generate the function $y(t) = t$ and view the generation in earlier mentioned **Scope** – this is the problem statement.

The function is a straight line passing through the origin and has a unity slope which is also known as the ramp. There is a block by the name **Ramp** (appendix B) in SIMULINK which generates t data by default. That means it is a source block as depicted in figure 1.6(f) or 1.6(g). Bring one **Ramp** and one **Scope** blocks in a new SIMULINK model file and connect the two blocks as shown in figure 1.7(c). Run the model by clicking the start simulation icon ▶ at icon bar of the model and doubleclick the **Scope** to view figure 1.7(d) presented curve which displays our wanted straight line plot. Horizontal and vertical axes of the **Scope** correspond to t and $y(t)$ data respectively so the **Ramp** generated the $y(t)$ and **Scope** just displayed that. You can inspect that the line of figure 1.7(d) is passing through the points (0,0) and (5,5) confirming the generation with correct slope.

✦ **Tutorial three**

If we differentiate the ramp $y(t) = t$ with respect to t, we should get $\frac{dy(t)}{dt} = 1$. We intend to simulate this mathematical operation.

The reader is familiar with **Derivative** block from subsection 1.2.3. The block differentiates any input signal to its input port and returns the derivative signal to its output port both in continuous sense. Insert the **Derivative** block between the **Ramp** and **Scope** blocks of figure 1.7(c) so that we have the model in figure 1.7(e). Select the **Ramp** or **Scope** block of the model in figure 1.7(c), use the left or right arrow key from the keyboard so that the space is enough between the two blocks to accommodate the **Derivative**. When you bring the **Derivative**, drop the block keeping its input and output ports in line with connection line of **Ramp** and **Scope**. SIMULINK is so smart that it connects the **Derivative** block automatically like figure 1.7(e). On forming the model, click the start simulation icon ▶ at the icon bar of model and doubleclick the **Scope** to view the output like figure 1.7(f) which is essentially a straight line parallel to horizontal axis and located at 1 in the vertical axis of **Scope**. In other words the vertical and horizontal axes of the **Scope** refer to $\frac{dy(t)}{dt}$ and t data respectively. That is what we expected from SIMULINK.

Note: Models in figures 1.7(a), 1.7(c), or 1.7(e) have connecting lines between various blocks. Data flowing through these lines is functional data not t data. It is the style of SIMULINK that t data in all these blocks is common. For example in figure 1.7(e), $y(t)$ data flows from **Ramp** to **Derivative** and $\frac{dy(t)}{dt}$ data flows from **Derivative** to **Scope** blocks.

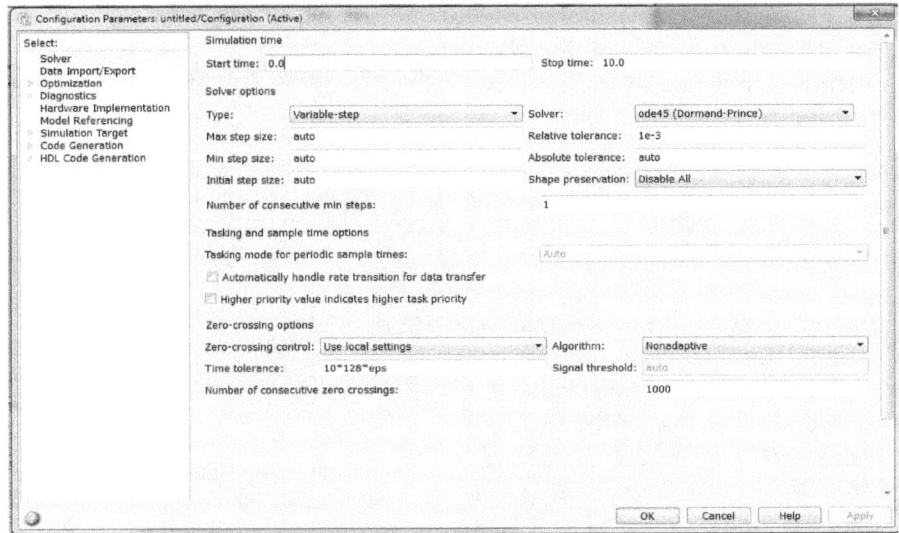

Figure 1.7(g) Simulation parameter window of SIMULINK

♦ ♦ Interval entering or t information in SIMULINK

In tutorial 2 the **Scope** (figure 1.7(d)) shows $y(t)=t$ versus t graph and in tutorial 3 the **Scope** (figure 1.7(f)) shows $\frac{dy(t)}{dt}=1$ versus t graph. We mentioned that horizontal axis of both **Scope**s represents t variation. The t variation seen in both **Scope**s is between 0 and 10 because that is default setting. Mathematically $y(t)$ or $\frac{dy(t)}{dt}$ is graphed over the interval $0 \le t \le 10$. In SIMULINK term the lower and upper bounds of interval $0 \le t \le 10$ are called the **Start time** and **Stop time** respectively.

What if we wish to enter other interval for example $-3 \le t \le 7.5$ instead of default $0 \le t \le 10$? Just mentioned **Start time** and **Stop time** then become -3 and 7.5 respectively which need to be entered. There is a window called **Configuration parameters** which keeps provision for entering the interval description, differential equation solver type (like stiff or nonstiff), step size of the computation (like adaptive or fixed), etc. The next question is how to reach to that window? If you click the **Simulation** menu (figure 1.3(b)), you find one submenu by the name **Model Configuration Parameters** in pulldown menu. Click the **Configuration Parameters** and the action lets you see the simulation parameter window of figure 1.7(g) or use **Ctrl+E** key. In upper portion of the window in figure 1.7(g) there are two slots for **Start time** and **Stop time** respectively. There we enter −3 and 7.5 from keyboard deleting default values for the lower and upper bounds of interval $-3 \le t \le 7.5$ respectively.

If the lower bound is 0 and upper bound is other than 10, you do not even need to open parameter window. In the untitled model file of figure 1.3(b), there is a slot for **Stop time** as indicated by the arrow. For example if the interval is $0 \le t \le 15$, we just enter 15 at **Stop time** slot of figure 1.3(b) without opening parameter window.

Note: There are many parameters in the simulation parameter window of figure 1.7(g) whose discussions are beyond the scope of the text. *We suggest that you do not change any other parameter unless you know about it.* SIMULINK

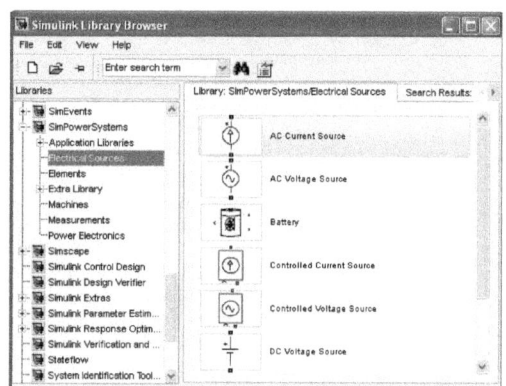

Figure 1.8(a) **SimPowerSystems** library of SIMULINK for electric circuit analysis

approach of modeling any dynamic problem is completely numerical. Also computer always works on discrete data instead of continuous one even though we do simulation in continuous sense.

1.2.7 Electric circuit library in MATLAB and SIMULINK

Electric circuit library is included both in MATLAB and SIMULINK. Built-in functions of MATLAB and blocksets of SIMULINK are devised to handle robust electrical circuit problems for professional use. We selected functions and blocks whose executions or operations are closely related to

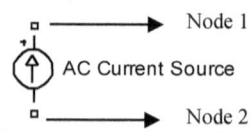

Figure 1.8(b) Two nodes of a SimPowerSystems block

introductory electric/electronic circuit text terms. In figure 1.3(a) you find SIMULINK library browser window. If you scroll down using the mouse in left half of the window, you find a library by the name **SimPowerSystems** which is what we will be utilizing most. Doubleclick the **SimPowerSystems** and wait for a while and find the subfamilies of **SimPowerSystems** as shown in figure 1.8(a). For example under the subfamily **Electrical Sources** you find **AC Current Source, AC Voltage Source,** etc. **Note that** you would not find it if you or your sponsor did not purchase the power system library.

♦ ♦ **Important difference between SIMULINK and SimPowerSystems blocks**

From subsections 1.2.4 and 1.2.6 we know that SIMULINK blocks receive or return functional data keeping t information common which makes blocks as one node element regardless of input or output port. On the contrary **SimPowerSystems** blocks are two node elements. This is arising from the fact that electric circuit elements are polarity intense. If you bring the **AC Current Source** block of figure 1.8(a) in an untitled model file, you find the block like figure 1.8(b). Figure 1.8(b) also shows two nodes of the

block. Earlier mentioned **Derivative** or **Ramp** block can not be connected to either node in figure 1.8(b), only can we connect any two blocks of **SimPowerSystems.** In terms of symbol, input or output port of SIMULINK block holds >

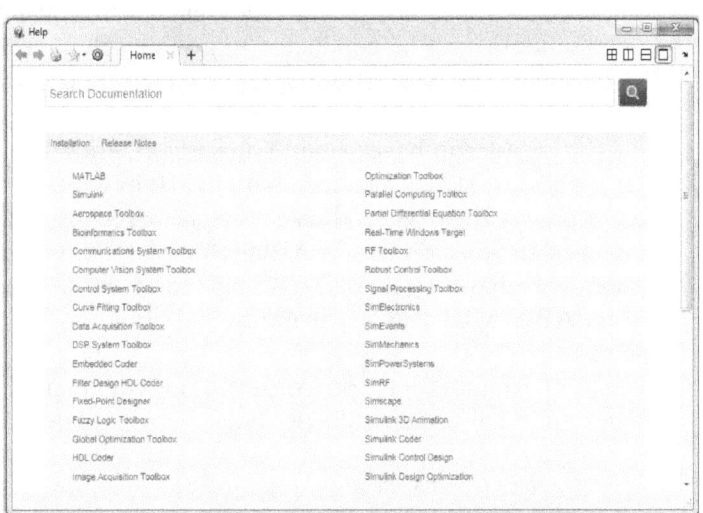

Figure 1.8(c) General Help window for MATLAB and SIMULINK

indication whereas **SimPowerSystems** block has the node indication ☐.

In MATLAB library of electrical or electronic circuits you find directory **powersys** or **powerdemo** which we will address later.

1.3 How to get help?

Help facilities in MATLAB are plentiful. One can access to information about a MATLAB function or SIMULINK block in a variety of ways. Command help finds the help of a particular function file. You are familiar with the function sin(x) from earlier discussion and can have command prompt help regarding the sin(x) as follows:

>>help sin ↵   ← Function name without the argument

-25-

> sin   Sine of argument in radians.
> sin(X) is the sine of the elements of X.
>
> See also asin, sind.
> ⋮

One disadvantage of this method is the user has to know the exact file name of a function. For a novice this facility may not be appreciative.

Casually we know partial name of a function or try to check whether any function exists by that name. Suppose we intend to see whether any function by name **simpowersys** (the rationale is this is the library name in SIMULINK) exists. Machine does that by intermediacy of command **lookfor** (no space gap between **look** and **for**):

> \>>lookfor simpowersys ↵
> power_analyze - Analyze a circuit built with SimPowerSystems.
> power_ltiview - (SYS) opens a graphical user interface used to link SimPowerSystems
> ⋮

Above return is showing all possible functions bearing the file name **simpowersys** or having the file name **simpowersys** partly. Now the command **help** can be conducted to go through a particular one for example the first one is **power_analyze** and execute **help power_analyze** to see its description at the command prompt.

In order to have window form help, click Help icon (i.e. ?) of figure 1.1(a) and MATLAB responds with the Help window of figure 1.8(c). As the figure shows, help is available content-based or index-based. If you have some search word on MATLAB or SIMULINK, you can search that through Search of figure 1.8(c). This help form is better when one navigates MATLAB/SIMULINK's capability not looking for a particular function/ block.

If you execute **powerdemo** in conjunction with **help**, you find names of many electrical/electronic circuit related MATLAB/SIMULINK embedded files:

> \>>help powerdemo ↵
> SimPowerSystems Demos
>
> General Demos library
>   power_filter   - Steady-state operation of linear circuit filter
>   power_transient   - Transient analysis of a linear circuit
> ⋮

MATLAB exhibits a long list of embedded file names which you see above. Take the first one as example, execute **help power_filter** at the command prompt to see its description.

Hidden algorithm or mathematical expression is often necessary whose assistance we can have through the search option from MathWorks Website provided that your PC is connected to internet.

However we close the introductory discussion on MATLAB and SIMULINK with this.

Chapter 2

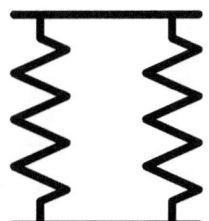

Basic Circuit Modeling

A direct current (DC) circuit is mainly composed of resistors, ideal voltage sources, and ideal current sources. Alternating current (AC) circuit mainly brings inductor and capacitor onboard. SIMULINK blocks are designed to handle robust electrical circuits yet appropriate parameters of SIMULINK blocks help us simulate conventional circuit terms in an easeful way. The chapter is all about implementations on fundamentals of electric circuits taught in freshman or sophomore level course. In order to provide a degree of familiarity, our approach has been "we have a circuit – want to simulate it" and tried to outline:

♦ ♦   Resistor modeling and DC/AC circuit construction
♦ ♦   Simulation of electric voltage and current in circuits
♦ ♦   Implementations on electric circuit measurements

2.1 How to model a resistor?

A direct current or DC circuit is the combination of resistors connected in various circuital forms. In order to analyze any DC circuit, the first question appears how we model a resistor. In SIMULINK we have a block by name **Series RLC Branch** which in general models resistance-inductance-capacitance or R-L-C. Let us bring the **Series RLC Branch** block (appendix B) in a new SIMULINK model file (subsection 1.2.2). Figure 2.1(a) presents the block appearance. Following doubleclick on the block, we see the parameter window as shown in figure 2.1(b) and find **Branch type** under the **Parameters** in the window. By default the setting is RLC meaning suited for modeling resistance-inductance-capacitance. If we click **RLC**, we find a popup showing different element options. From the popup we choose **R** for resistor modeling. Suppose we intend to model a

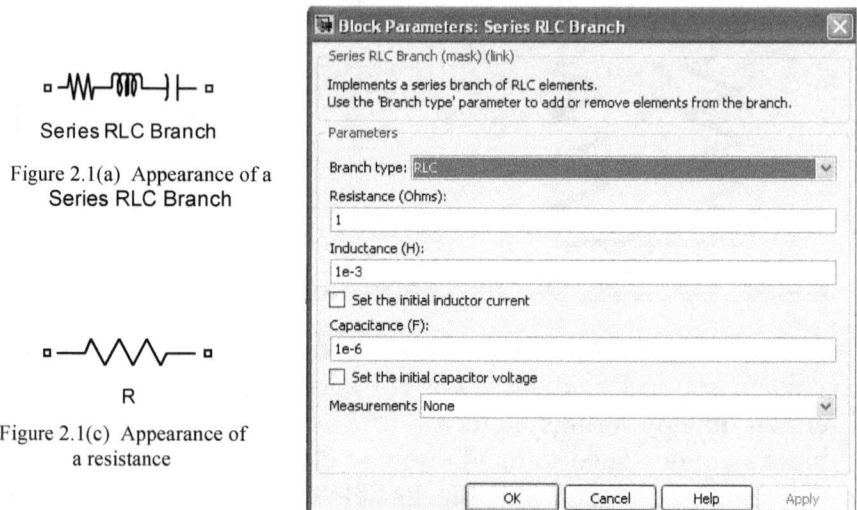

Series RLC Branch

Figure 2.1(a) Appearance of a
Series RLC Branch

R

Figure 2.1(c) Appearance of
a resistance

Figure 2.1(b) Block parameter window of
Series RLC Branch

resistance of $R = 5\Omega$. In figure 2.1(b), we enter **5** in the slot of **Resistance (Ohms)** and click OK. Appearance of the **Series RLC Branch** is element adaptive. Rename (subsection 1.2.3) the **Series RLC Branch** as R. In doing so, we see the block appearance of figure 2.1(c) therefore we modeled the resistor.

Resistance of different units is often seen in electrical circuits. For example the resistances of value 25 mΩ, 4.7 KΩ, and 33 MΩ are entered by writing **25e-3, 4.7e3**, and **33e6** respectively where **e** stands for **10** to the power (appendix A).

2.2 How to model a DC circuit?

A DC circuit is composed of resistors. Resistors are connected in various circuital forms such as series, parallel, Y-Δ, or combination of them. First of all we have to have a circuit diagram we intend to model. Let us go through the following examples in this regard.

✦ Example 1

We wish to model the circuit of figure 2.2(a) in which two resistors are connected in series. The procedure we need to model the circuit in figure 2.2(a) is the following:

⇒ open a new SIMULINK model file (section 1.2.2),

⇒ bring a **Series RLC Branch** block in the model file (appendix B), doubleclick it, select **R** from **Branch type** popup, enter **Resistance** as 3 (section 2.1), and rename the block as **R=3**,

⇒ copy the block **R=3** by clicking the copy icon in the model file, paste it in the same model file, rename the copied block as **R=7**, doubleclick the block, and change the **Resistance** value from previous 3 to 7, and

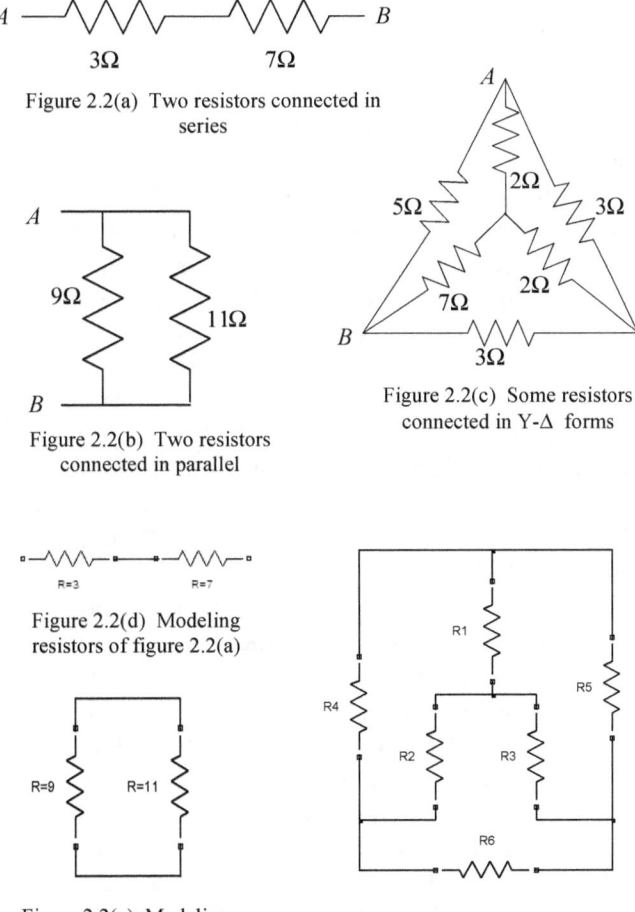

Figure 2.2(a) Two resistors connected in series

Figure 2.2(b) Two resistors connected in parallel

Figure 2.2(c) Some resistors connected in Y-Δ forms

Figure 2.2(d) Modeling resistors of figure 2.2(a)

Figure 2.2(e) Modeling resistors of figure 2.2(b)

Figure 2.2(f) Modeling resistors of figure 2.2(c)

⇒ place the blocks R=3 and R=7 side by side and connect them as shown in figure 2.2(d) (subsection 1.2.3).

Therefore we modeled the two resistors of figure 2.2(a) resembling to SIMULINK model in figure 2.2(d).

◆ **Example 2**

In this example we intend to model the parallel resistors of figure 2.2(b). We can model the circuit starting from the model (figure 2.2(d)) of example 1. Delete the connection line between the resistor blocks R=3 and R=7, rename the block R=3 as R=9, doubleclick the block R=9 (figure 2.1(b)), change its **Resistance** value to 9 keeping the others unchanged in the parameter window, rightclick the R=9, and click the **Rotate Block** under the popup **Format** which makes the resistance R=9 vertical. Figure 2.2(e) is model of the parallel resistors. In a similar fashion model the R=11 of figure 2.2(e) and finally connect them.

◆ **Example 3**

Referring to figure 2.2(c), there are several resistors connected in Y-Δ forms which we wish to model.

You find the resistor alignment of figure 2.2(c) inclined. To date we can model only the horizontal and vertical resistors in SIMULINK. As long as a circuit is connection-wise correct, there is no harm using horizontal and vertical resistors to model the circuit. Not to mention, resistor placement in the model is user-defined. Our resistor placement happened like figure 2.2(f). One important point needs to be mentioned – *no two blocks in a SIMULINK model should have the same name*. For this reason we named resistors of figure 2.2(c) as R1, R2, etc in model 2.2(f) instead of R=2 as done before. There are five vertical and one horizontal resistors in the figure 2.2(f) and their correspondences with regard to figure 2.2(c) are R1=2, R2=7, R3=2, R4=5, R5=3, and R6=3 (all in Ω). Let us go through the following procedure to model the circuit:

⇒ open a new SIMULINK model file, bring a **Series RLC Branch** in the model, rename the block as R1, rightclick the block R1, click the **Rotate Block** under **Format** popup to model a vertical resistor, doubleclick the R1, select R from the popup, enter the value as 2 (section 2.1) in doing so the R1 modeling is finished,

⇒ select the R1, copy it in the icon bar of SIMULINK, paste it four times to generate the other four vertical resistors (that is R2, R3, R4, and R5) of figure 2.2(f) and doubleclick each of the four resistor blocks and enter its **Resistance** value in the parameter window thereby modeling all vertical resistors, and

⇒ copy and paste **R1**, find **R6**, rightclick the **R6**, click the **Rotate Block** under **Format** popup to model the horizontal resistor, and enter its **Resistance** value as 3.

Finally place all blocks relatively like figure 2.2(f) and connect them to end modeling of Y-Δ form resistors in figure 2.2(c).

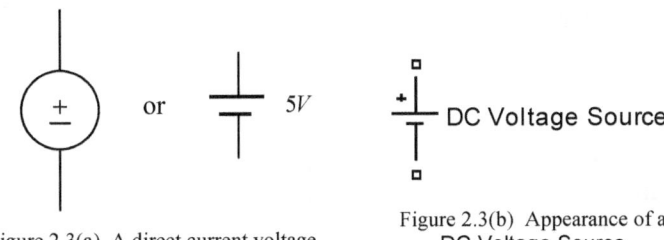

Figure 2.3(a) A direct current voltage source of value 5 volts

Figure 2.3(b) Appearance of a DC Voltage Source

2.3 How to model a source element in DC circuits?

There are two types of source element seen in basic electrical circuits namely direct voltage and current sources. Figure 2.3(a) presents the schematic symbol of a $5V$ direct current voltage source which we wish to model. Open a new SIMULINK model file and bring the block **DC Voltage Source** in the model file (subsection 1.2.2 and appendix B). Figure 2.3(b) shows the appearance of the **DC Voltage Source** which simulates an ideal voltage source in any direct current electrical circuit. If you

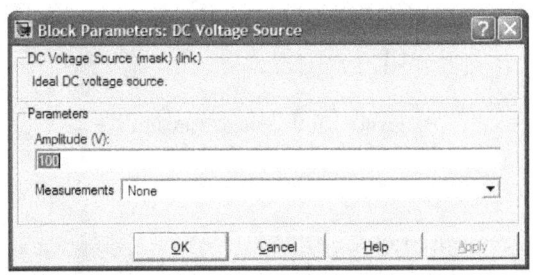

Figure 2.3(c) Block parameter window of a firstly opened DC Voltage Source

doubleclick the block, you see the firstly opened parameter window like figure 2.3(c) in which slot for entering voltage value lies below the word **Amplitude (V):**. In order to model the ideal voltage source of figure 2.3(a), we delete the default amplitude 100 from the slot, type 5 from the keyboard, and click OK. In the case of other unit voltages, employ the scale factor for example voltages $3.3mV$ and $23.2KV$ are entered by writing **3.3e-3** and **23.2e3** respectively where **e** stands for 10 to the power (appendix A).

An ideal current source modeling is slightly different from that of the voltage one. The main reason is SIMULINK blocks are designed to handle robust and professional circuits. Figure 2.3(e) shows the schematic representation of an ideal direct current source. We model an ideal current source by using block **Controlled Current Source** whose appearance is

Mohammad Nuruzzaman

seen in figure 2.3(f). As you see, there are two nodes (□) and one input port in **Controlled Current Source**. Open a new SIMULINK model file and bring the block in the model. Doubleclicking the block displays firstly opened parameter window of figure 2.3(d). Whatever function we apply to input port (90 degree rotated s) of the block transforms as the current source value. If we apply constant 3 to the input port of the block, the block becomes a current source of 3 amperes; again if we apply a sine wave, the same sine wave becomes a current source, and so on. In the

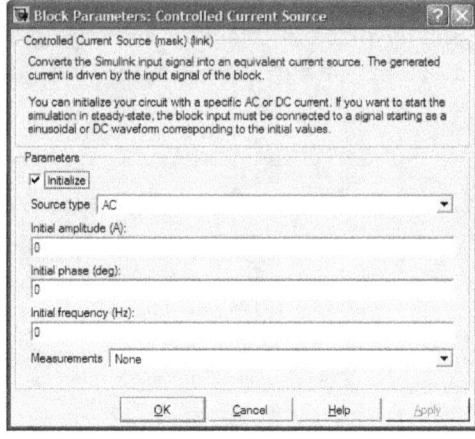

Figure 2.3(d) Block parameter window of a firstly opened **Controlled Current Source**

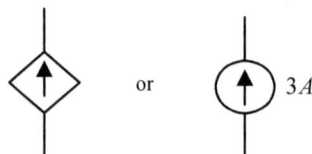

Figure 2.3(e) A direct current source of value 3 amperes

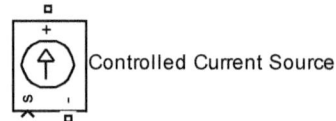

Figure 2.3(f) Appearance of a **Controlled Current Source**

parameter window of figure 2.3(d), we change the **Source type** from AC to DC and uncheck the **Initialize** (the parameter window is parameter-adaptive and you see that the moment you act). Bring a **Constant** in the same model file, doubleclick the block to see its parameter window like figure 2.3(g), and change its **Constant** value from the default 1 to 3 (which

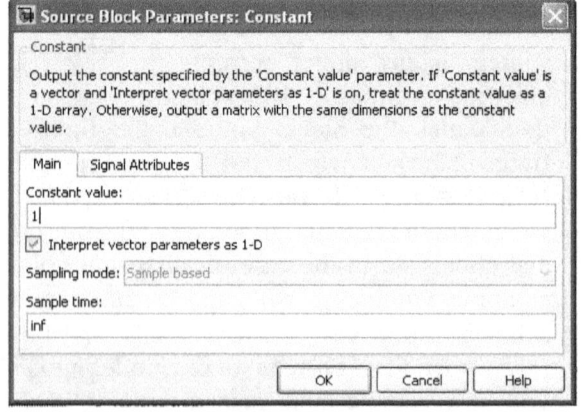

Figure 2.3(g) Block parameter window of a firstly opened **Constant**

is ideal current source current value in amperes). Connect the **Constant** block with **Controlled Current Source** like figure 2.3(h) which essentially models the current source of figure 2.3(e). We may have other unit currents for example $3.6mA$ and $34.7\mu A$ which need writing **3.6e-3** and **34.7e-6** respectively where **e** stands for 10 to the power. In order to model a $3.6mA$ ideal current source what we need is doubleclick the **Constant** and enter **3.6e-3** at the **Constant** slot of parameter window in figure 2.3(g).

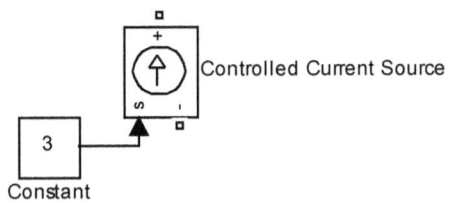

Figure 2.3(h) Modeling ideal current source of figure 2.3(e)

2.4 How to find the DC circuit voltage in a model?

Practically DC circuit voltages are determined by the use of a voltmeter. We know that a voltmeter is always connected in parallel to measure the voltage between any two terminals or nodes of an electrical circuit. The notion applies in SIMULINK model as well. The block **Voltage Measurement** (appendix B) simulates the action of a voltmeter whose schematic representation is shown in figure 2.4(a) but to show the value of voltmeter voltage, we need a **Display** in conjunction with the block. Depicted figure 2.4(b) is the model simulating the action of voltmeter in figure 2.4(a). A DC voltmeter

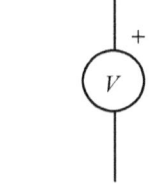

Figure 2.4(a) Schematic symbol of a DC voltmeter

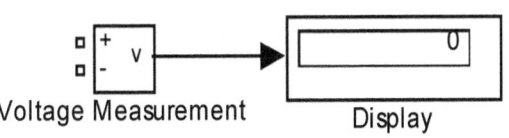

Voltage Measurement   Display

Figure 2.4(b) Modeling a DC voltmeter

is polarity sensitive meaning it has + and − polarities, so is the **Voltage Measurement** (appears as + and − in the block appearance of figure 2.4(b)). We cite one modeling example on voltage determining in the following.

◆ **Example**

It is given that voltage between the nodes A and B of bridge circuit in figure 2.4(c) is 0.6533 mV which we wish to determine.

The circuit needs one ideal voltage source and five resistors (sections 2.1-2.3) modeling as depicted in figure 2.4(d), which is the model of this example. Complete procedure to model the circuit is the following:

⇒ open a new SIMULINK model file (subsection 1.2.2), bring one Series RLC Branch block (appendix B) in the model, doubleclick the block, select R from the Branch type popup, enter the Resistance as 3 in the parameter

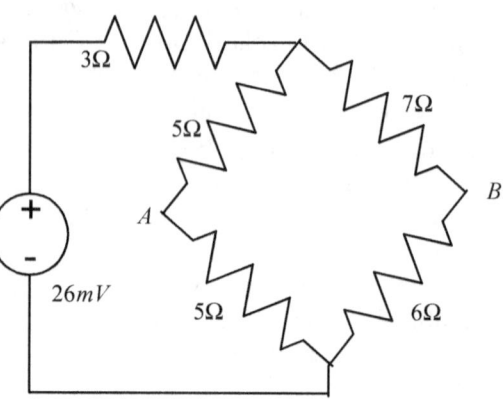

Figure 2.4(c) An electric bridge circuit

window thereby modeling the 3Ω horizontal resistor of figure 2.4(c),

⇒ rename the Series RLC Branch as R1, copy the R1 in the clipboard, paste it in the same model (appears as R2), rightclick on the R2, click the Rotate Block through Format popup to form a vertical resistor, doubleclick R2, change its Resistance value to 5, copy R2 in the clipboard, paste it three times to make R3, R4, and R5 appear, doubleclick each one of the last three resistors and set the Resistance as 7, 5, and 6 respectively,

⇒ bring one DC Voltage Source, one Voltage Measurement, and one Display blocks in the model, doubleclick the DC Voltage

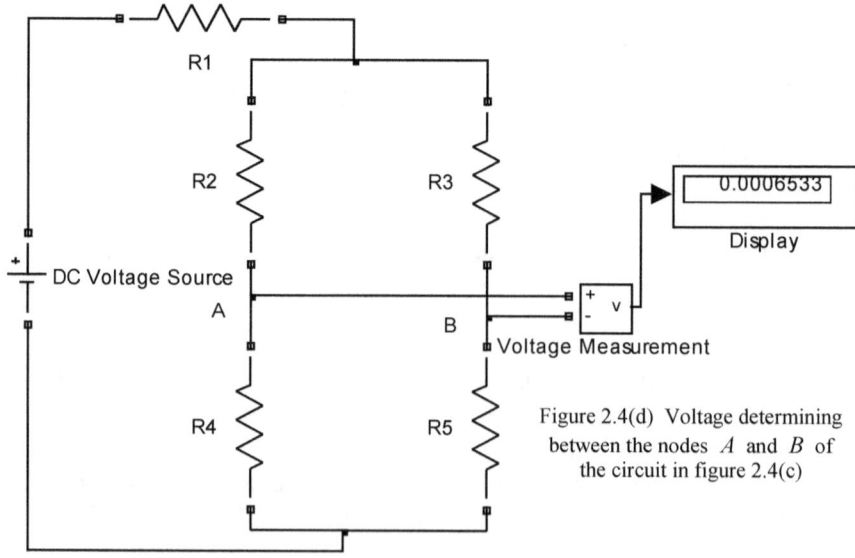

Figure 2.4(d) Voltage determining between the nodes A and B of the circuit in figure 2.4(c)

Source, and enter 26e-3 as Amplitude in parameter window for 26 *mV* voltage source,

⇒ place so mentioned blocks relatively and connect them as presented in figure 2.4(d), and

⇒ doubleclick near point A of the model in figure 2.4(d) and type A at the blinking cursor to indicate the node A, and do so near the point B.

Finally click the simulation icon ▶ at the icon bar to run the model and find the voltage between terminals A and B as 0.0006533 (in V) in Display.

2.5 How to find the DC circuit current in a model?

We know that an ammeter whose schematic symbol is shown in figure 2.5(a) measures the current in any practical electrical circuit. The current measurement in an electrical circuit is somewhat tricky because of disconnection involvement. In the voltage measurement we directly connect a voltmeter to circuit nodes without any disconnection on the contrary an ammeter is always connected in series to measure a current through the circuit branch. This necessitates that we disconnect any given electrical circuit branch and place the ammeter in series with correct polarity in the branch. However the block **Current Measurement** (appendix B) simulates the action of an ammeter and a **Display** block shows value of the direct current flowing through the branch therefore figure 2.5(b) displayed model is equivalent to the current

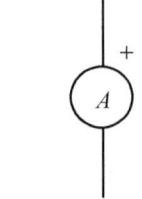

Figure 2.5(a) Schematic symbol of a DC ammeter

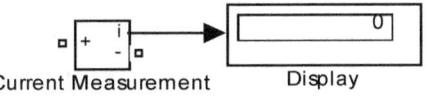

Current Measurement   Display

Figure 2.5(b) Modeling a DC ammeter

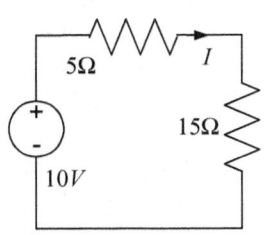

Figure 2.5(c) A series circuit with two resistors

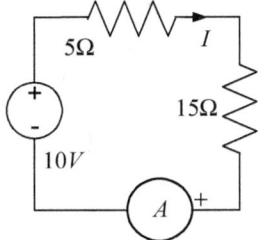

Figure 2.5(d) Series circuit of figure 2.5(c) with a DC ammeter

measurement ammeter of figure 2.5(a). A DC ammeter is polarity sensitive, so is the **Current Measurement** block. The arrangement of **Current**

Measurement should be such that the user-required current enters to +ve terminal of the block. Let us go through the following examples on direct current finding.

✦ Example 1

Figure 2.5(c) shows two resistors connected in series and we wish to find the current I in the circuit. Clearly we connect an ammeter in series with the resistor of figure 2.5(c) according to polarity as shown in figure 2.5(d). Value of the current I or ammeter reading should be 0.5 amperes. This is what we expect from the circuit simulation.

In order to model the circuit of figure 2.5(d), we proceed with the following:

⇒ open a new SIMULINK model (subsection 1.2.2), bring one DC Voltage Source, one Series RLC Branch, one Current Measurement, and one Display blocks in the model,

⇒ doubleclick the DC Voltage Source and enter 10 as Amplitude (V) in the parameter window to model the source element (section 2.3),

Figure 2.5(e) Modeling the circuit of figure 2.5(d)

⇒ doubleclick the Series RLC Branch, select R from the Branch type popup, enter the Resistance as 5 (section 2.1) in the parameter window, and rename the block as R=5 to model horizontal resistor of figure 2.5(d),

⇒ copy the R=5 in the clipboard, paste it in the same model, rename the block as R=15, doubleclick the block, enter 15 as the Resistance in the parameter window, rightclick the mouse on the R=15, and click the Rotate Block through Format popup to model vertical resistor of figure 2.5(d),

⇒ rightclick each of the Display and Current Measurement blocks and click the Flip Block through Format popup to orient the ammeter of figure 2.5(d), and

⇒ place so mentioned blocks relatively and connect them as presented in figure 2.5(e).

After all these we click simulation icon ▶ at the icon bar to run the model and find the required current I as 0.5 ampere in Display of figure 2.5(e).

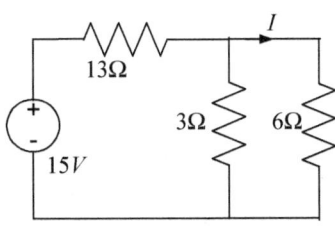
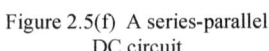

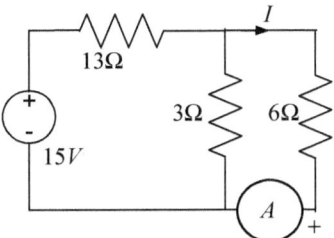

Figure 2.5(f) A series-parallel DC circuit

Figure 2.5(g) Series-parallel DC circuit of figure 2.5(f) connected with an ammeter

✦ Example 2

This example needs us to find current I through the 6Ω resistor of figure 2.5(f). It is given that the current I is 0.3333 amperes by using the equivalent resistance and current-division analyses – which we intend to obtain from SIMULINK circuit simulation.

The simulation requires that we connect an ammeter in the branch whose current is to be determined and the ammeter connection is shown in figure 2.5(g). Depicted in figure 2.5(h) is the model of figure 2.5(g). In order to simulate the circuit of the figure 2.5(g), we carry out the following:

⇒ open a new SIMULINK model file and bring one **DC Voltage Source**, one **Series RLC Branch**, one **Current Measurement**, and one **Display** blocks in the model,

⇒ doubleclick the **DC Voltage Source** and enter **15** as the **Amplitude (V)** in the parameter window for modeling the source element,

⇒ doubleclick the **Series RLC Branch** block, select **R** from the **Branch type** popup, enter the **Resistance** as **13** in the parameter window, and rename the block as **R=13** to model the horizontal resistor of figure 2.5(g),

⇒ copy the **R=13** in the clipboard, paste it in the same model,

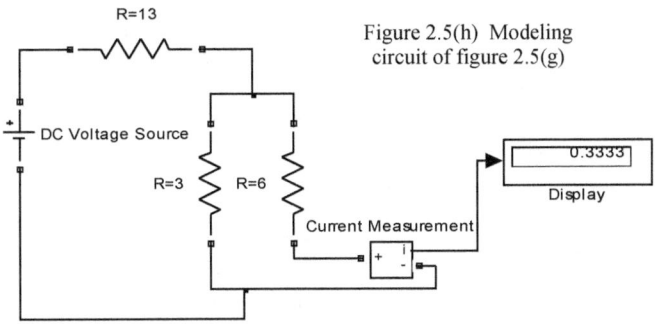

Figure 2.5(h) Modeling circuit of figure 2.5(g)

rename the block as **R=3**, doubleclick the block, enter **3** as the Resistance, rightclick on the **R=3**, and click the **Rotate Block** through **Format** popup to model the left vertical resistor of figure 2.5(g),

⇒ copy the **R=3** in the clipboard, paste it in the same model file, rename the block as **R=6**, doubleclick the block, and enter **6** as the **Resistance** in the parameter window to model the second vertical resistor of figure 2.5(g), and

⇒ place various blocks relatively and connect them as presented in figure 2.5(h).

At the end we click the simulation icon ▶ at icon bar to run the model and find the required current *I* as 0.3333 amperes in **Display** of figure 2.5(h).

2.6 How to find an equivalent resistance of DC circuits?

Once we construct a DC circuit employing series, parallel, or Y-Δ form resistors, it is no complicated task to find equivalent resistance of the circuit without any calculation. We are going to cite one approach to finding the equivalent resistance of a DC circuit. There may be other approaches. A circuital trick is

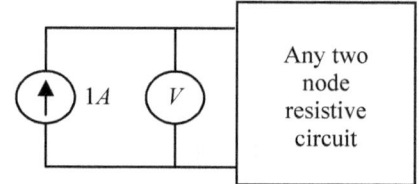

Figure 2.6(a) Circuit concept on equivalent resistance between any two node resistive network

exercised here. If we pass 1*A* direct current to any two node resistive circuits, voltage drop in the circuit is the equivalent resistance. Figure 2.6(a) depicts the concept behind equivalent resistance finding in which the *V* is a voltmeter and its reading becomes the equivalent resistance of circuit. Sections 2.3 and 2.4 describe ideal current source and voltmeter modelings respectively. Let us go through the following examples on equivalent resistance.

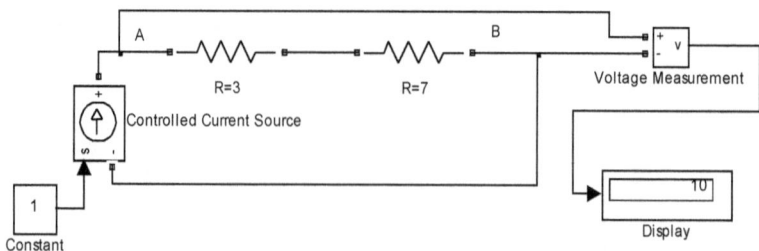

Figure 2.6(b) Model for equivalent resistance measurement of circuit in figure 2.2(a)

❖ Example 1

Equivalent resistance of series circuit in figure 2.2(a) is 10 Ω which we wish to find. To obtain the resistance, we connect 1A direct current source to the nodes *A* and *B* and measure the voltage drop across the nodes *A* and *B*. This voltage is actually our equivalent resistance. As a complete procedure, we need to carry out the following:

⇒ open a new SIMULINK model file (subsection 1.2.2), bring one **Constant**, one **Controlled Current Source**, two **Series RLC Branch**, one **Voltage Measurement**, and one **Display** blocks (appendix B) in the model,

⇒ doubleclick the **Controlled Current Source**, change its Source type from **AC** to **DC** in the parameter window, and uncheck the **Initialize** in the parameter window as well,

⇒ doubleclick the first **Series RLC Branch**, select R from the **Branch type** popup, enter the **Resistance** as **3** in the parameter window (section 2.1), rename the block as **R=3**, and similarly model the second **Series RLC Branch** as **R=7**, and

⇒ place the blocks relatively and connect them as shown in figure 2.6(b) – model for this example.

Finally click simulation icon ▶ at the icon bar to run the model and find the equivalent resistance 10 (in Ω) in **Display**.

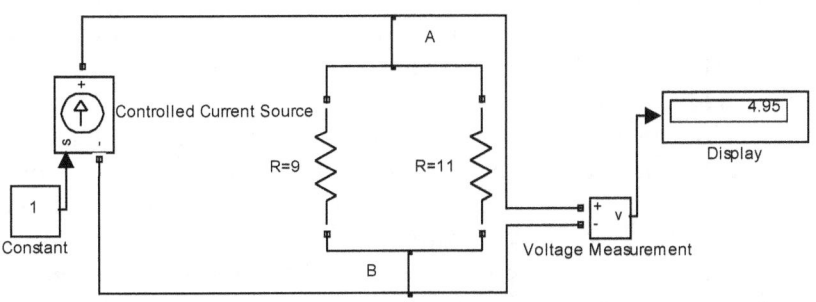

Figure 2.6(c) Determining equivalent resistance of parallel circuit in figure 2.2(b)

❖ Example 2

The parallel circuit in figure 2.2(b) has the equivalent resistance 4.95 Ω between the nodes *A* and *B* which we wish to find.

Example 1 mentioned procedure is equally applicable for any two node resistive circuits. The circuit of figure 2.2(b) is modeled in example 2 of section 2.2. In addition to the circuit, we need example 1 mentioned 1A current source and voltmeter to nodes *A* and *B*. With all these, figure 2.6(c) presents SIMULINK model for finding equivalent resistance. On constructing the model, click simulation icon ▶ at the icon bar to run the model and find equivalent resistance 4.95 (in Ω) in Display.

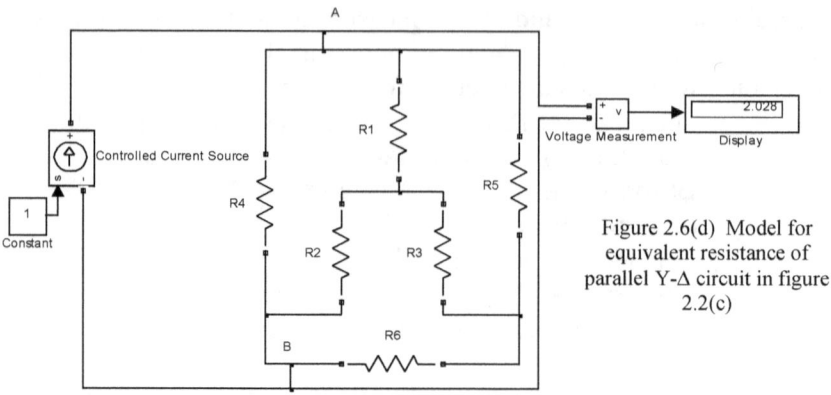

Figure 2.6(d) Model for equivalent resistance of parallel Y-Δ circuit in figure 2.2(c)

✦ Example 3

It is given that the Y-Δ circuit of figure 2.2(c) has an equivalent resistance 2.028Ω between the nodes *A* and *B* which we wish to find.

Example 3 of section 2.2 describes modeling of the Y-Δ form resistors. All we need is connect aforementioned 1*A* current source and voltmeter simulator to nodes *A* and *B*. Figure 2.6(d) depicts the complete SIMULINK model for equivalent resistance finding. Click the simulation icon ▶ at the icon bar to run the model and find the equivalent resistance 2.028 (in Ω) in Display.

We hope these three examples will help the reader find the equivalent resistance of many complicated resistive circuits. In all three examples, you find the node labels A and B in the model. Just doubleclick mouse pointer at the desired position and find the cursor blinking. Type A at the blinking position and click mouse in other area of the model. We did not rename the Display. Modeling could have been better if the Display were renamed as Equivalent Resistance.

2.7 How to find a node voltage in DC circuits?

In electric circuit analysis the node voltage is a special voltage which can also be measured by using the Voltage Measurement of section 2.4. In node voltage analysis of DC electric circuits, one discerning point is we keep a common or reference voltage node which is set to 0*V*. SIMULINK block Ground (appendix B) simulates a reference node whose appearance is presented in figure 2.7(b). In order to find any node voltage, we connect −ve port of Voltage Measurement to the Ground and the +ve port of Voltage Measurement is connected to the node whose voltage is to be determined. If we have two or more nodes, so **Voltage Measurement** blocks are required and each of the negative ports of the blocks needs a **Ground**.

When we need to determine many node voltages, presence of several **Voltage Measurement** blocks makes the circuit modeling clumsy. There is another block called **Neutral** which marks and numbers a node in any electrical circuit and whose appearance is shown in figure 2.7(c). We place the **Voltage Measurement** and **Neutral** blocks outside the main circuit area to model a circuit neatly. For every node voltage, we need two **Neutral** blocks – one connects to the circuit required node point and the other to the +ve port of the **Voltage Measurement**. The default node number of **Neutral** is 10 but you can doubleclick the **Neutral** and change the node number to any other for example 1, 2, 3, etc in its

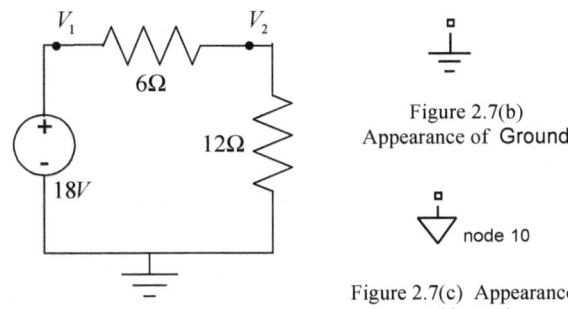

Figure 2.7(b)
Appearance of **Ground**

Figure 2.7(c) Appearance of **Neutral**

Figure 2.7(a) Two node voltages in a simple series circuit

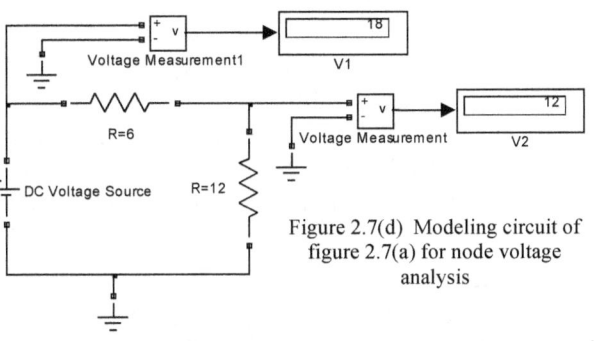

Figure 2.7(d) Modeling circuit of figure 2.7(a) for node voltage analysis

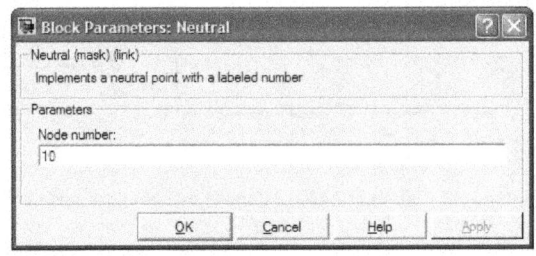

Figure 2.7(e) Block parameter window of **Neutral**

parameter window of figure 2.7(e). While using the **Neutral**, a **Ground** needs to be connected to the –ve port of **Voltage Measurement** for every node voltage.

We present some examples on node voltage simulation in the sequel.

Mohammad Nuruzzaman

◆ Example 1

The two resistor series circuit of figure 2.7(a) has two node voltages $V_1=18\,V$ and $V_2=12\,V$. Our objective is to find these two node voltages through SIMULINK circuit simulation which needs the following steps:

⇒ open a new SIMULINK model file (subsection 1.2.2), bring one Series RLC Branch, one DC Voltage Source, two Voltage Measurement, three Ground (one for the circuit and two for the Voltage Measurement), and one Display blocks in the model,

⇒ doubleclick the DC Voltage Source and enter 18 as the Amplitude (V) in parameter window to model the DC source,

⇒ doubleclick the Series RLC Branch, select R from the Branch type popup, change its Resistance value to 6 in parameter window (section 2.1), rename the block as R=6, copy the R=6 in clipboard, paste it once, rename the pasted block as R=12, doubleclick the R=12, change its value to 12, rightclick on the R=12, and click the Rotate Block through Format popup to turn the resistor to a vertical one,

⇒ rename the Display as V1, copy it in the clipboard, and paste it to see another Display by the name V2, and

⇒ place the blocks relatively and connect them as shown in figure 2.7(d) that is the model for this example.

Finally we click simulation icon ▶ at the icon bar to run the model and see the required node voltages V_1 and V_2 in V1 and V2 blocks respectively.

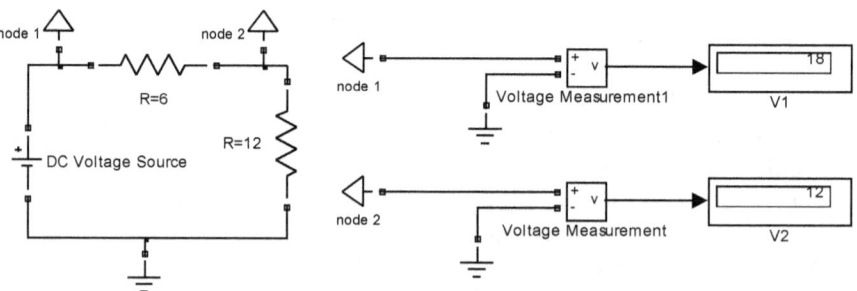

Figure 2.7(f) Alternate modeling for node voltage of circuit in figure 2.7(a) using Neutral

◆ Example 2

As an alternative we find the node voltages of circuit in figure 2.7(a) employing earlier mentioned Neutral. Figure 2.7(f) shows the necessary model. Let us perceive how it is different from the model in figure 2.7(d). Referring to the figure, the node 1 and node 2 correspond to nodes V_1 and V_2 of given circuit in figure 2.7(a) respectively. When you bring the Neutral

-42-

in the model, you find appearance of the block like figure 2.7(c). Doubleclick the block, change its node number from default 10 to 1 in the parameter window, rightclick on the block, and click the **Flip Block** through the **Format** popup to model **node 1** of figure 2.7(f). Copy the **node 1** in the clipboard and paste it to see another **node 1** block. Rightclick the latter **node 1** and click **Rotate Block** through the **Format** popup to model the horizontal **node 1** connected to **Voltage Measurement1** in figure 2.7(f). Similarly we model the **node 2**. Finally we run the model to see the two node voltages in the blocks **V1** and **V2** respectively.

2.8 How to find a loop current in DC circuits?

Loop current seen in basic electrical circuits is also a special current which we determine by the same **Current Measurement** block of section 2.5. Actually we determine the branch current of an electrical circuit and that current serves the purpose of loop current. Loop current has only theoretical importance because it is a means of determining the branch current. We present one example on loop current in the following.

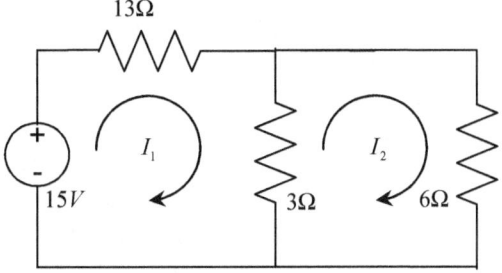

✦ Example

Figure 2.8(a) shows a simple series-parallel DC circuit with two loop currents I_1 and I_2. It is given that $I_1 = 1\ A$ and $I_2 = 0.3333\ A$ and our objective is to find these two loop currents through circuit simulation.

Figure 2.8(a) A series-parallel DC circuit with two loop currents

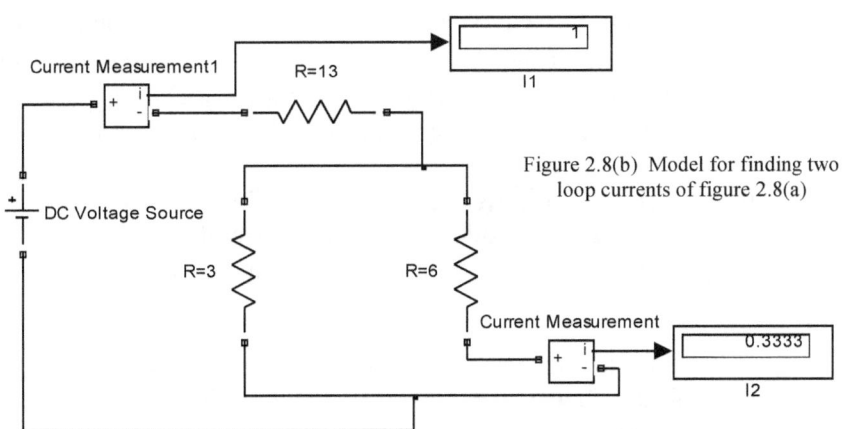

Figure 2.8(b) Model for finding two loop currents of figure 2.8(a)

Currents I_1 and I_2 are the branch currents through the resistors 13Ω and 6Ω of circuit in figure 2.8(a) respectively so if we determine these two currents employing two ammeters in two branches, we solve the problem. The main circuit is modeled as example 2 in section 2.5 (figure 2.5(h)) and modified circuit for the loop current determining is shown in figure 2.8(b). We copy the **Current Measurement** of figure 2.5(h) in the clipboard, paste it to see the **Current Measurement1**, copy the **Display** of figure 2.5(h) in the clipboard, paste it to see the **Display1**, rename the **Display** and **Display1** as **I2** and **I1** respectively. On connecting so mentioned blocks like figure 2.8(b), we click simulation icon ▶ at the icon bar and find the required loop currents I_1 and I_2 in **I1** and **I2** blocks respectively.

2.9 How to find the Thévenin equivalent of DC circuits?

Any two terminal DC electric circuit has its Thévenin equivalent. The equivalent is composed of two components – one ideal voltage source V_{eq} and one equivalent series resistance R_{eq}. Figure 2.9(a) shows the Thévenin equivalent between any two nodes A and B of an electrical circuit. In finding the Thévenin equivalent, following steps are adopted:

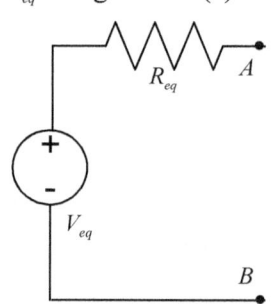

(a) we connect a voltmeter (i.e. **Voltage Measurement-Display** set) to given nodes (section 2.4) A and B and take the voltmeter reading (i.e. **Display** reading) as the ideal voltage source equivalent V_{eq} and

Figure 2.9(a) Thévenin equivalent between any two nodes A and B of an electrical circuit

(b) we find the equivalent resistance R_{eq} employing section 2.6 mentioned technique (i.e. by including **Constant-Controlled Current Source-Voltage Measurement-Display** set to circuit) but turning all voltage sources to $0V$ and all current sources to $0A$.

Both simulations use **Voltage Measurement-Display** set which must be connected to the given two nodes therefore the use of one set is enough. Let us go through the following examples on Thévenin equivalent.

◆ **Example 1**

We intend to find the Thévenin equivalent between nodes A and B of circuit in figure 2.4(c). It is given that the equivalent related to figure 2.9(a) is $V_{eq} = 0.6533\,mV$ (A is of +ve polarity) and $R_{eq} = 5.734\Omega$. Our objective is to find these two values through SIMULINK circuit simulation.

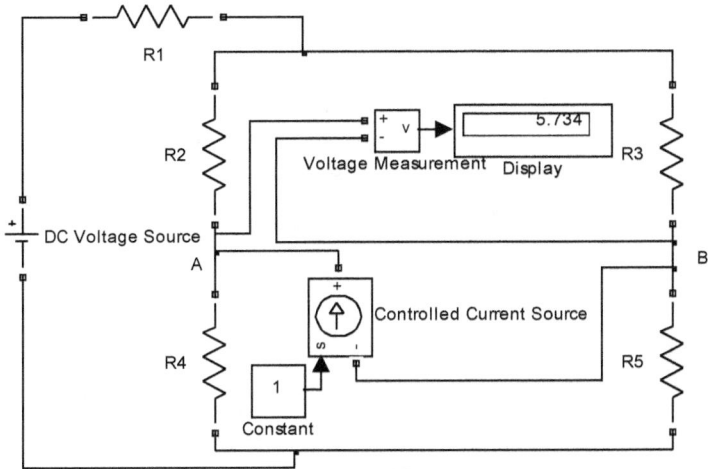

Figure 2.9(b) Modeling circuit of figure 2.4(c) for Thévenin equivalent

For V_{eq} :

This circuit is modeled in section 2.4. We also determined the voltage between the nodes A and B in that example actually that is the V_{eq} we are seeking for. However figure 2.9(b) depicts the model with additional block elements to determine both components of the Thévenin equivalent. Note that the **Voltage Measurement – Display** set of figure 2.4(d) is rearranged.

For R_{eq} :

The very next step is to connect a 1 A current source to the required A and B nodes by using **Controlled Current Source** and **Constant** blocks as shown in figure 2.9(b).

For the equivalent:

Each equivalent determining takes place one at a time. For the V_{eq} measurement, we doubleclick the **Constant** block, enter the Constant value as 0 in the parameter window, click the simulation icon ▶ to run the model, and find the V_{eq} value in **Display**. Again we doubleclick the **DC Voltage Source**, set its **Amplitude** as 0 in the parameter window to deactivate the source, enter the **Constant** value as 1, click the simulation icon ▶ to run the model,

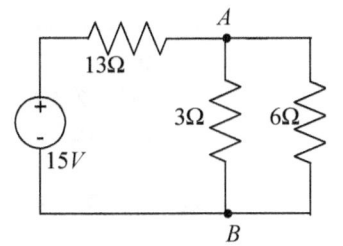

Figure 2.9(c) A series-parallel DC circuit

and find the R_{eq} value in **Display**.

✦ Example 2

Determine Thévenin equivalent between nodes A and B for the circuit in figure 2.9(c). It is provided that the equivalent related to figure 2.9(a) is $V_{eq} = 2\,V$ (A is of +ve polarity) and $R_{eq} = 1.7333\,\Omega$ which we intend to obtain from SIMULINK circuit simulation.

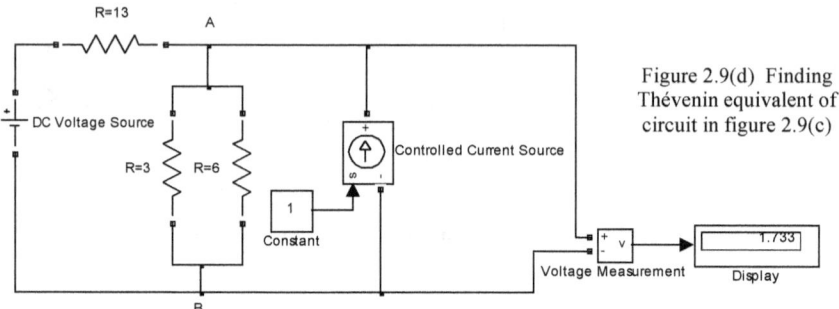

Figure 2.9(d) Finding Thévenin equivalent of circuit in figure 2.9(c)

For the circuit modeling:

We described modeling of circuit in figure 2.9(c) as example 2 of section 2.5 whose model is seen in figure 2.5(h). Modify the model in figure 2.5(h) to obtain the model in figure 2.9(d) which determines Thévenin equivalent for this example hence we delete the **Current Measurement** of figure 2.5(h) and bring one **Constant**, one **Voltage Measurement**, and one **Controlled Current Source** blocks in the model. Connect the blocks as shown in figure 2.9(d).

For the equivalent:

For the V_{eq} simulation, we doubleclick **Constant** of figure 2.9(d), enter the **Constant** value as 0 in the parameter window, click the simulation icon ▶ to run the model, and find the V_{eq} value in **Display**. Then we doubleclick the **DC Voltage Source**, set its **Amplitude** as 0 in parameter window to deactivate the source, enter **Constant** value as 1 on doubleclicking it, click the simulation icon ▶ to run the model, and find the R_{eq} value in

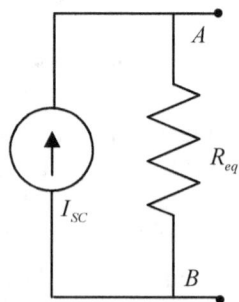

Figure 2.10(a) Norton equivalent between any two nodes A and B of an electrical circuit

Display.

2.10 How to find the Norton equivalent of DC circuits?

Like Thévenin equivalent, any two terminal electrical circuit has its Norton equivalent. There are two elements in the equivalent – one ideal current source I_{SC} and one parallel resistor R_{eq}. Figure 2.10(a) shows Norton equivalent between any two nodes A and B of an electrical circuit. In order for finding the Norton equivalent, followings are exercised:

(a) place a short circuit or an ideal ammeter to given two nodes A and B and take ammeter reading as the ideal current source equivalent I_{SC} and

(b) find the equivalent resistance R_{eq} employing section 2.6 mentioned technique but turning all voltage sources to $0V$ and current sources to $0A$ which are present in given circuit. This is exactly the resistance component of Thévenin equivalent discussed in last section. The only difference is the same resistance is now in parallel with I_{SC}.

In section 2.5 we elaborately explained how one determines the current through any branch of an electrical circuit by using the **Current Measurement – Display** set and apply here the same set to determine the

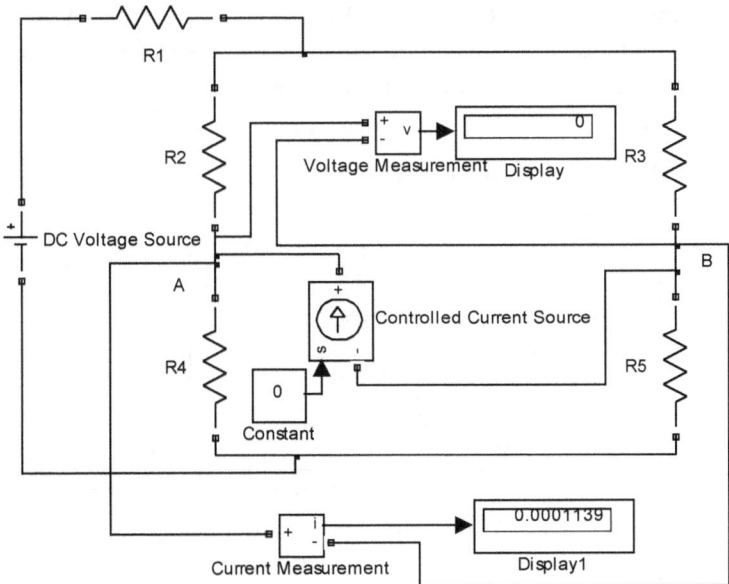

Figure 2.10(b) Model for finding Norton equivalent of circuit in figure 2.4(c)

I_{SC}. As done in the last section, we connect one **Constant – Controlled Current Source – Voltage Measurement – Display** blocks set for R_{eq} finding at given nodes A and B. Let us go through the following examples on Norton equivalent.

♦ Example 1

We wish to find the Norton equivalent for circuit in figure 2.4(c). It is given that the equivalent between nodes A and B of circuit in figure 2.4(c) is $I_{SC}=0.1139\,mA$ (direction is from A to B) and $R_{eq}=5.734\Omega$ related to figure 2.10(a). Our aim is to have these.

For the same circuit we found Thévenin equivalent as example 1 of last section. Additional blocks starting from the model in figure 2.9(b) we need are one **Current Measurement** and one **Display1** for I_{SC} as shown in figure 2.10(b) – model for this example, which are to be connected to the nodes A and B. Note that overlapping connection line between the node A and +ve of **Current Measurement** does not mean short circuit in the figure 2.10(b).

Equivalent finding:

Once again the **Constant-Controlled Current Source-Voltage Measurement-Display** blocks set of figure 2.10(b) is only for finding the equivalent resistance. For determining the I_{SC}, we doubleclick the **Constant** block, enter its **Constant** value as 0 in the parameter window keeping the **DC Voltage Source** value $26\,mV$ or **26e-3**, click the simulation icon ▶ to run the model, and find the I_{SC} value as **0.0001139** A in **Display1**. The R_{eq} finding is identical with that of example 1 of last section.

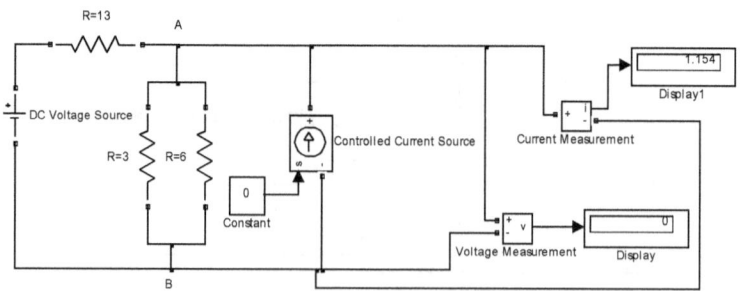

Figure 2.10(c) Model for finding Norton equivalent of circuit in figure 2.9(c)

♦ Example 2

Determine the Norton equivalent between the nodes A and B for circuit in figure 2.9(c). It is provided that the equivalent is $I_{SC}=1.154\,A$ (direction is from A to B) and $R_{eq}=1.7333\Omega$ which we aim to obtain.

Example 2 of last section explains the model (figure 2.9(d)) for Thévenin equivalent to which one **Current Measurement** and one **Display1** set is connected to the nodes A and B as shown in figure 2.10(c) – model for this example.

Equivalent finding:

We doubleclick the **Constant**, enter the **Constant** value as 0 in the parameter window, doubleclick the **DC Voltage Source**, enter the **Amplitude** value as **15**, click the simulation icon ▶ to run the model, and find the I_{SC} value as 1.154 in **Display1**. For R_{eq}, see the example 2 of last section because it is the same.

2.11 How to find a DC circuit power?

DC electric power dissipated by any resistor is basically the product of voltage across and current through the resistor. There is no specific block for DC electrical power simulation that is why we apply the **Product** (appendix B) block

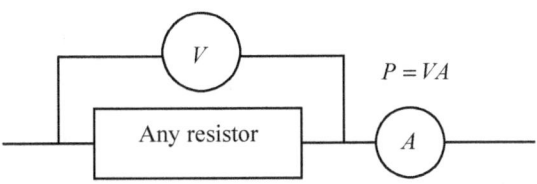

Figure 2.11(a) Voltmeter-Ammeter method for measurement of electrical power in a DC circuit

to multiply the voltage across and current through any resistor. Figure 2.11(a) presents general voltmeter-ammeter method for measurement of DC electrical power. The voltmeter and ammeter readings we obtain by using **Voltage Measurement** and **Current Measurement** respectively. Emphasis should be given to the connection that the **Voltage Measurement** and **Current Measurement** must be connected in parallel and series to the resistor whose power is to be determined respectively.

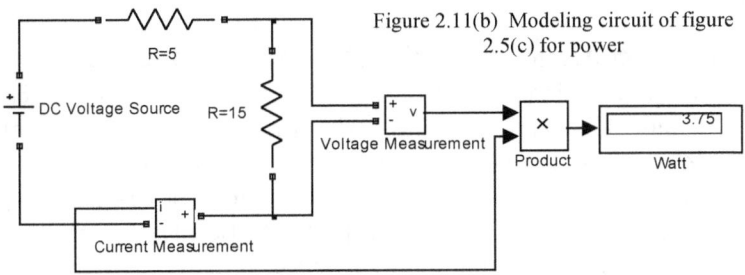

Figure 2.11(b) Modeling circuit of figure 2.5(c) for power

✦ **Example 1**

Figure 2.5(c) mentioned circuit says that the electrical power absorbed by the resistor 15Ω is 3.75 watts which we wish to obtain.

The circuit is modeled as example 1 in section 2.5 (figure 2.5(e)). In that model we bring one **Voltage Measurement** and one **Product** blocks and rename the **Display** block as **Watt** to be consistent with the modeling. We connect additional blocks like figure 2.11(b) – model for this example, click the simulation icon ▶ to run the model, and find the electrical power value 3.75 in **Watt** block. Note that the overlapping connection emerging from the i of **Current Measurement** in figure 2.11(b) does not mean a short circuit.

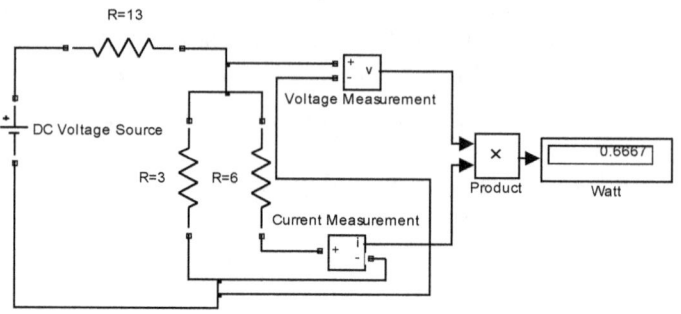

Figure 2.11(c) Electric power determining for circuit in figure 2.5(f)

✦ **Example 2**

The 6Ω resistor of circuit in figure 2.5(f) dissipates 0.6667 watts electrical power which we intend to obtain through SIMULINK circuit simulation.

The circuit is modeled as example 2 in section 2.5 – figure 2.5(h) is model for the circuit. Additionally we need one **Voltage Measurement** and one **Product** blocks in the model and rename the **Display** as **Watt**. Figure 2.11(c) presents the modified model starting from the one in figure 2.5(h). On connecting the additional blocks, we click the simulation icon ▶ at the icon bar to run the model and find the electrical power value 0.6667 in **Watt** block.

2.12 How to model an inductor and a capacitor?

An alternating current or AC circuit load elements are composed of resistors, inductors, and capacitors which are connected in various circuital forms. In section 2.1 we introduced the **Series RLC Branch** which in general models resistance-inductance-capacitance or R-L-C. Figure 2.1(b) shown parameter window has the property **Branch type**, from the popup you can select **L** for inductor and **C** for capacitor.

We intend to model an inductor of value $L=3\,H$ so we bring a **Series RLC Branch** block (appendix B) in a

L

Figure 2.12(a) Appearance of an inductor

C

Figure 2.12(b) Appearance of a capacitor

-50-

new SIMULINK model file (figure 2.1(a)). Doubleclick the block, select **L** from the **Branch type** popup, enter its **Inductance** as **3**, rename (subsection 1.2.3) the **Series RLC Branch** as **L** therefore we find the modeled inductor like figure 2.12(a).

Again in order to model a capacitor of $4.7\,F$, we bring a **Series RLC Branch** block in a new SIMULINK model file, doubleclick the block, select **C** from the **Branch type** popup, enter its **Capacitance** as **4.7**, rename the **Series RLC Branch** as **C** therefore we find the modeled capacitor like figure 2.12(b). Scale factor can be used to model the elements other than default values. For instance the elements of values $2.5\,mH$, $5.6\,\mu H$, $5.7\,\mu F$, $6.8\,pF$, and $2.3\,nF$ are entered by writing **2.5e-3, 5.6e-6, 5.7e-6, 6.8e-12,** and **2.3e-9** in parameter window slots respectively (appendix A).

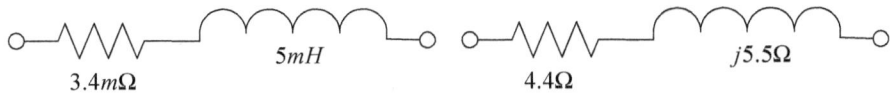

Figure 2.12(c) An AC circuit consisted of R - L load in absolute units	Figure 2.12(d) An AC circuit consisted of R - L load in reactance form

2.13 How to model an AC circuit?

Primarily an AC circuit other than source elements comprises with resistors, inductors, and capacitors. Since inductors and capacitors have frequency dependency, given circuit specification can be in absolute unit or in reactance form. We address some elementary AC circuit modeling in the following.

Figure 2.12(e) Model of AC circuit in figure 2.12(c)

♦ **Example 1**

We wish to model the circuit of figure 2.12(c) in which one resistor and one inductor are connected in series. Sections 2.1 and 2.12 explain the modeling of resistor only and inductor only respectively. The reader may say why do not we set the **Resistance, Inductance, and Capacitance** values as **3.4e-3, 5e-3,** and **inf** in parameter window of **Series RLC Branch** respectively? This option is okay if we do not need to connect any voltmeter between the junction point or node of the resistor and inductor. If we need the node to be connected with some other node, it is compulsory that we model the resistor and inductor separately. Considering separate modeling, we carry out the following to model the circuit of figure 2.12(c):

⇒ open a new SIMULINK model file (subsection 1.2.2), bring a **Series RLC Branch** in the model (appendix B), rename the block as **R**, doubleclick the **R**, select **R** from the **Branch type** popup, and enter its **Resistance** as 3.4e-3 to model only the resistor,

⇒ bring another **Series RLC Branch** in the model, rename the block as L, doubleclick the L, select L from the **Branch type** popup, and enter its **Inductance** as 5e-3 in the parameter window to model only the inductor, and

⇒ place the two blocks relatively and connect them like figure 2.12(e) to finish modeling of AC circuit in figure 2.12(c).

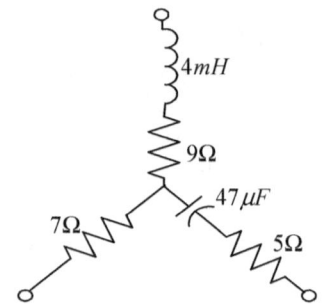

Figure 2.12(f) A three phase AC circuit

❖ **Example 2**

Referring to figure 2.12(d), the two elements are now given in reactance form for which knowing the value of circuit frequency is mandatory say 60 *Hz* so values of the **Resistance** and **Inductance** we enter as **4.4** and **5.5/2/pi/60** in the parameter window of example 1 respectively (because $X_L = 2\pi f L$, appendix A).

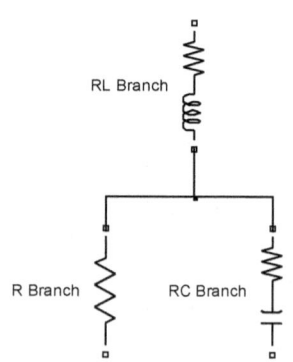

Figure 2.12(g) Model of three phase circuit in figure 2.12(f)

❖ **Example 3**

A three phase AC circuit is shown in figure 2.12(f). Although capacitor of the circuit is inclined, we model it vertically in SIMULINK model. The two inclined resistors of figure 2.12(f) are also modeled vertically however we conduct the following steps to perform the modeling (assuming that the resistors in the inductor and capacitor branches are lumped):

⇒ open a new SIMULINK model file, bring a **Series RLC Branch** block in the model, copy and paste the block twice through the clipboard to see other two blocks by names **Series RLC Branch1** and **Series RLC Branch2**, rename **Series RLC Branch** as RL Branch, doubleclick the RL Branch, select RL from **Branch type** popup, enter its **Resistance** as 9 and **Inductance** as 4e-3, rightclick on RL Branch, and click **Rotate Block** under Format popup to model it vertically,

⇒ rename the **Series RLC Branch1** as R Branch, doubleclick the block, select R from **Branch type** popup, enter its **Resistance** as

7, rightclick on **R Branch**, and click the **Rotate Block** under **Format** popup to model the block vertically,

⇒ rename the **Series RLC Branch2** block as **RC Branch**, doubleclick the block, select **RC** from **Branch type** popup, enter its **Resistance** as **5** and **Capacitance** as **47e-6**, rightclick on the **RC Branch**, and click the **Rotate Block** under **Format** popup to model the block vertically, and

⇒ finally place so designed blocks relatively and connect them like figure 2.12(g) to finish the modeling.

2.14 How to model an AC circuit source element?

An elementary AC circuit mostly contains two ideal sources – voltage and current. Figures 2.13(a) and 2.3(e) present the schematic symbols of the voltage and current sources respectively.

The block AC Voltage Source simulates the voltage source of figure 2.13(a)

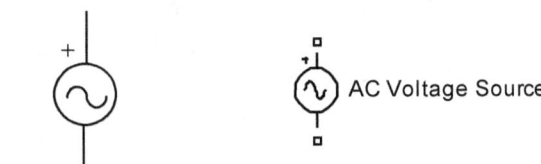

Figure 2.13(a) Schematic symbol of an ideal AC voltage source

Figure 2.13(b) Model of ideal AC voltage source

whose icon appearance is shown in figure 2.13(b). Open a new SIMULINK model file (section 1.2), bring the **AC Voltage Source** in the model (appendix B), and doubleclick the block to see its parameter window like figure 2.13(c). In the parameter window you find the slots for **Peak amplitude** in V, **Phase** in degrees, and **Frequency** in Hz of voltage source to be modeled.

Suppose we intend to model an AC voltage source with specifications peak amplitude $50\,V$ and frequency $50\,Hz$ therefore we enter **50** and **50** in slots of the **Peak amplitude** and **Frequency** deleting default values and leaving other settings unchanged in parameter window of figure 2.13(c) respectively.

Frequently the AC voltage source values are given in terms of the phasor form for instance $75V\angle-30^0$ we enter by writing **75**

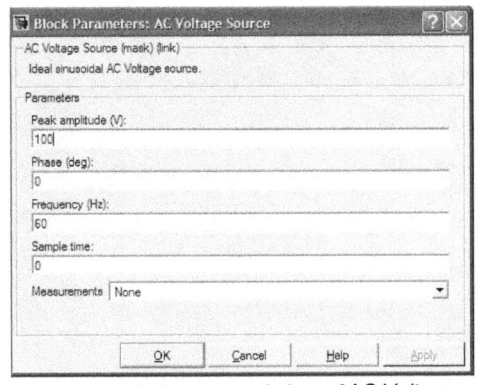

Figure 2.13(c) Parameter window of AC Voltage Source

and **−30** in the slots of **Peak amplitude** and **Phase** in parameter window of figure 2.13(c) respectively. If the given phasor value is in RMS form, we multiply the phasor by $\sqrt{2}$ to turn that to peak value therefore we enter **75*sqrt(2)** as the **Peak amplitude** when the $75V\angle-30^0$ is in RMS form. In the case of other unit than volt we use scale factor for instance peak amplitudes $75mV\angle-30^0$ and $75KV\angle-30^0$ are entered by writing **75e-3** and **75e3** respectively.

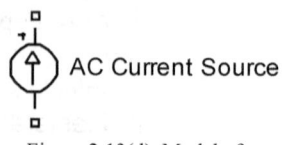 AC Current Source

Figure 2.13(d) Model of an ideal AC current source

Figure 2.3(e) shows the schematic symbol of an ideal AC current source whose SIMULINK counterpart is **AC Current Source** and whose icon appearance is presented in figure 2.13(d). Open a new SIMULINK model file, bring the **AC Current Source** in the model, and doubleclick the block to see its parameter window like figure 2.13(e). The current source parameter entering is similar to that of the **AC Voltage Source** which we just discussed. The point is we

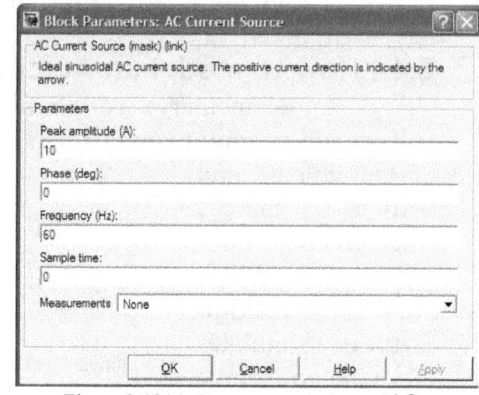

Figure 2.13(e) Parameter window of AC Current Source

enter all parameter specifications for a voltage in the **AC Voltage Source** but now we do the same for the current.

We wish to terminate the chapter with the discussion on AC source elements.

Exercises

1. Model each following electrical circuit in SIMULINK: (a) figure E2.1(a) with $R_1 = 7\Omega$ and $R_2 = 9\Omega$ (b) figure E2.1(b) with $R_1 = 7\Omega$, $R_2 = 5\Omega$, $R_3 = 2\Omega$, and $R_4 = 9\Omega$ (c) figure E2.1(c) based on indicated resistor values in the figure (d) figure E2.1(d) based on indicated resistor values in the figure.

2. In question (1) determine the equivalent resistance for each part using SIMULINK modeling.

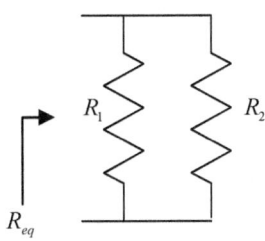

Figure E2.1(a) Two resistors in parallel

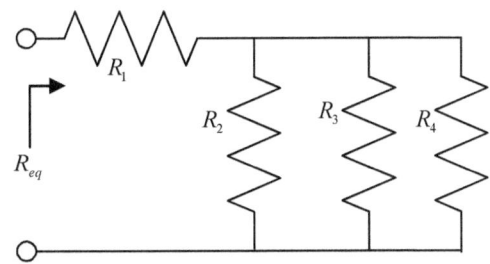

Figure E2.1(b) Three parallel resistors in series with another resistor

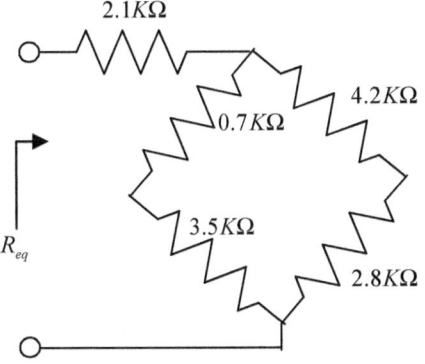

Figure E2.1(c) A bridge network of resistors

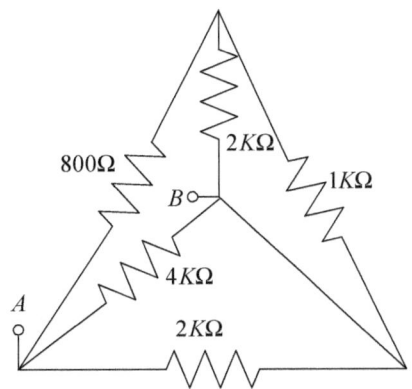

Figure E2.1(d) A network of interconnected resistors

3. Model a DC voltage source of $20\,V$ in SIMULINK. Do the same for $50\,mV$. Model a direct current source of $10\,A$ in SIMULINK. Do the same for $30\,\mu A$.

4. Model the single voltage source circuit of figure E2.2(a) in SIMULINK. Determine the following voltages in SIMULINK: (a) V_{ab} (b) V_{ca} (c) V_{cb}.

5. In the circuit of figure E2.2(b), a one ampere direct current source is connected to nodes A and B. Apply SIMULINK modeling to verify that the voltages between various nodes are the following: $V_{AB} = -2.338\,V$, $V_{CA} = 1.743\,V$, $V_{CE} = 0.533\,V$, $V_{DE} = 0.307\,V$, $V_{CF} = 1.336\,V$, $V_{BE} = 1.129\,V$, $V_{EA} = 1.210\,V$, $V_{BF} = 1.931\,V$, $V_{BD} = 0.821\,V$, $V_{DF} = 1.110\,V$, and $V_{AF} = -0.408\,V$.

6. Figure E2.3(a) shows some branch currents. Verify employing SIMULINK technique that these branch currents have the values as follows: $I_1 = 39.73\,mA$, $I_2 = -0.1648\,mA$, $I_3 = -0.01\,A$, and $I_4 = 10.16\,mA$.

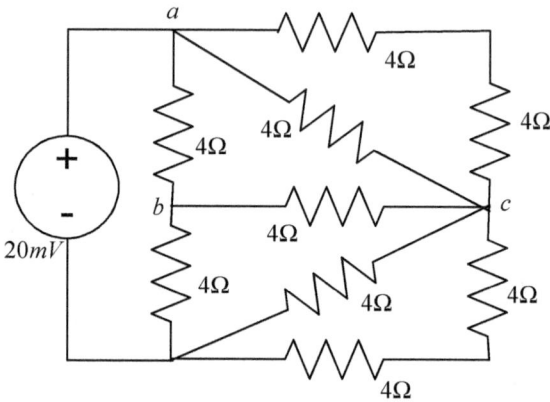

Figure E2.2(a) An electrical circuit with a single voltage source

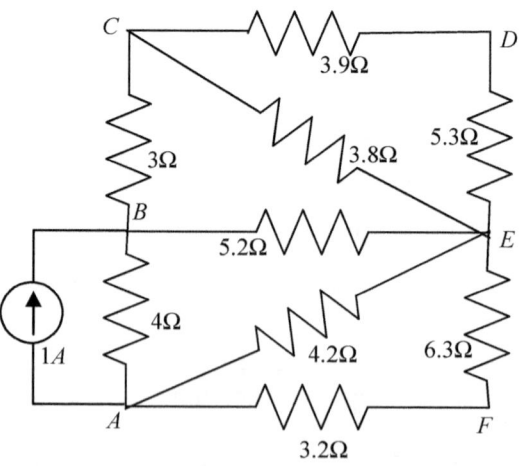

Figure E2.2(b) Circuit with a DC current source

7. Referring to the circuit of figure E2.3(b), the four node voltages are as follows: $V_1 = -1.857\,V$, $V_2 = -76.857\,V$, $V_3 = -76.429\,V$, and $V_4 = -102.429\,V$. Employ SIMULINK modeling to verify these voltages.

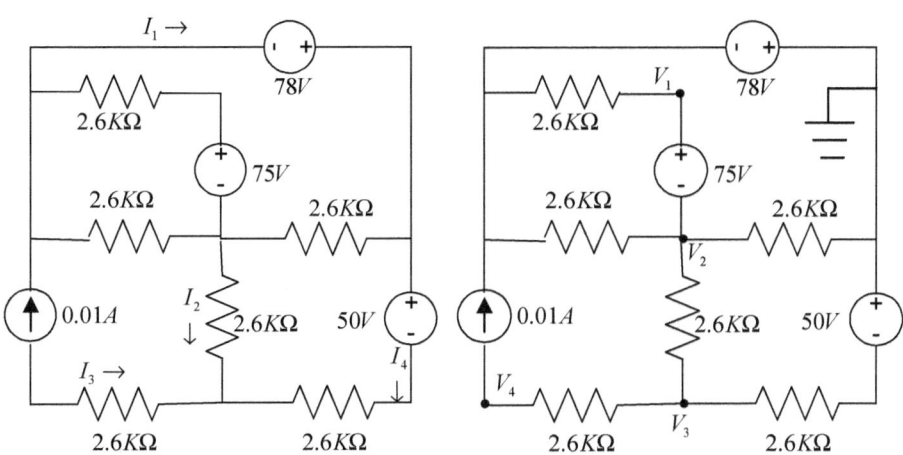

Figure E2.3(a) A DC circuit with three voltage and one current sources

Figure E2.3(b) A DC circuit with four node voltages

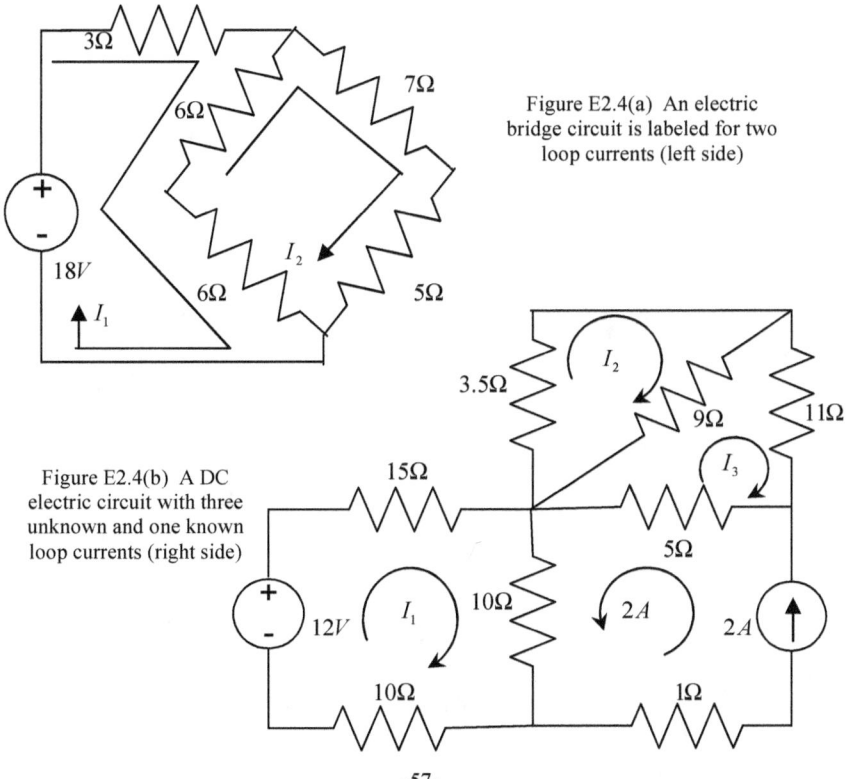

Figure E2.4(a) An electric bridge circuit is labeled for two loop currents (left side)

Figure E2.4(b) A DC electric circuit with three unknown and one known loop currents (right side)

8. Model the following DC circuits in SIMULINK and verify the loop currents:

   (a) circuit of figure E2.4(a) has the loop currents $I_1 = 2\,A$ and $I_2 = 1\,A$ and

   (b) the three loop currents in circuit of figure E2.4(b) are $I_1 = -0.2286\,A$, $I_2 = -0.3888\,A$, and $I_3 = -0.54\,A$.

9. Following DC circuits are labeled according to nodal voltage theory. Model each circuit in SIMULINK and verify the node voltages: (a) the circuit of figure E2.5(a) has the nodal voltages $V_1 = 2.9189\,V$, $V_2 = 12.5676\,V$, $V_3 = 1.1351\,V$, and $V_4 = -2.4324\,V$ and (b) the

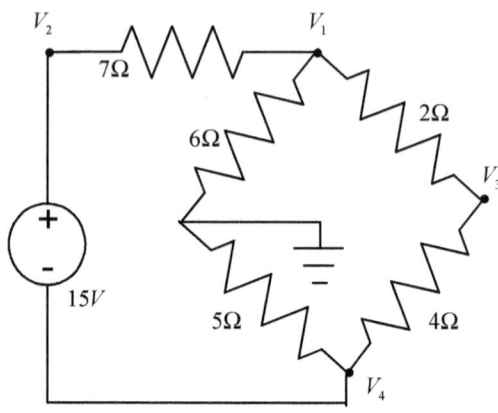

Figure E2.5(a) An electric bridge circuit is labeled for four node voltages

circuit of figure E2.5(b) has the seven nodal voltages as $V_1 = 6.5676\,V$,

$V_2 = 4.6216\,V$, $V_3 = -21.0541\,V$, $V_4 = -12.0541\,V$, $V_5 = -9.0811\,V$, $V_6 = -2.0811\,V$, and $V_7 = -2.4324\,V$.

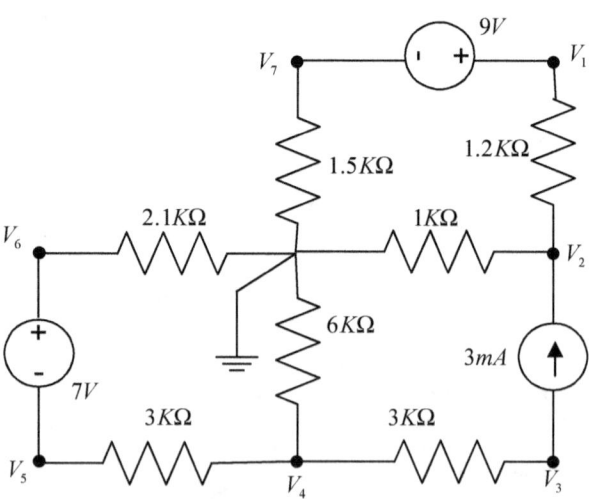

Figure E2.5(b) A multi-resistor and multi-source DC electric circuit with seven node voltages

10. Determine the Thevenin equivalent between the nodes C and F employing SIMULINK model for the circuit in figure E2.2(b).

11. Determine the Thevenin equivalent between the nodes V_1 and V_2 employing SIMULINK model for the circuit in figure E2.5(b).

12. Determine the Norton equivalent between the nodes C and F employing SIMULINK model for the circuit in figure E2.2(b).

13. Determine the Norton equivalent between the nodes V_1 and V_2 employing SIMULINK model for the circuit in figure E2.5(b).

14. Employing SIMULINK model, determine the electrical power dissipated by the resistor between the nodes C and D of circuit in figure E2.2(b).

15. Employing SIMULINK model, determine the electrical power dissipated by the resistor between the nodes V_1 and V_2 of circuit in figure E2.5(b).

16. Model the AC circuit of figure E2.5(c) in SIMULINK.

17. Subject to a frequency $60Hz$, model the AC circuit of figure E2.5(d) in SIMULINK.

18. Model an AC voltage source of peak amplitude $110V$ and frequency $60Hz$ in SIMULINK.

19. Model an alternating current (AC) source of peak amplitude $7A$ and frequency $60Hz$ in SIMULINK.

20. Model each following AC source in SIMULINK: (a) $110V\angle-60^0$ (b) $50mV\angle40^0$ (c) $8A\angle60^0$ (d) $4.6\mu A\angle75^0$.

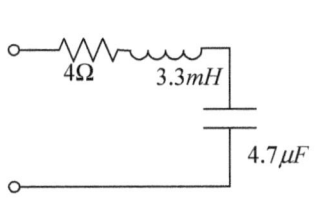

Figure E2.5(c) A series AC circuit

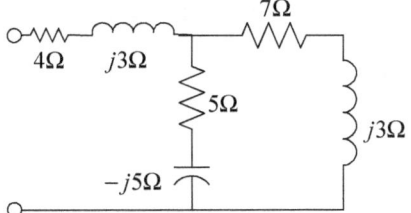

Figure E2.5(d) A series-parallel AC circuit

Answers:

(1) Hint: sections 2.1 and 2.2

(2) (a) $R_{eq} = 3.9375\Omega$ (b) $R_{eq} = 8.2329\Omega$ (c) $R_{eq} = 4.725\ K\Omega$ (d) $R_{eq} = 698.4127\Omega$ Hint: section 2.6

(3) Hint: section 2.3

(4) (a) $V_{ab} = 10\ mV$ (b) $V_{ca} = -10\ mV$ (c) V_{cb} = extremely small number or $0\ mV$ Hint: section 2.4

(5) Hint: section 2.4

(6) Hint: section 2.5

(7) Hint: section 2.7

(8) Hint: section 2.8

(9) Hint: section 2.7

(10) $V_{TH} = 1.336\ V$ with C +ve and F −ve and $R_{TH} = 4.354\Omega$ Hint: section 2.9

(11) $V_{TH} = 1.947\ V$ with V_1 +ve and V_2 −ve and $R_{TH} = 810.8\Omega$ Hint: section 2.9

(12) $I_{SC} = 0.3068\ A$ flowing C to F and $R_{TH} = 4.354\Omega$ Hint: section 2.10

(13) $I_{SC} = 2.4\ mA$ flowing V_1 to V_2 and $R_{TH} = 810.8\Omega$ Hint: section 2.10

(14) $P = 0.01311\ W$ Hint: section 2.11

(15) $P = 3.156\ mW$ Hint: section 2.11

(16) Hint: section 2.13

(17) Hint: section 2.13

(18) Hint: section 2.14

(19) Hint: section 2.14

(20) Hint: section 2.14

Chapter 3

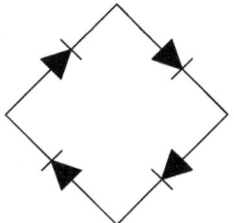

Modeling Diode Circuits

In this chapter we mainly simulate PN junction semiconductor diode circuits of elementary level. It is well known that a PN junction diode is a one-way device offering low resistance when forward-biased and behaving almost as an insulator when reverse-biased. As a two terminal device, the simulation is conducted mostly on ideal volt-ampere characteristic of the diode. Embedded SIMULINK blocks help us model the following:

- ❤ ❤ PN junction semiconductor diode and its basic circuits
- ❤ ❤ Diode based clipping and capacitive filtering circuits
- ❤ ❤ Zener diode circuits and rectification of AC signals
- ❤ ❤ Simulation on the measurement of diode circuits

3.1 How to model a diode in SIMULINK?

Diode is a frequently seen component in electronic circuits which we model by SIMULINK block **Diode**. Link of the block is found in appendix B. Figure 3.1(a) shows the schematic symbol of a diode D. Open a new SIMULINK model file (subsection 1.2.2) and bring the **Diode** in the model

Mohammad Nuruzzaman

file. Figure 3.1(b) shows appearance of the block. On doubleclicking the Diode, we see the parameter window as presented in figure 3.1(d). In electronic circuitry a diode has user-supplied forward bias voltage V_γ which we enter at the **Forward voltage (Vf) V** slot in figure 3.1(d). For an ideal diode this forward voltage is zero. Values of the **Snubber resistance Rs (Ohms)** and **Snubber capacitance Cs (F)** should be infinity (has MATLAB code **inf**) and 0 for an ideal diode but numerical nature of modeling sometimes requires some finite values respectively. For numerical computation reason let us put some small finite value in the slot for example the default one is 0.001Ω.

Assume that the forward biased voltage V_γ of a diode is $0.5\,V$ then we enter the **Forward voltage (Vf) V**, **Snubber capacitance Cs (F)**, and **Snubber resistance Rs** in figure 3.1(d) as **0.5**, **0**, and **inf** deleting the default values respectively. Since the **Diode** is value adaptive, figure 3.1(c) shows appearance of the block following the values' entering therefore figure 3.1(c) represents the model of diode seen in figure 3.1(a) with the same polarity.

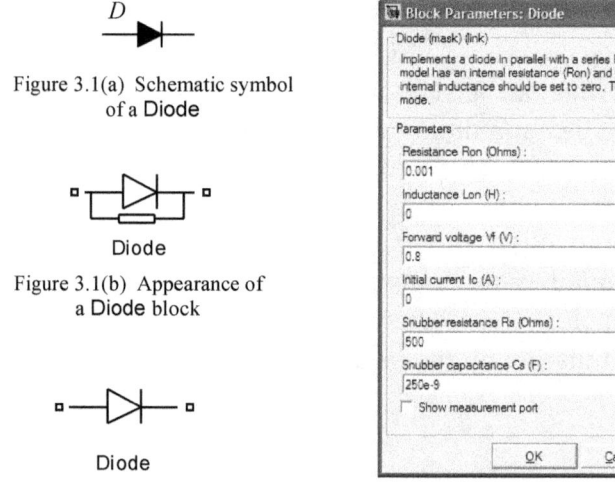

Figure 3.1(a) Schematic symbol of a Diode

Figure 3.1(b) Appearance of a Diode block

Figure 3.1(c) Adapted appearance of Diode

Figure 3.1(d) Block parameter window of Diode

3.2 Modeling a basic diode circuit

A basic diode circuit is composed of one AC source, one diode, and one resistor as seen in figure 3.2(a). Modelings of a resistor and an AC voltage source are addressed in sections 2.1 and 2.14 respectively.

It is given that diode D of figure 3.2(a) is having $0V$ forward bias voltage V_γ and $R=100\Omega$. When an input AC voltage V_i as seen in figure 3.2(b) is applied to the diode circuit, the output voltage V_0 across the R should be like figure 3.2(c). We wish to implement this circuit behavior.

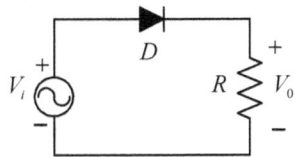

Figure 3.2(a) A basic Diode circuit

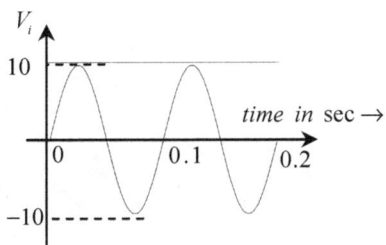

Figure 3.2(b) Plot of input AC source voltage in figure 3.2(a)

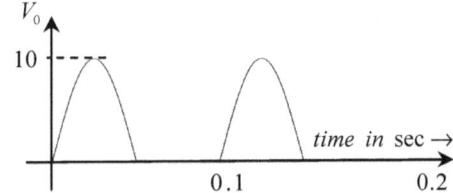

Figure 3.2(c) Output of diode circuit in figure 3.2(a)

Figure 3.2(d) shows model for the basic diode circuit of figure 3.2(a). In order to obtain the model, let us carry out the following:

⇒ open a new SIMULINK model file (subsection 1.2.2) and

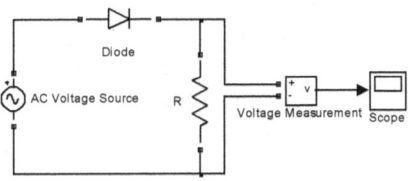

Figure 3.2(d) Model of basic Diode circuit

⇒ get one AC Voltage Source (for V_i), one Diode (for D), one Series RLC Branch (for R), one Voltage Measurement (for V_0), and one Scope blocks in the model (appendix B).

Figure 3.2(b) says that the V_i is having amplitude $10V$ and time period $T=0.1$sec. Frequency of the input sine wave should be $f=\dfrac{1}{T}=10\,Hz$ so to model the V_i do the following:

⇒ doubleclick the AC Voltage Source and enter the values of Peak amplitude (V) and Frequency (Hz) as 10 and 10 in parameter window respectively.

Given resistor value is 100Ω whose modeling takes place as follows:

⇒ rename the Series RLC Branch as R (subsection 1.2.3) to make sense with given element, doubleclick the R, select R from the Branch type popup, enter the Resistance as 100 in parameter window, rightclick on the R, and click the **Rotate Block** under the **Format** popup to model a vertical resistor in doing so the R modeling is finished.

For the ideal diode modeling we carry out the following:

⇒ doubleclick the **Diode** and enter the values of **Forward voltage (Vf) V**, **Snubber resistance Rs (Ohms)**, and **Snubber capacitance Cs (F)** as **0**, **inf**, and **0** in parameter window respectively (section 3.1).

A diode is a nonlinear device and its model needs stiff differential equation solver so click the menu **Simulation** (section 1.2.1) down **Configuration parameters** in model file. Prompt window of figure 1.7(g) appears and select **ode23s** (which solves stiff differential equations) as model **Solver**. Figure 3.2(b) says that the V_i duration is 0.2 secs or the interval is $0 \leq t \leq 0.2$ secs therefore we enter the stop time as **0.2** in slot of **Stop time** of window in figure 1.7(g). After that

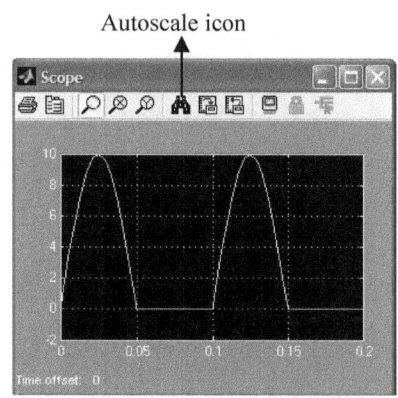

Figure 3.2(e) **Scope** output for diode circuit in figure 3.2(a)

⇒ place the blocks relatively like figure 3.2(d), connect them as shown in the figure, and click the simulation icon ▶ to run the model and

⇒ then doubleclick the **Scope** to see voltage output with autoscale setting (run time axis and default axis settings might be different that is why autoscale setting is essential) as shown in figure 3.2(e).

Certainly figure 3.2(e) displayed output corresponds to the V_0 wave shape of figure 3.2(c).

The simulated circuit behavior illustrates an ideal diode implementation that is why the voltage in figure 3.2(e) is from 0 to $10V$.

What if we have the diode D of figure 3.2(a) with a forward biased voltage $V_\gamma = 1V$? Under this condition we expect the wave shape output should be from 0 to $9V$. In order to

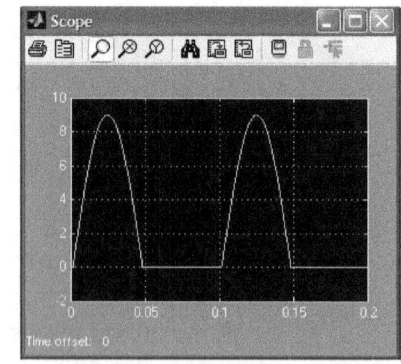

Figure 3.2(f) **Scope** output for diode circuit of figure 3.2(a) with forward bias voltage

include the forward bias voltage we doubleclick the **Diode** of figure 3.2(d) and enter the **Forward voltage (Vf)** V as 1 in the parameter window. Run the model again by clicking the simulation icon ► and the **Scope** output with autoscale setting is shown in figure 3.2(f). Now the **Scope** output is from 0 to $9V$ as expected.

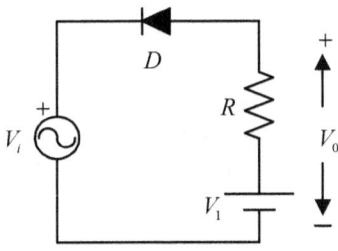

Figure 3.3(a) A clipping circuit

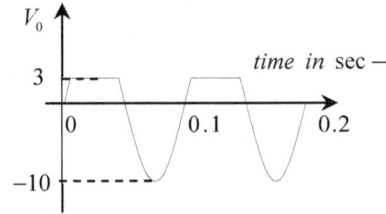

Figure 3.3(b) Plot of output V_0 of clipping circuit in figure 3.3(a)

3.3 Clipping circuit modeling

Clipping circuits transform any arbitrary wave shape (voltage or current) variation to user-defined reference level. The clipping operation can be above or below the reference level. These circuits are also referred to as voltage or current limiters. SIMULINK helps us easily implement the clipping circuit behavior for which following two examples are demonstrated.

✦ Example 1

Figure 3.3(a) shows a single level clipping circuit. In the circuit suppose forward bias voltage V_γ of diode is $0V$ and input V_i to the circuit is the wave shape of figure 3.2(b). Circuit analysis says that V_0 wave shape output from the circuit in figure 3.3(a) follows the variation as in figure 3.3(b) when $R = 1\Omega$ and $V_1 = 3V$ — we wish to simulate this circuit response.

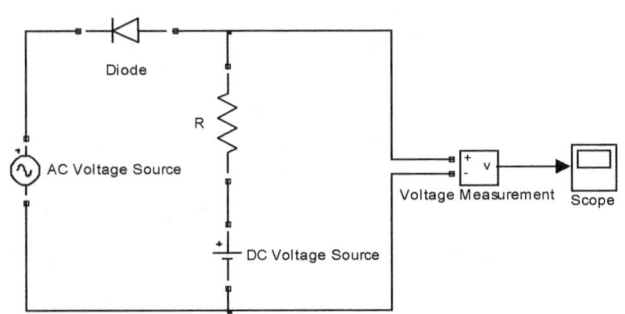

Figure 3.3(c) Model of clipping circuit in figure 3.3(a)

Model of figure 3.3(c) shows the implementation of clipping circuit. In order to construct the model we go through the following steps:

⇒ open a new SIMULINK model file (section 1.2.2),

⇒ get one AC Voltage Source (for V_i), one DC Voltage Source (for V_1), one Diode (for D), one Series RLC Branch (for R), one Voltage Measurement (for V_0), and one Scope blocks in the model file (appendix B),

⇒ rename the Series RLC Branch as R (subsection 1.2.3) to make sense with given element, doubleclick the R, select R from the Branch type popup, enter its Resistance as 1 in parameter window (section 2.1), rightclick on the R, and click the Rotate Block under the Format popup to model a vertical resistor,

⇒ doubleclick the Diode and enter values of the Resistance Ron (Ohms), Forward voltage (Vf) V, Snubber resistance Rs (Ohms), and Snubber capacitance Cs (F) as 1e-10, 0, inf, and 0 in the parameter window respectively (section 3.1) – the Ron resistance must be very small because the series resistance is 1Ω and we chose 10^{-10} which has the code 1e-10,

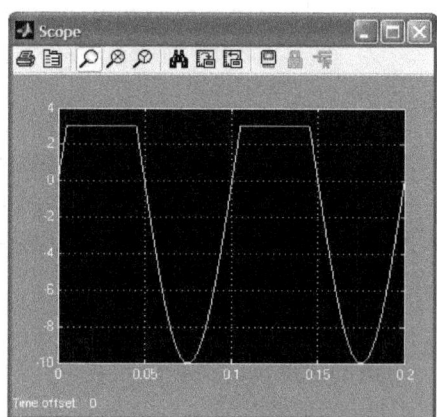

Figure 3.3(d) Scope return for the clipping circuit of figure 3.3(a)

⇒ rightclick on the Diode and click the Flip Block under the Format popup to model the diode orientation of figure 3.3(a),

⇒ doubleclick the AC Voltage Source and enter values of the Peak amplitude (V) and Frequency (Hz) as 10 and 10 in parameter window (section 2.14, time period and frequency of input V_i in figure 3.2(b) are 0.1sec and 10 Hz) respectively,

⇒ doubleclick the DC Voltage Source and enter value of Amplitude (V) as 3 in parameter window for entering V_1,

⇒ click the menu Simulation down the Configuration parameters at the menu bar, select ode23s as the model Solver, and enter the stop time as 0.2 at the slot of Stop time (see last section),

⇒ place the blocks relatively like figure 3.3(c) and connect them as shown in the figure and click the simulation icon ▶ to run the model, and

⇒ finally doubleclick the **Scope** to see the voltage output with autoscale setting (figure 3.2(e)) as shown in figure 3.3(d).

As expected, the figure 3.3(d) shown output confirms the V_0 wave shape of figure 3.3(b).

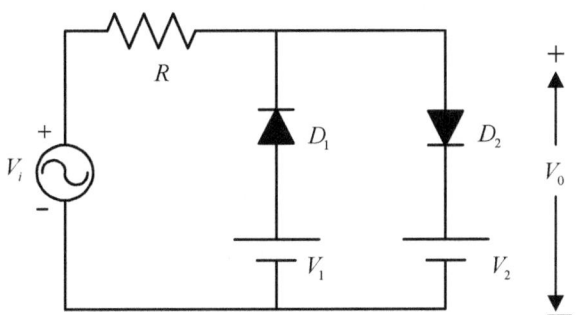

Figure 3.3(e) A diode circuit with two clipping levels

✦ **Example 2**

In example 1 we demonstrated one level clipping. As an example of two level clipping, diode circuit of figure 3.3(e) can be applied.

Diodes D_1 and D_2 of figure 3.3(e) are ideal. The R, V_1, and V_2 in the circuit are 1Ω, $2V$, and $6V$ respectively. When input V_i to the circuit is the wave shape of figure 3.2(b) and $V_2 > V_1$, the V_0 wave takes the shape of figure 3.3(f) which we intend to verify by SIMULINK circuit modeling and do so by the following:

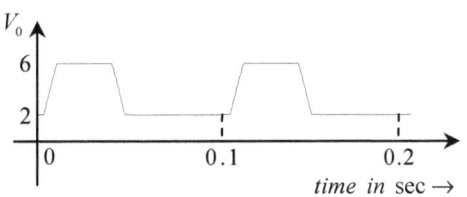

Figure 3.3(f) Output V_0 from clipping circuit in figure 3.3(e)

⇒ open a new SIMULINK model file,

⇒ get one **AC Voltage Source** (for V_i), one **DC Voltage Source** (for V_1), one **Diode** (for D_1), one **Series RLC Branch** (for R), one **Voltage Measurement** (for V_0), and one **Scope** blocks in the model file,

⇒ rename the **Series RLC Branch** block as R, doubleclick the R, select R from the **Branch type** popup, and enter its **Resistance** as 1 in the parameter window,

⇒ doubleclick the **Diode**, enter the values of the **Resistance Ron (Ohms)**, **Forward voltage (Vf) V**, **Snubber resistance Rs (Ohms)**, and **Snubber capacitance Cs (F)** as 1e-10, 0, inf, and 0 in the parameter window respectively, rename the **Diode** as D1, rightclick on the D1, click the **Flip Block** under the

Format popup, rightclick again on the D1, and click the Rotate Block under the Format popup (last two actions for the orientation of D_1 in figure 3.3(e)),

⇒ copy just modeled D1 in the clipboard, paste it to see the D2, rightclick on the D2, and click the Flip Block under Format popup to model D_2 of the circuit,

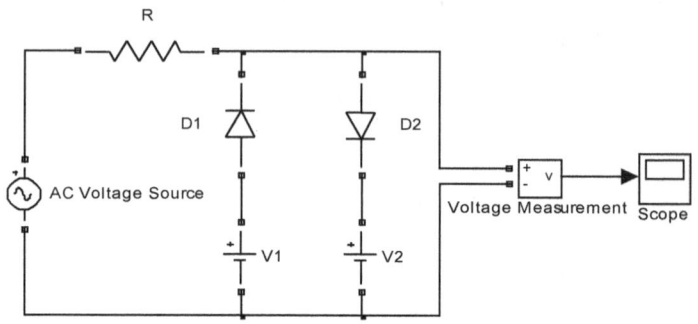

Figure 3.3(g) Model of clipping circuit in figure 3.3(e)

⇒ rename the DC Voltage Source as V1, doubleclick the V1, enter its Amplitude (V) as 2 for V_1 in the parameter window, copy just modeled V1 in the clipboard, paste it to see V2, doubleclick the V2, and enter its Amplitude (V) as 6 for V_2 in the parameter window,

⇒ doubleclick the AC Voltage Source and enter the values of the Peak amplitude (V) and Frequency (Hz) as 10 and 10 in the parameter window for V_i respectively, and

Figure 3.3(h) Scope return for clipping circuit of figure 3.3(e) corresponding to output V_0

⇒ click the menu Simulation down Configuration parameters at the menu bar, select ode23s as the model Solver, and enter the stop time as 0.2 in slot of the Stop time.

Figure 3.3(g) is the model we are looking for circuit in figure 3.3(e). Place the blocks relatively and connect them as shown in figure 3.3(g) and click the simulation icon ▶ to run the model. Then doubleclick the Scope to see the voltage output with autoscale setting as shown in figure 3.3(h). Obviously the return in figure 3.3(h) confirms the expected V_0 shape of figure 3.3(f).

In a similar fashion any other diode-applied clipping circuit can easily be implemented.

3.4 Filter circuit modeling

A diode circuit output is usually not a perfect DC frequently fluctuating one. It is a common practice that we use a capacitor in conjunction with the diode circuit output to perform filtering operation. In this section we explain the modeling of a filter circuit by taking the following example.

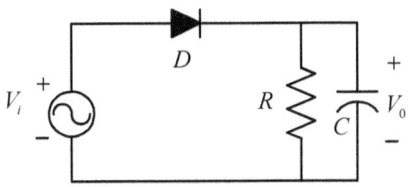

Figure 3.4(a) Filtering output of basic Diode circuit

Referring to basic diode circuit of figure 3.2(a), a capacitor of value $C=1\ mF$ is connected in parallel with the resistor R as shown in figure 3.4(a). Given that due to presence of the capacitor, wave shape of V_0 takes the form of figure 3.4(b) which we wish to simulate.

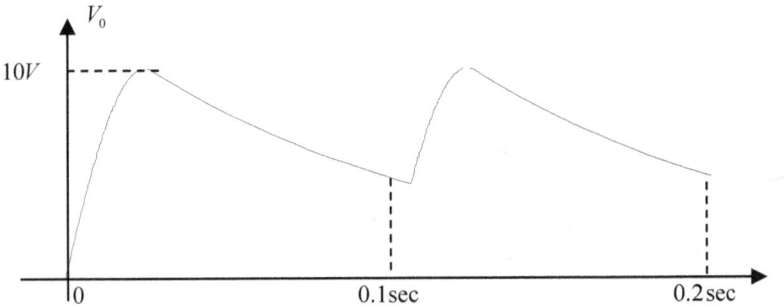

Figure 3.4(b) Output V_0 of filtering circuit in figure 3.4(a)

All element values of circuit are taken from section 3.2. This modeling starts from the model in figure 3.2(d). Delete the connection lines of **Voltage Measurement** in figure 3.2(d), copy and paste the **R** through the clipboard to see **R1**, rename the **R1** as **C**, doubleclick the **C**, select **C** from the **Branch type** popup (section 3.1), and enter its **Capacitance** value as **1e-3** in the parameter window (appendix A). Figure 3.4(c) shows model of the filter circuit in figure 3.4(a) therefore we connect the **C** as in the figure. Click the simulation icon ▶ at the icon bar to run the model and doubleclick **Scope** to see the voltage output with autoscale setting (figure 3.2(e)) as shown in figure 3.4(d). It goes without saying that the return in figure 3.4(d) is equivalent to the expected V_0 shape of figure 3.4(b).

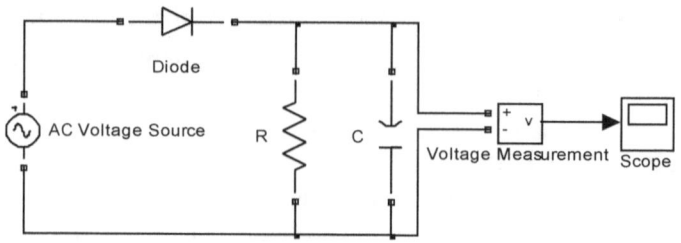

Figure 3.4(c) Model of filter circuit in figure 3.4(a)

Nonetheless input and output wave shapes together are often sought for comparison reason. It is given that the input V_i and output V_0 of circuit in figure 3.4(a) take functional variation as shown in figure 3.4(e). We wish to simulate this combined variation as well.

We need two **Voltage Measurement** blocks – one for V_i and the other for V_0. This modeling starts from the model in figure 3.4(c). Copy and paste the **Voltage Measurement** through the clipboard to see **Voltage Measurement1** in model of figure 3.4(c) – this will determine the V_i. On a common time axis, we combine two similar voltage signals by the block **Mux** (appendix B). Get the **Mux** in the model file.

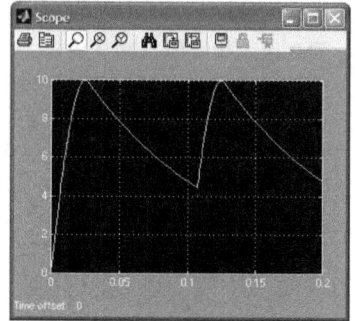

Figure 3.4(d) **Scope** output for filtering circuit of figure 3.4(c) corresponding to V_0

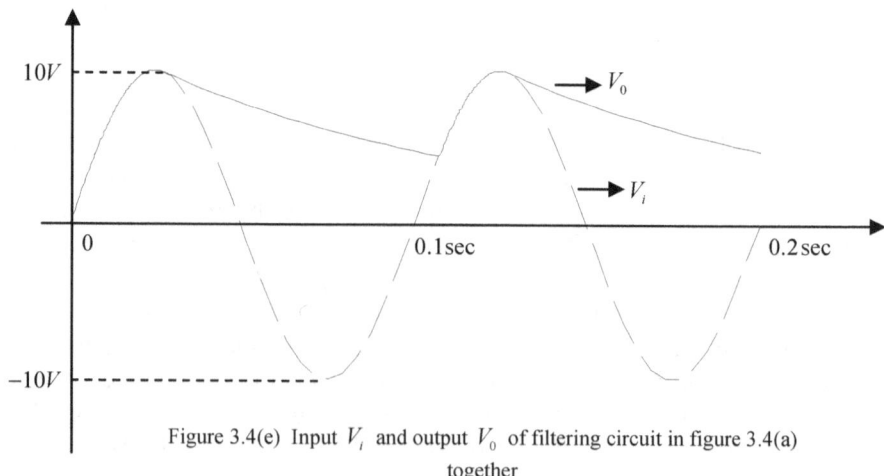

Figure 3.4(e) Input V_i and output V_0 of filtering circuit in figure 3.4(a) together

Figure 3.4(f) is the sought model for circuit in figure 3.4(a).

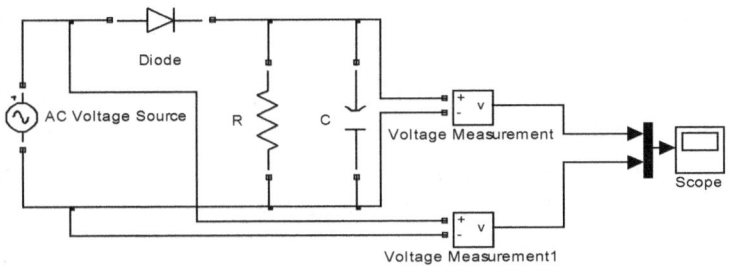

Figure 3.4(f) Model of filter when input and output voltages are required

Place the blocks relatively and connect them as shown in figure 3.4(f) and click the simulation icon ▶ to run the model. On doubleclicking the **Scope** followed by autoscale setting, we view the turnout as shown in figure 3.4(g). The **Scope** output in figure 3.4(g) is the one which is required variation of figure 3.4(e).

3.5 Zener diode modeling

A Zener diode has the schematic symbol as shown in figure 3.5(a). Since there is no dedicated block for a Zener diode to date, we model the diode by employing two **Diode** blocks of section 3.1. In elementary level a Zener diode usually has two parameters namely forward bias voltage V_γ and reverse break voltage V_Z assuming that the reader is familiar with the theory. One **Diode** block models the V_γ and the other does V_Z.

Figure 3.4(g) **Scope** return from model in figure 3.4(f)

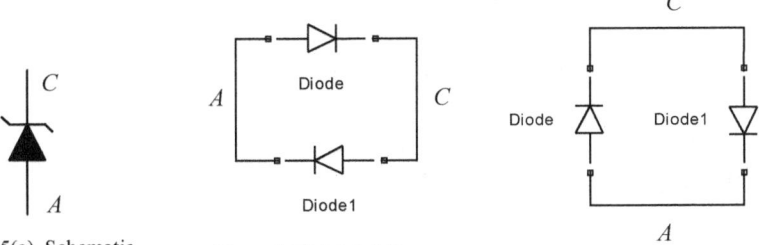

Figure 3.5(a) Schematic symbol of a Zener diode

Figure 3.5(b) Modeling a Zener diode horizontally

Figure 3.5(c) Modeling a Zener diode vertically

Open a new SIMULINK model file (subsection 1.2.2) and bring the **Diode** in the model (appendix B). We intend to model a Zener diode like figure 3.5(a) which has $V_\gamma=0.7V$ and $V_z=12V$. Doubleclick the Diode and enter its Forward voltage (Vf) V, Snubber capacitance Cs (F), and Snubber resistance Rs as 0.7, 0, and inf in the parameter window respectively thus modeling only the V_γ. Copy and paste the Diode through the clipboard to see Diode1, doubleclick the Diode1, enter its Forward voltage (Vf) V as 12 in the parameter window for modeling only the V_z, rightclick on the Diode1, and click the Flip Block under Format popup. Connect the two Diodes like figure 3.5(b) that depicts the model of Zener diode seen in figure 3.5(a). Polarity consistency of the two nodes i.e. A and C is also shown in the figures 3.5(a) and 3.5(b).

Model of figure 3.5(b) shows horizontal orientation of the Zener diode. Circuit connection off and on requires vertical placement that happens by the model in figure 3.5(c). In order to do so, we need to rotate the Diode by ninety degrees before connecting the lines. The rotation of a block in SIMULINK always takes place 90° clockwise but the polarity of the one in figure 3.5(a) requires otherwise. For this reason after getting the Diode in the model file and setting its parameters, we rightclick on the Diode, click the Rotate Block under the Format popup, again rightclick on the Diode, and click the Flip Block under the Format popup thus modeling the Diode of figure 3.5(c). Copy and paste the Diode through the clipboard to see Diode1, rightclick on Diode1, and click the Flip Block under the Format popup for Diode1 of figure 3.5(c). Make sure that the values of V_γ and V_z are properly set as the Forward voltage (Vf) V in the parameter windows of Diode and Diode1 respectively that defines the polarity of figure 3.5(a). We plan to include one example on Zener diode simulation in the following.

◆ Example

Given that with $R=10\Omega$, $R_L=1K\Omega$, $V_\gamma=0.7V$ for the diode D, $V_\gamma=0.7V$ and $V_z=4V$ for the Zener diode, and V_i as in section 3.2, the voltage output of circuit in figure 3.5(d) follows the functional variation of figure 3.5(e). Our objective is to simulate this circuit behavior by applying SIMULINK modeling.

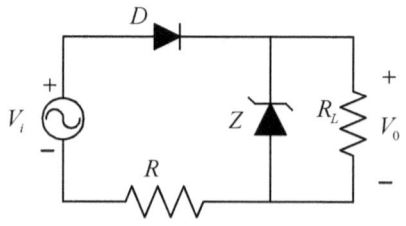

Figure 3.5(d) A Zener Diode circuit

Figure 3.5(f) is the model for simulating the problem and the entire modeling procedure is cited below:

⇒ open a new SIMULINK model file,

⇒ get one **AC Voltage Source** (for V_i) in the model file, doubleclick the **AC Voltage Source**, and enter values of the **Peak amplitude (V)** and **Frequency (Hz)** as **10** and **10** in parameter window respectively,

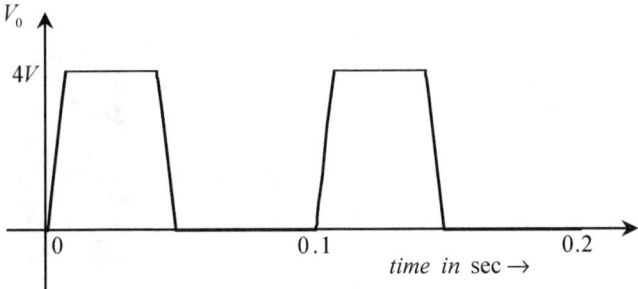

Figure 3.5(e) Output V_0 from the Zener diode circuit in figure 3.5(d)

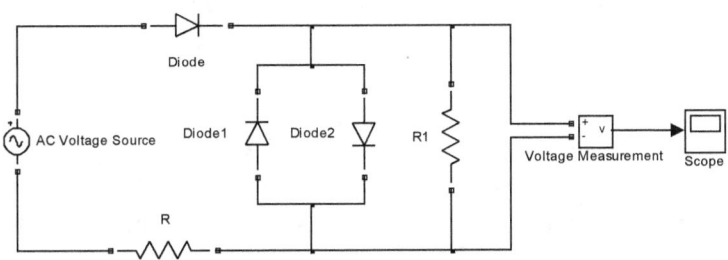

Figure 3.5(f) Model of the Zener diode circuit in figure 3.5(d)

⇒ get one **Series RLC Branch** (for R) in the model file, rename the **Series RLC Branch** as **R**, doubleclick **R**, select **R** from the **Branch type** popup (section 2.1), and enter its **Resistance** as **10** in parameter window,

⇒ copy just modeled **R** in clipboard, paste it to see **R1** (which models R_L), doubleclick **R1**, enter its **Resistance** as **1e3** in parameter window for $R_L = 1\ K\Omega$, rightclick on **R1**, and click **Rotate Block** under the **Format** popup to model it vertically,

⇒ get one **Diode** block in the model file for D, doubleclick **Diode**, enter values of the **Resistance Ron (Ohms)**, **Forward voltage (Vf) V**, **Snubber resistance Rs (Ohms)**, and **Snubber capacitance Cs (F)** as **1e-10**, **0.7**, **inf**, and **0** in parameter window respectively,

⇒ copy and paste **Diode** through clipboard to see **Diode1** (its forward bias voltage is the same as **Diode's**), rightclick on **Diode1**, click **Flip Block** under the **Format** popup, rightclick again on the **Diode1**, click the **Rotate Block** under the **Format** popup (**Diode1** of figure 3.5(f) respresents V_γ of Zener diode),

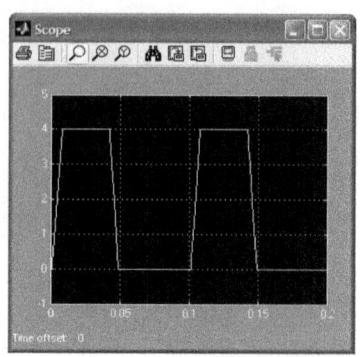

copy and paste the **Diode1** through clipboard to see **Diode2**, rightclick on **Diode2**, click **Flip Block** under the **Format** popup (**Diode2** of figure 3.5(f) represents V_z of Zener diode), doubleclick **Diode2**, and enter the value of the **Forward voltage (Vf) V** as 4 in parameter window for V_z,

Figure 3.5(g) **Scope** output for Zener diode circuit of figure 3.5(d) corresponding to V_0

⇒ get one **Voltage Measurement** and one **Scope** blocks in the model file for V_0,

⇒ click the menu **Simulation** down **Configuration parameters** at the menu bar, select **ode23s** as the model **Solver**, and enter stop time as **0.2** in slot of the **Stop time** in the window, and

⇒ finally place the blocks relatively and connect them as shown in figure 3.5(f), click the simulation icon ▶ to run the model, and doubleclick **Scope** to see the voltage V_0 with autoscale setting (figure 3.2(e)) as shown in figure 3.5(g).

Needless to mention, the return in figure 3.5(g) is in accordance with the expected V_0 shape of figure 3.5(e).

3.6 Rectifier circuit modeling

Rectification means converting an alternating current signal to direct one. Section 3.2 illustrated diode circuit is basically half wave rectification of sinusoidal AC input. There is another by name full wave rectification whose schematic circuit is seen in figure

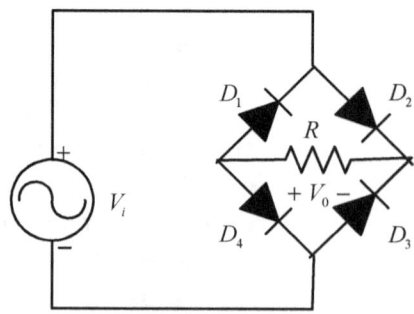

Figure 3.6(a) A full wave rectifier bridge circuit

3.6(a). The diodes D_1, D_2, D_3, and D_4 of circuit in figure 3.6(a) are ideal. Input V_i to the circuit is the same as in section 3.2. When $R = 1\ K\Omega$, the voltage output V_0 across the R follows functional variation of figure 3.6(b) which we intend to obtain through SIMULINK modeling.

There are two options to construct this model. The reader may employ four **Diode** blocks of section 3.1 and connect the circuit like figure 3.6(a). The other option is employ the **Universal Bridge** block which keeps provision for full wave rectification based on diode circuits. We plan to implement the second option i.e. D_1 - D_2 - D_3 - D_4 of figure 3.6(a) is implemented by the **Universal Bridge**.

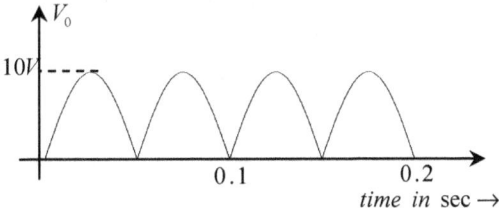

Figure 3.6(b) Full wave rectified sine wave

Open a new SIMULINK model file (subsection 1.2.2) and get the block in the model file (appendix B). Its first appearance is shown in figure 3.6(c) and doubleclick the block to see its parameter window as in figure 3.6(e). The **Universal Bridge** is devised for handling different power electronic devices and for three phase in general. We set parameters of the **Universal Bridge** so that it behaves only as a single phase rectifier.

In figure 3.6(e) there is a popup menu by name **Number of bridge arms**. The default number is **3**. Click the popup and select it to be 2 for the single phase diode circuit. Again there is another popup which is **Power Electronic device**. Default popup is **Thyristors** and select it to be **Diodes**

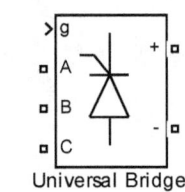

Universal Bridge

Figure 3.6(c) Appearance of
Universal Bridge

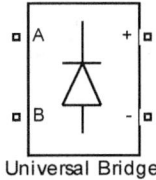

Universal Bridge

Figure 3.6(d) Adapted **Universal
Bridge**

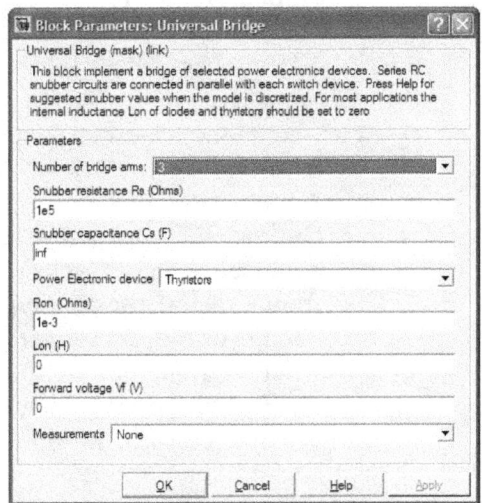

Figure 3.6(e) Block parameter window of
Universal Bridge

by using the mouse. In doing so the **Universal Bridge** adapts to the icon of figure 3.6(d). This is the block which will act as the full wave bridge rectifier on diode circuit in figure 3.6(a) or you can say the D_1 - D_2 - D_3 - D_4 portion of figure 3.6(a) is implemented by the icon of figure 3.6(d). In the icon nodes **A** and **B** indicate connecting nodes of the input V_i and the **+ve** and **−ve** indicate output V_0 of the circuit in figure 3.6(a).

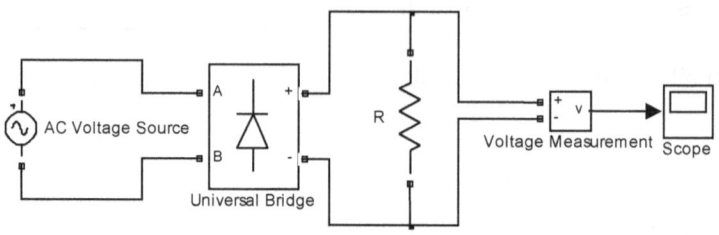

Figure 3.6(f) Model of diode bridge circuit in figure 3.6(a)

Figure 3.6(f) shows complete model of the circuit in figure 3.6(a). For the other elements,

⇒ get one **AC Voltage Source** (for V_i) in the model file, doubleclick **AC Voltage Source**, and enter values of the **Peak amplitude (V)** and **Frequency (Hz)** as **10** and **10** in parameter window respectively,

⇒ get one **Series RLC Branch** (for R) in the model file, rename the **Series RLC Branch** as **R**, doubleclick the **R**, select **R** from the **Branch type**

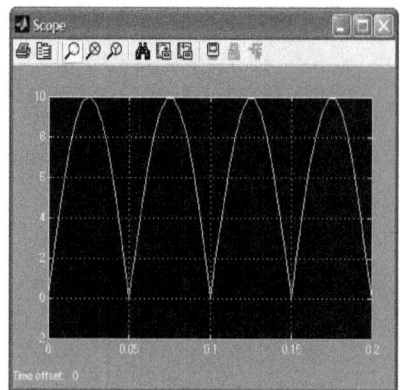

Figure 3.6(g) **Scope** output for full wave rectified bridge circuit of figure 3.6(a) corresponding to V_0

popup (section 2.1), enter its **Resistance** as **1e3** in the parameter window, rightclick on the **R**, and click the **Rotate Block** under the **Format** popup to model it vertically,

⇒ get one **Voltage Measurement** and one **Scope** blocks in the model file for V_0,

⇒ click the menu **Simulation** down **Configuration parameters** of the model file, select **ode23s** as model **Solver** in the prompt window (figure 1.7(g)), and enter stop time as 0.2 in the slot of **Stop time**, and

⇒ finally place the blocks relatively and connect them as shown in figure 3.6(f), click simulation icon ▶ to run the model, and doubleclick **Scope** to see voltage output V_0 with autoscale setting (figure 3.2(e)) as in figure 3.6(g) which obliges the variation of wave in figure 3.6(b).

The example considers all diodes ideal. Should the reader need to enter the forward bias voltage, that option is also provided in parameter window of figure 3.6(e) assuming that all four diodes are identical.

3.7 Some measurement simulations on a diode circuit

The reader might need to know some values such as DC component, AC component, etc present in a wave once a diode circuit has been modeled. Built-in blocks of SIMULINK easily help us obtain those for which following simulations on measurement are demonstrated.

✦ DC or average value of a signal

In section 3.2 we simulated the basic diode circuit which actually performs half wave rectification. If $v(t)$ is any periodic signal over period

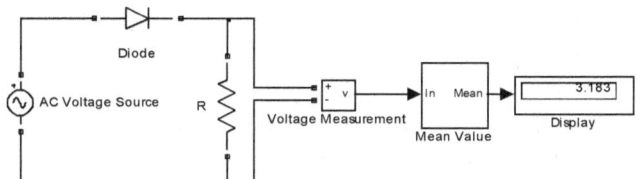

Figure 3.7(a) Model for finding the average value of a diode circuit output

T, the average value is $\frac{1}{T}\int_{t=0}^{T} v(t)dt$. We know that an ideal half rectified wave has DC or average value $\frac{V_m}{\pi}$ when V_m is amplitude of the associated sine wave. Section 3.2 cited amplitude is $10V$ so we should expect $\frac{V_m}{\pi}=3.1831$ as the average value if the wave is half rectified.

SIMULINK block **Mean Value** (appendix B) determines the average or DC value of a signal over the user-defined time period. In section 3.2 we described modeling of the circuit completely and figure 3.2(d) is the model. In the model we delete the **Scope** and get one **Mean Value** and one **Display** blocks. The **Scope** is for viewing a wave shape but the **Display** for viewing numerical values. Connect the **Mean Value** like figure 3.7(a). Since the sine wave has time period 0.1 secs, doubleclick the **Mean Value** and enter the **Averaging period (s)** as 0.1 in the parameter window. The time interval information should also be consistent with the period. As **Stop time** you can enter any multiple of the period for example 0.1, 0.2, 0.3, etc (figure 1.7(g)) however run the model and value in Display confirms the average value of half rectified sine wave.

✦ Root mean square or RMS value of a signal

Root mean square or RMS value of the periodic $v(t)$ is defined as $\sqrt{\dfrac{1}{T}\displaystyle\int_{t=0}^{T}v^2(t)dt}$. We know that an ideal half rectified wave has the RMS value $V_m/2$. For the ongoing example the RMS value should be 5 which we intend to verify.

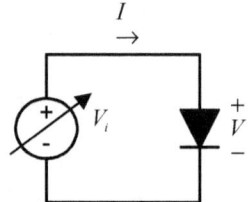

Figure 3.7(b) Model for finding the RMS value of a diode circuit output

Block RMS helps us verify that. Figure 3.7(b) depicts the model for RMS value finding. Unlike the Mean Value, the parameter window of RMS requires frequency in Hz so doubleclick the RMS and enter Fundamental frequency as 10 in the parameter window. Make sure Stop time is multiple of the period 0.1 like Mean Value. Run the model and find the RMS value 5 in Display.

Figure 3.7(c) A variable voltage applied to a diode

3.8 I/V characteristic of a diode/its circuit

Figure 3.7(c) depicts a variable voltage V_i connected to a diode. We wish to see I versus V for the diode (i.e. diode current versus diode voltage). The variable voltage we simulate by a ramp function i.e. $V_i=t$ which is exercised by a Ramp block (appendix B) in SIMULINK. Recall that DC Voltage Source simulates constant voltage (section 2.3) and there is no direct way to change it without programming. The best approach can be the use of a Controlled Voltage Source which is controlled by the Ramp. The time interval (figure 1.7(g)) will dictate the V_i variation for instance if we choose $0 \le t \le 5\sec$, the V_i variation is going to be $0 \le t \le 5V$. Regarding the diode, it can be ideal or non ideal as you decide.

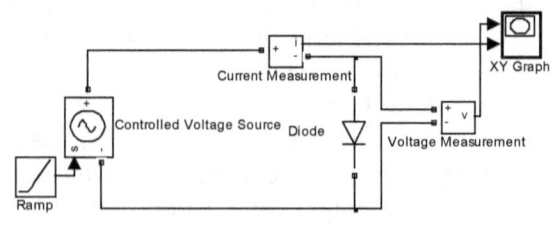

Figure 3.7(d) Model for diode I-V characteristic

Concerning section 3.1 if you change the parameters of parameter window, certainly the diode characteristics are going to change. Let us choose the **Resistance Ron (Ohms), Forward voltage (Vf) V, Snubber resistance Rs (Ohms), and Snubber capacitance Cs (F)** of figure 3.1(d) as $10^3\Omega$, $0.7V$, $10^7\Omega$, and $0F$ respectively. Say the V_i changes over $0 \le V_i \le 7V$, we wish to see the I-V characteristic for the diode of figure 3.7(c).

Figure 3.7(d) depicts complete model for the characteristic. In order to construct the model, open a new SIMULINK model file (subsection 1.2.2) and get one **Ramp**, one **Controlled Voltage Source**, one **Voltage Measurement**, one **Current Measurement**, and one **Diode** blocks in the model file. Enter last paragraph quoted quantities as **1e3, 0.7, 1e7, and 0** respectively in parameter window of figure 3.1(d), leave the other parameters as default. Rightclick on the **Diode** and click the **Rotate Block** under the **Format** popup for vertical orientation. Block **XY Graph** shows the plot of y versus x type data which will be used for I-V characteristic plotting so get the block in the model. Connect the blocks as shown in figure 3.7(d). Click the menu **Simulation** down **Configuration parameters** of the model file, select **ode23s** as model **Solver** in prompt window (figure 1.7(g)) for diode's presence, and enter the **Stop time** as **7** for $0 \le V_i \le 7V$.

The **XY Graph** has two input ports as seen in figure 3.7(d) – lower and upper, which represent y or vertical and x or horizontal data respectively. The **XY Graph** has some default axis setting which needs changing. Doubleclick the **XY Graph** and enter **x-min** and **x-max** as 0 and 7 respectively for $0 \le V_i \le 7V$. For the diode current we are not sure. Click the simulation icon ▶ to run the model. The diode current is so small that it aligns with the horizontal axis and not visible hence doubleclick the **XY Graph**, enter **y-min** and **y-max** as 0 and 0.01 respectively, and run the model again. The grapher shows the diode I-V characteristic as presented in figure 3.7(e). Horizontal and vertical axes labeling indicates the I-V link of the diode. The graph starts from 0.7, why is that? The answer is forward bias voltage chosen is

$I \uparrow$

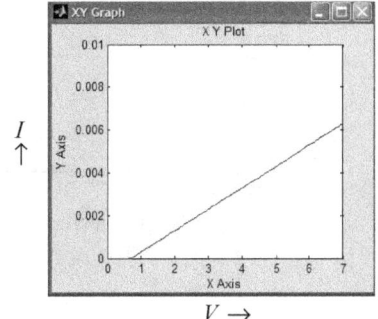

$V \rightarrow$

Figure 3.7(e) I-V characteristic of a diode

$0.7V$. Here we considered a single diode in fact you may apply this technique to a load resistor if it is necessary for instance half rectification circuit of figure 3.2(a).

That brings the end of this chapter.

Exercises

1. Model the ideal diode of figure 3.1(a) in SIMULINK. Do the same if the diode has a forward biased voltage $V_\gamma = 1V$.

2. Model the basic diode circuit of figure 3.2(a) in SIMULINK assuming ideal diode where the V_i is a sinusoid with amplitude variation $\pm 12V$, frequency $60\ Hz$, duration over $0 \le t \le \frac{1}{30} secs$, and $R = 75\ \Omega$. Do the same if the D has forward bias voltage $V_\gamma = 0.7V$. Obtain the V_0 waveshapes in both cases.

3. Obtain each of the following diode circuit output wave shapes by applying SIMULINK model:
   (a) regarding figure E3.1(a), the V_i is a sinusoid which has amplitude variation $\pm 15V$, frequency $60\ Hz$, and duration over $0 \le t \le \frac{1}{30} secs$. The D has forward bias voltage $V_\gamma = 0.7V$ also $R = 75\Omega$ and $V_1 = 4V$,
   (b) regarding figure E3.1(b), all elements in the circuit have the values as in part (a) only difference is the reversed polarity of V_1,
   (c) regarding figure E3.1(c), all elements in the circuit have the values as in part (a), and

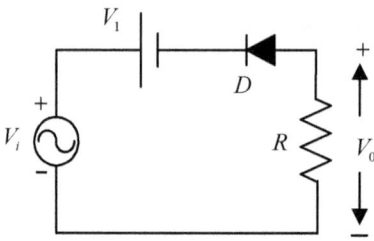

Figure E3.1(a) A clipping circuit

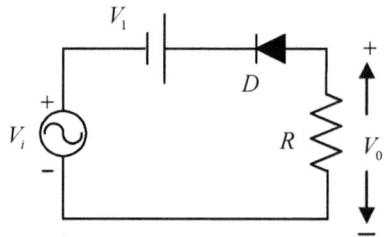

Figure E3.1(b) A clipping circuit

(d) regarding figure E3.1(d), elements in the circuit have the values as follows: $R = 1\ K\Omega$, V_γ of $D_1 = 0.4V$, V_γ of $D_2 = 0.7V$, V_i as in part a, $V_1 = 4V$, and $V_2 = 8V$.

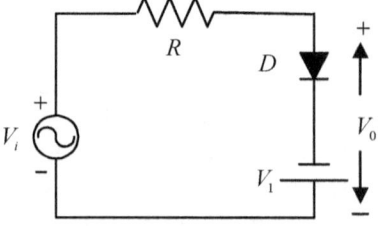

Figure E3.1(c) A clipping circuit

4. Simulate the capacitively filtered wave shapes of the following circuits by applying SIMULINK model:

(a) regarding figure 3.4(a), the V_i is a sinusoid which has amplitude variation $\pm 50\,V$, frequency $50\,Hz$, and duration over $0 \leq t \leq \frac{1}{25}\sec s$. The D is a silicon diode that has forward bias voltage $V_\gamma = 0.6\,V$ also $R = 10\,K\Omega$ and

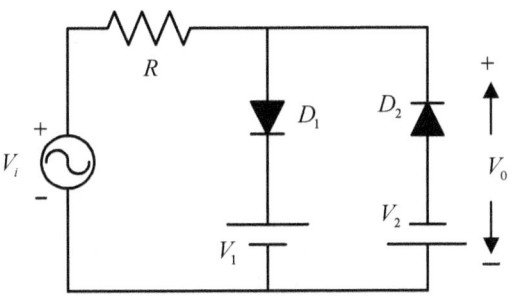

Figure E3.1(d) A two level clipping circuit

$C = 25\,\mu F$. The V_0 versus t is required over the given interval.

(b) regarding figure E3.2(a), the V_i is a sinusoid which has amplitude variation $\pm 75\,V$, frequency $60\,Hz$, and duration over $0 \leq t \leq \frac{1}{30}\sec s$. Each of the $D_1 - D_2 - D_3 - D_4$ in the figure is a silicon diode which has forward bias voltage $V_\gamma = 0.7\,V$ also $R = 1\,K\Omega$ and $C = 47\,\mu F$. The V_0 versus t with and without C are required as a single plot over the given interval.

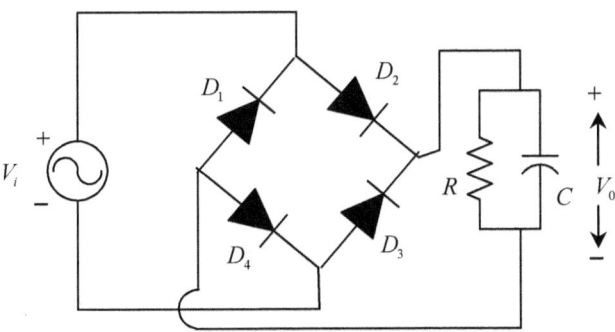

Figure E3.2(a) Filtering output of a full wave rectified
bridge circuit by using a capacitor

5. Concerning the Zener diode circuit of figure 3.5(d), we have $R = 20\Omega$, $R_L = 3.5\,K\Omega$, $V_\gamma = 0.8\,V$ for the diode D, $V_\gamma = 0.6\,V$ and $V_z = 12\,V$ for the Zener diode Z, and V_i is a sinusoid which has amplitude variation $\pm 75\,V$, frequency $60\,Hz$, and duration over $0 \leq t \leq \frac{1}{30}\sec s$. Obtain the voltage output V_0 waveshape from the Zener diode circuit by SIMULINK modeling.

6. Determine the average value of each following signal in SIMULINK: (a) V_0 waveshape of problem (2) considering forward bias (b) V_0 waveshape of problem 3(c) (c) V_0 waveshape of problem 4(b) without and with the capacitor.

7. Determine the RMS value of each signal of question 6 in SIMULINK.

8. A nonideal diode has the following parameters: on resistance, forward bias voltage, snubber resistance, and snubber capacitance are $1.5K\Omega$, $0.85V$, $10^8\Omega$, and $1\,nF$ respectively. Obtain the I-V characteristic of the diode in the forward bias zone if the voltage across the diode changes from 0 to $5V$.

9. Figure E3.2(b) depicts a voltage doubler circuit using two diodes where V_i is a sinusoid which has amplitude variation $\pm 10V$, frequency $60\,Hz$, and duration over $0 \le t \le \frac{1}{15} \sec s$. Obtain the voltage output V_0 waveshape by SIMULINK modeling subject to $C_1 = 1\,mF$, $C_2 = 1\,mF$, and V_γ of each diode being $0.7V$.

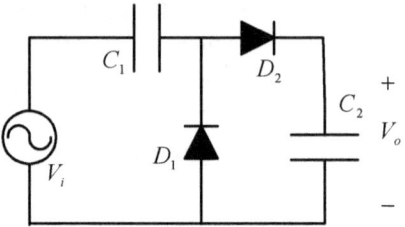

Figure E3.2(b) Voltage doubler circuit using diode

Answers:

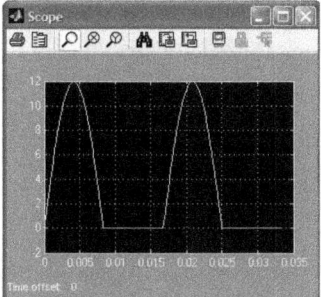

Figure A3.1(a) V_0 of problem
2 without forward bias

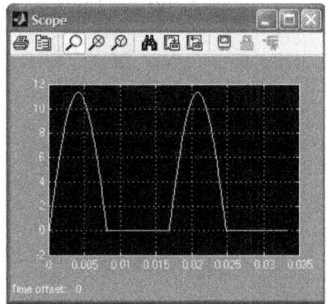

Figure A3.1(b) V_0 of problem
2 with forward bias

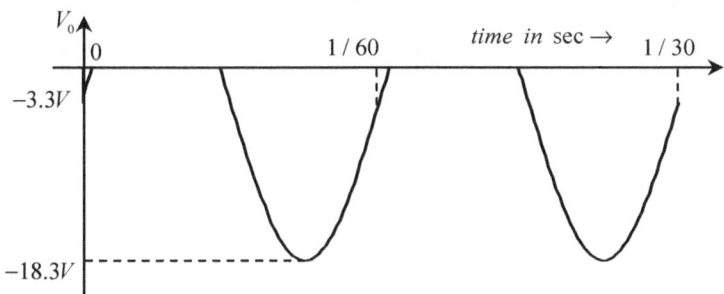

Figure A3.2(a) Output V_0 of clipping circuit in figure E3.1(a)

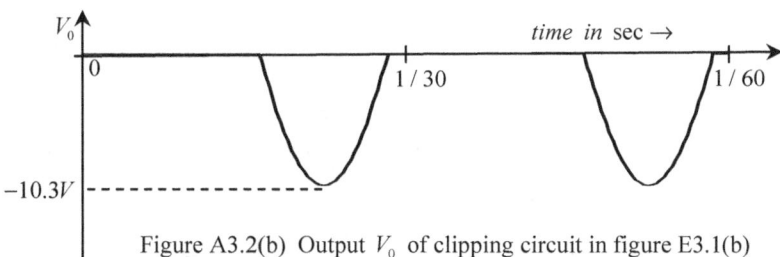

Figure A3.2(b) Output V_0 of clipping circuit in figure E3.1(b)

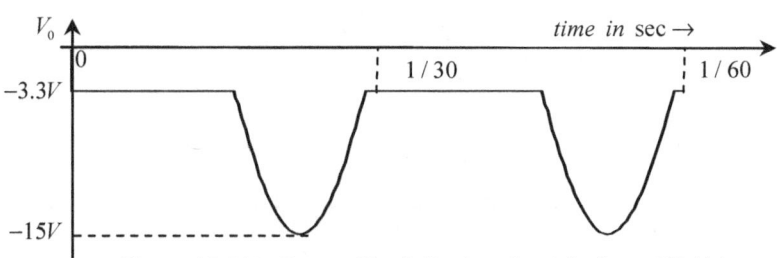

Figure A3.2(c) Output V_0 of clipping circuit in figure E3.1(c)

(1) Hint: section 3.1
(2) Figures A3.1(a) and A3.1(b).　　　　Hint: section 3.2
(3) (a) Figure A3.2(a)　　　　(b) Figure A3.2(b)　　　　(c) Figure A3.2(c)
　　　(d) Figure A3.2(d)　　　　　　　　Hint: section 3.3

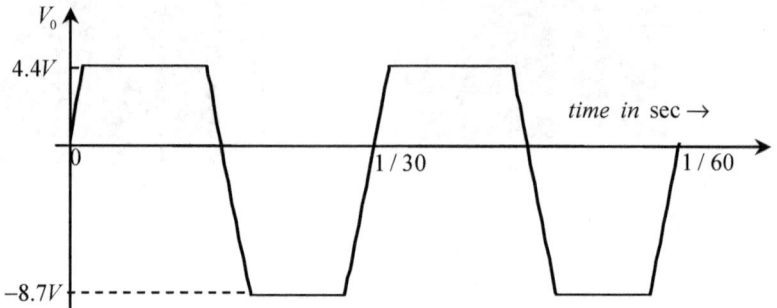

Figure A3.2(d)　Output V_0　of clipping circuit in figure E3.1(d)

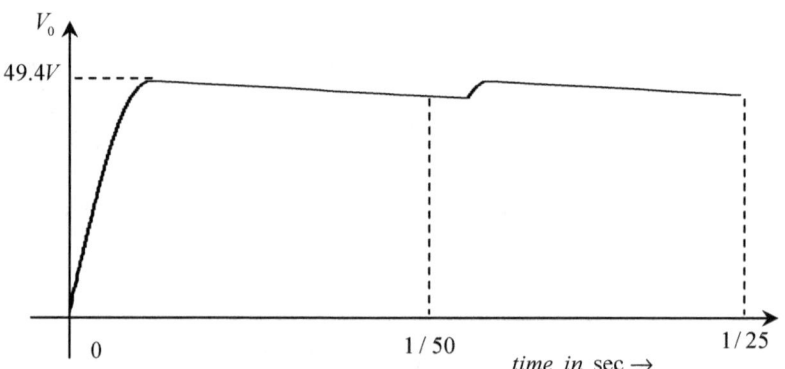

Figure A3.3(a)　Output V_0　from filter diode circuit of figure 3.4(a)

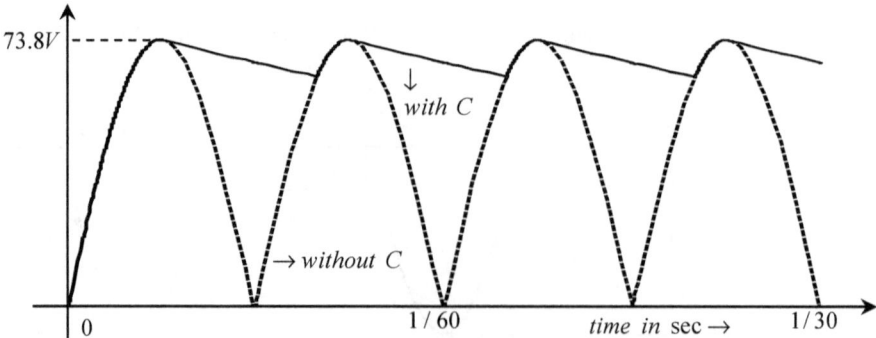

Figure A3.3(b)　Output V_0　from filter diode circuit of figure E3.2(a)

(4) (a) Figure A3.3(a)   (b) Figure A3.3(b). The model is shown in figure
    A3.3(c).
    Hint: sections 3.4 and 3.6
(5) Figure A3.4(a)
    Hint: section 3.5

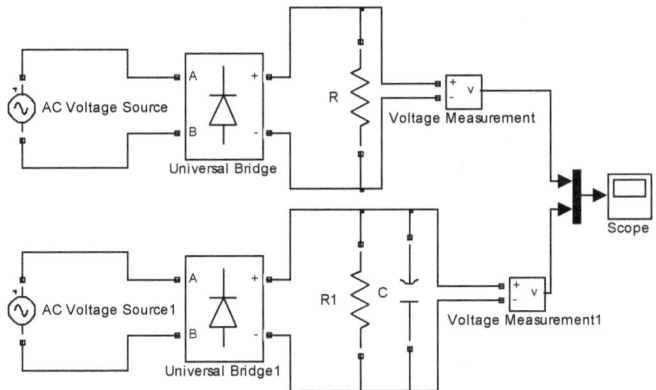

Figure A3.3(c)   Model of full rectified bridge circuit in figure E3.2(a)
with and without the capacitor

(6) (a) 3.479 V (b) –6.557 V (c) 46.43 V without   capacitor   and
    68.77 V with capacitor
    Hint: section 3.7
(7) (a) 5.565 V (b) 7.902 V (c) 51.74 V without   capacitor   and
    68.84 V with capacitor
    Hint: section 3.7
(8) Figure A3.4(b)   Hint: section 3.8
(9) Figure A3.4(c) subject to on resistance, snubber resistance, and snubber
    capacitance as 0.001 Ω , 1$K\Omega$, and 0 F respectively
    Hint: sections 2.12 and 3.1-3.2

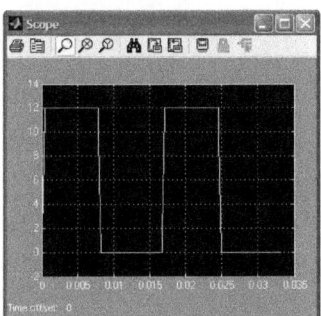

Figure A3.4(a) Scope output
from Zener diode circuit of
problem (5)

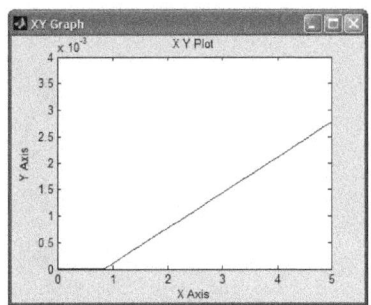

Figure A3.4(b) I/V characteristic of
diode in problem (8)

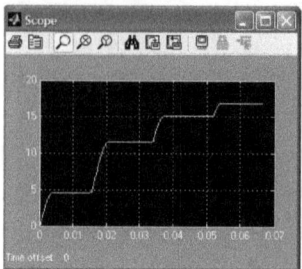

Figure A3.4(c) Voltage output
of voltage doubler circuit

Chapter 4

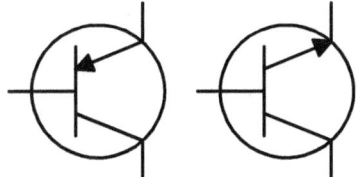

Modeling Transistor Circuits

Bipolar junction transistor (BJT) has special importance in electronics in the sense that it slowly replaced vacuum tubes of electronic systems and paved the way for integrated circuits or ICs which are at the heart of today's consumer electronics. Principal focus of this chapter is to implement the structure and operation of BJTs in MATLAB/ SIMULINK platform. Since a BJT is mostly nonlinear in its I-V characteristic, unified model is difficult to exercise however our study material selection is on the following:

    ❖❖ Bipolar junction transistor modeling – NPN and PNP
    ❖❖ Load line and I-V characteristics of a transistor
    ❖❖ Small and large signal models of transistors
    ❖❖ Voltage gain, current gain, and input/output resistance

4.1 Transistor as a voltage controlled current source

    A bipolar junction transistor can be considered as a voltage controlled current source which we demonstrate in this section. Figure 4.1(a) shows the symbol of an NPN bipolar junction transistor where C, E, and B refer to collector, emitter, and base of the transistor respectively. Depicted in figure 4.1(b) is model of the transistor which is basically a voltage controlled current source $I_s e^{\frac{V_{BE}}{V_T}}$ where I_s=reverse saturation current in ampere, V_{BE}=base to emitter voltage in volt, and V_T=thermal threshold in volt. To

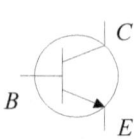

Figure 4.1(a) Symbol
of an NPN bipolar
junction transistor

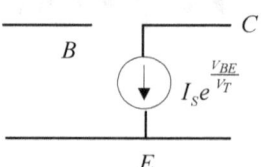

Figure 4.1(b) Transistor
as a voltage controlled
current source

mention about SIMULINK, the transistor of figure 4.1(a) we model by **Controlled Current Source** (section 2.3). The block has three ports **s**, plus, and minus which correspond to B, E, and C of the transistor in figure 4.1(a) respectively.

Controlled Current Source

Figure 4.1(c) Linking transistor terminology to SIMULINK block

The voltage V_{BE} is a variable and user-supplied. A **Constant** block generates the V_{BE}. Say we have V_{BE}=600 mV, I_s= $10\times10^{-15} A$, and V_T=26 mV for a transistor, then **Value** of the **Constant** should be **10e-15*exp(600e-3/26e-3)** for entering $I_s e^{\frac{V_{BE}}{V_T}}$ (appendix A). The plus and minus ports of **Controlled Current Source** generate the current $I_s e^{\frac{V_{BE}}{V_T}}$. Let us see the following examples.

◆ **Example 1**

Figure 4.1(d) depicts an NPN bipolar transistor circuit with R=1 $K\Omega$, I_s=5.1×10$^{-16} A$, V_{BE}=752 mV, and V_T= 26 mV. Determine the current I_C and voltage V_C of the transistor.

Figure 4.1(e) presents the SIMULINK model of transistor in figure 4.1(d). In order to build the model, following steps are carried out:

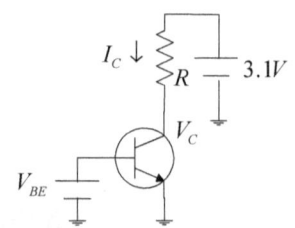

Figure 4.1(d) A transistor circuit

⇒ open a new SIMULINK model file (subsection 1.2.2),

⇒ get one **Constant** block in the model and enter its **Value** as **5.1e-16*exp(752e-3/26e-3)**, and enlarge the block to see its contents,

⇒ get one **Controlled Current Source** block in the model, rightclick on the block, and click the **Flip Block** under the **Format** popup to make sense with collector-emitter vertical position,

⇒ get one **Series RLC Branch** block in the model file, rename it as **R** (subsection 1.2.3), doubleclick **R**, select **R** from the popup (section 2.1), enter its **Resistance** as **1e3**, rightclick on the block, and click the **Rotate Block** under the **Format** popup to turn the resistor vertical,

⇒ get one **Current Measurement** (for I_c), one **Voltage Measurement** (for V_c), and two **Display** blocks in the model,

⇒ rightclick on the **Current Measurement**, click the **Flip Block** under the **Format** popup due to I_c polarity, and do the same for the **Display1**,

⇒ get one **DC Voltage Source** in the model, doubleclick it, and enter value of **Amplitude (V)** as **3.1** in parameter window,

⇒ include annotations **E** and **C** to make sense for emitter and collector respectively (subsection 1.2.3), and

⇒ get one **Ground** block in the model, copy it to the clipboard, and paste it two times to see the other two blocks.

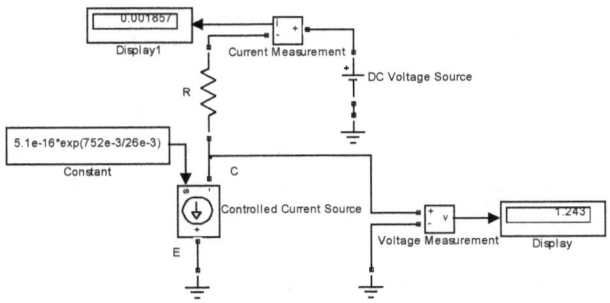

Figure 4.1(e) Model of the transistor circuit in
figure 4.1(d)

Place the blocks relatively and connect them like figure 4.1(e) and eventually click the simulation icon ▶ at the icon bar to run the model and find the V_c and I_c in **Display** and **Display1** which read out as $1.243\,V$ and $0.001857\,A$ respectively.

✦ Example 2

A PNP transistor circuit is shown in figure 4.1(f) which includes an audio voltage V_{audio}. The circuit parameters are $I_S = 1.5 \times 10^{-16} A$ and $V_T = 26 \, mV$. The V_{audio} changes from 0 to 5 mV. Determine the current I_C and voltage V_R for both voltage limits of V_{audio}.

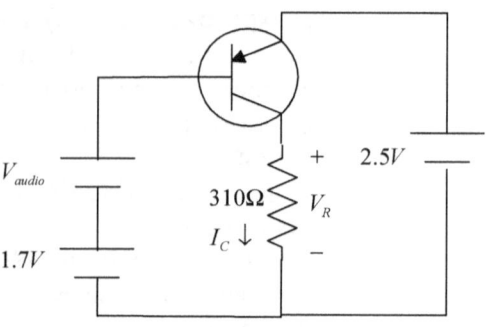

Figure 4.1(f) A PNP transistor circuit

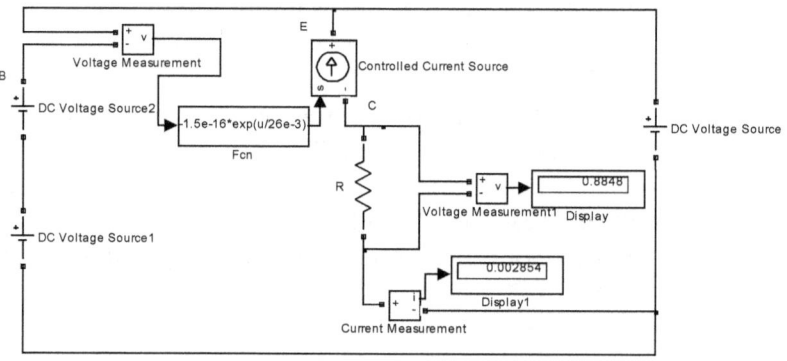

Figure 4.1(g) Model of transistor circuit in figure 4.1(f)

Shown in figure 4.1(g) is the model of this example. This circuit is different from the one in example 1. The resistor is in collector circuit and there is no ground in the circuit. The transistor collector current is now $I_S e^{\frac{V_{EB}}{V_T}}$. The V_{EB} is basically the voltage between emitter of the transistor (which is up in figure 4.1(f)) and + ve node of V_{audio}. How do we get this voltage difference collected? The answer is through **Voltage Measurement** block. Output of the **Voltage Measurement** can be fed to s port or base of **Controlled Current Source** but passing through a functional block **Fcn** which defines $I_S e^{\frac{V_{EB}}{V_T}}$. Let us construct the model by following steps:

⇒ open a new SIMULINK model file,
⇒ get three **DC Voltage Source** blocks in the model (for 1.7V, 2.5V, and V_{audio}), doubleclick the **DC Voltage Source** and **DC Voltage Source1**, enter the values of **Amplitude (V)** as **2.5** and

1.7 in parameter window respectively, and leave the **DC Voltage Source2** for V_{audio},

⇒ get one **Controlled Current Source**, one **Series RLC Branch**, one **Current Measurement**, two **Voltage Measurement**, one **Fcn**, and two **Display** blocks in the model,

⇒ doubleclick the **Series RLC Branch**, rename it as **R**, doubleclick **R**, select **R** from the popup, enter its **Resistance** as **310**, rightclick on the block, and click the **Rotate Block** under the **Format** popup to turn the resistor vertical,

⇒ doubleclick the **Fcn**, enter its code as **-1.5e-16*exp(u/26e-3)**, and enlarge the block to see its contents (the negative sign is for the transistor being PNP and the property of **Fcn** is input port signal must be defined as **u** which is here V_{EB}),

⇒ include annotations **E** and **C** to make sense for emitter and collector identification respectively, and

⇒ place the blocks relatively and connect them like figure 4.1(g).

Now doubleclick the **DC Voltage Source2**, enter value of **Amplitude (V)** as **0** for V_{audio}=0, click the simulation icon ▶ at the icon bar to run the model, and find the V_R and I_C in **Display** and **Display1** which read out as $1.072\,V$ and $0.003459\,A$ respectively. Again enter value of **Amplitude (V)** as **5e-3** for V_{audio}=5 mV, click the simulation icon ▶ at the icon bar to run the model, and find the V_R and I_C in **Display** and **Display1** which read out as $0.8848\,V$ and $0.002854\,A$ respectively.

4.2 Transistor load line characteristic

Voltage developed at any of the three terminals of a transistor usually does not exceed the supply voltage. If the load resistance is changed, how the load voltage or other in an electronic circuit changes is basically load line characteristic of the transistor.

The circuit of figure 4.2(a) is actually the circuit of figure 4.1(d). Point is V_{CC} of figure 4.2(a) replaced the voltage $3.1\,V$ of figure 4.1(d). The circuit is modeled as example 1 in last section. Considering identical transistor characteristic of the

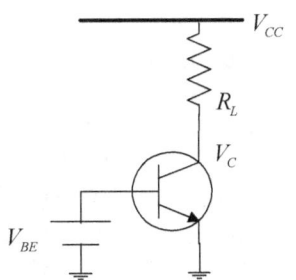

Figure 4.2(a) Transistor circuit with a load resistance R_L

example, we have V_{CC}=5 V and R_L changes from 0 to 2 $K\Omega$. We wish to see the V_C (not load voltage) versus R_L load line characteristic of the transistor.

The problem with SIMULINK block is definitive value is required to make the block operational. For instance the **R** block of figure 4.1(e) only works for a single known value which should not be variable, that is not so in this problem. In this type of problem MATLAB-SIMULINK combined strategy works fine. What we do is keep the **R** as variable in SIMULINK and pass the **R** from MATLAB to SIMULINK though programming means.

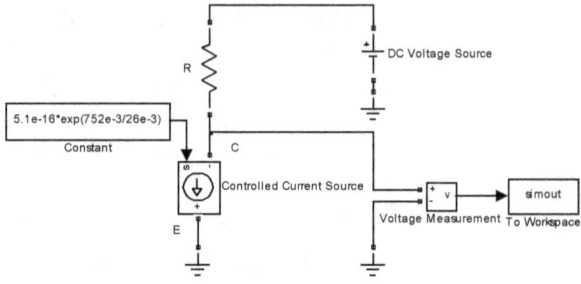

Figure 4.2(b) Model of transistor circuit in figure 4.2(a) for load line characteristic

The R_L changes from 0 to 2 $K\Omega$ but there is no step mentioned say 100 Ω. In MATLAB jargon (subsection 1.1.2) we write it as **RL=0:100:2e3;** where **RL** is a user-chosen variable representing R_L. There is another problem; when R_L =0, some SIMULINK blocks may not work so we put some negligible quantity epsilon which in MATLAB term is **eps** hence the modified **RL** is **RL=eps:100:2e3;**.

Figure 4.2(b) depicts model for load line characteristic of the transistor. Start from the model in figure 4.1(e). Delete the **Display1** and **Current Measurement** because we do not need them. Doubleclick the **DC Voltage Source** and enter the value of **Amplitude (V)** as **5** for V_{CC} =5V. Doubleclick the **R** and enter its **Resistance** as **R** for variable R_L. Get a **To Workspace** block which transports data from SIMULINK to MATLAB. Delete the **Display** and connect **To Workspace** in line to that for transporting V_C to MATLAB. Another point, the **To Workspace** functions for various kinds of data from which we need only matrix or array type hence doubleclick the **To Workspace** and select **Array** as **Save Format** in the parameter window. Save the model file by some name **test**.

The **To Workspace** has default variable **simout** which retains the output from **Voltage Measurement** or V_C but the problem is time dependency of SIMULINK block (subsection 1.2.4) causes many values of V_C to appear from which we need only the last value and which we get by **simout(end)**. Let us carry out the following two lines at the command prompt of MATLAB:

```
>>RL=eps:100:2e3; ↵
>>VC=[ ]; for R=RL, sim('test.mdl');VC=[VC simout(end)]; end ↵
```

Ignore the warnings. Clearly the first line is for R_L out of the two. We exercised data accumulation technique (appendix D.3) to hold the V_C values returned from SIMULINK (i.e. **VC=[VC simout(end)];**) where **VC** stands for V_C and is a user-chosen variable. By a for-loop (appendix D.4) every sample in RL is chosen by writing for R=RL. The command sim('test.mdl'); is equivalent to running the SIMULINK model test.mdl staying in MATLAB. Anyhow after all these R_L and V_C values are available to RL and VC respectively. Exercise plot of appendix E for the load line:

>>plot(RL,VC) ↵

Figure 4.2(c)

Include labeling (underscore is for the subscript) and grid by:

>>xlabel('R_L') ↵
>>ylabel('V_C') ↵
>>grid ↵

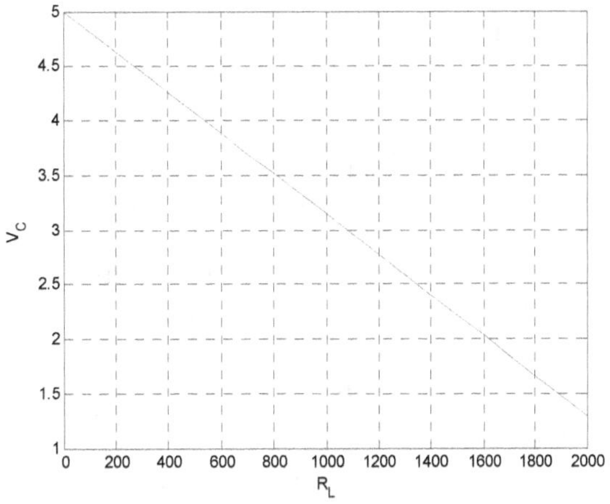

Figure 4.2(c) Load line characteristic of transistor

Figure 4.2(c) is the expected load line of the transistor.

4.3 Transistor V/V and I/V characteristics

A bipolar junction transistor has governing equations relating to its various voltages and currents. If the expressions for these quantities are given, how do we graph them? – that is the topic of this section. The best approach should be the use of embedded plotter for the expression. Take the example of NPN transistor voltages and currents which are given by:

$$I_C = \beta I_B,$$

$$I_C = I_S e^{\frac{V_{BE}}{V_T}},$$

$$I_B = \frac{I_S}{\beta} e^{\frac{V_{BE}}{V_T}},$$

$$I_E = I_C + I_B, \text{ and}$$

$$V_{CE} = V_{CC} - I_C R_L.$$

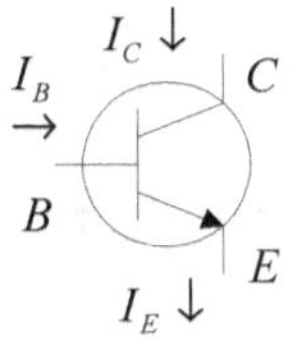

Figure 4.3(a) NPN transistor with current directions

The transistor with various currents is shown in figure 4.3(a) where β is common emitter current gain and the other symbols have previous section quoted meanings. Appendix E mentioned ezplot is exercised in any I/V characteristic plotting. In the case of multiple voltage-current interdependencies command hold can be exercised. Let us see the following examples on voltage-voltage or voltage-current interdependence plotting.

Example 1: I_C versus V_{BE}

Graph I_C versus V_{BE} of the NPN transistor in figure 4.3(a) for $I_S = 1.5 \times 10^{-16} A$ and $V_T = 26\, mV$ over $500mV \leq V_{BE} \leq 800mV$.

Carry out the following at the command prompt:

```
>>ezplot('1.5e-16*exp(Vbe/26e-3)',[0.5 0.8]) ↵
```

Figure 4.3(b) is the outcome from above execution in which the vertical or y axis is the I_C in Ampere. Clearly the **Vbe** (our chosen variable) stands for V_{BE}. The first input argument of **ezplot** is the code for $I_S e^{\frac{V_{BE}}{V_T}}$ under quote. Inside the **ezplot** all quantities are in standard unit.

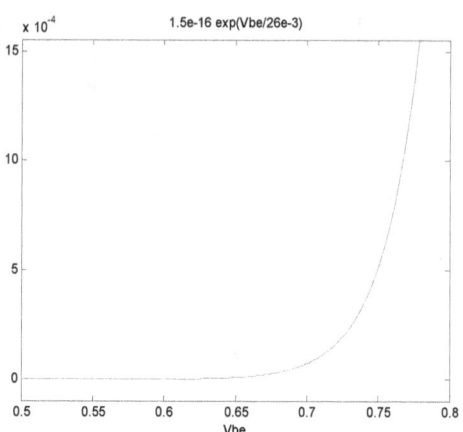

Figure 4.3(b) Graph of I_C versus V_{BE} of an NPN transistor

Example 2: I_B versus V_{BE}

The plot of I_B versus V_{BE} is similar to that of I_C versus V_{BE} just β needs to be considered. With the example 1 parameters and $\beta = 100$, the command would be:

```
>>ezplot('1.5e-16*exp(Vbe/26e-3)/100',[0.5 0.8]) ↵
```

The figure is not shown for the space reason.

Example 3: I_C versus V_{BE} for several I_S

In this example our objective is to graph I_C versus V_{BE} of example 1 transistor for $I_{S1} = 1 \times 10^{-16} A$, $I_{S2} = 2 \times 10^{-16} A$, and $I_{S3} = 5 \times 10^{-16} A$ over $500mV \leq V_{BE} \leq 800mV$.

Similar to example 1, let us plot the I_C versus V_{BE} for I_{S1} as follows:

```
>>ezplot('1e-16*exp(Vbe/26e-3)',[0.5 0.8]) ↵
```

We can retain the last graph by the command **hold**:

```
>>hold ↵
```

Whichever graph is plotted is superimposed now on the last one with the same axes setting. Knowing so, the other two graphs (for I_{S2} and I_{S3}) we include by:

```
>>ezplot('2e-16*exp(Vbe/26e-3)',[0.5 0.8]) ↵
>>ezplot('5e-16*exp(Vbe/26e-3)',[0.5 0.8]) ↵
```

Attach mark of distinction on each curve by the command **legend**:

```
>>legend('for Is=1','for Is=2','for Is=5') ↵
```

The input arguments of **legend** respectively indicate the graphing order which are completely user-chosen and under quote for instance the first graph has **'for Is=1'** for I_{S_1}. Figure 4.3(c) depicts the three variations of I_C.

All graphs plotted this way are indicated by the same color. Click Edit plot icon (figure 1.2(c)), bring mouse pointer on any curve, leftclick to select the curve, and rightclick the mouse to find a popup where you see line color or width option. From the popup choose any color for the curve. The indicatory box of figure 4.3(c) becomes color updated automatically. Since the text is written in black and white form, you do not see the curve demarcation in terms of color.

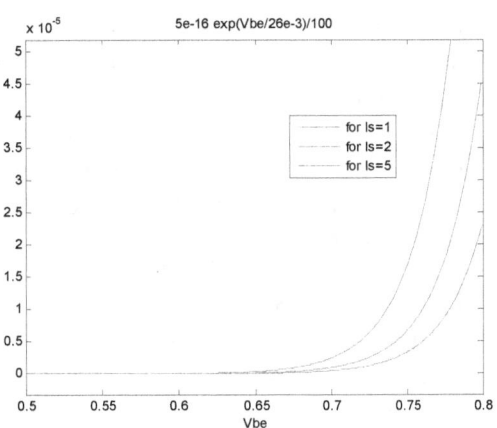

Figure 4.3(c) Graph of I_C versus V_{BE} for three I_S

Example 4: Constant I_C versus V_{CE} for several V_{BE}

If V_{CC} is not given, we can not link I_C to V_{CE} i.e. I_C does not depend on V_{CE} directly besides V_{CC} is circuit dependent. Say we have $V_{BE1}=0.7\,V$, $V_{BE2}=0.75\,V$, and $V_{BE3}=0.8\,V$ for the transistor of example 1 with $I_S=5\times10^{-16}A$ and wish to see I_C versus V_{CE}. In this sort of plot you may assume any reasonable variation of V_{CE} which will be on the horizontal or x axis say 1 to $2\,V$. Knowing so, the implementation is similar to that of the example 3:

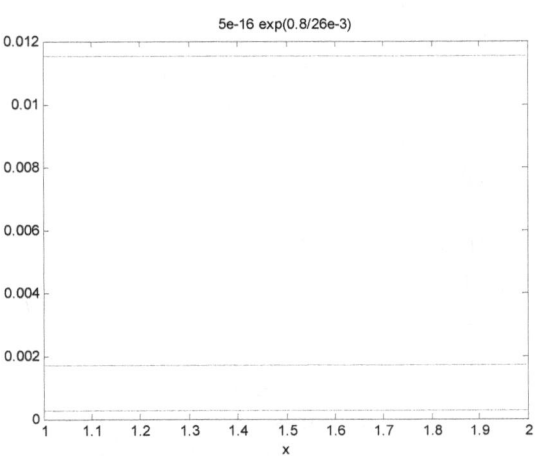

Figure 4.3(d) Graph of I_C versus V_{CE} for three V_{BE}

```
>>ezplot('5e-16*exp(0.7/26e-3)',[1 2]) ↵
>>hold ↵
```

```
>>ezplot('5e-16*exp(0.75/26e-3)',[1 2]) ↵
>>ezplot('5e-16*exp(0.8/26e-3)',[1 2]) ↵
```

The values of I_C are extremely small and so closely located that you find the three curves as one. The graph is not shown for the space reason. This problem can be solved if we set the I_C or vertical axis slightly more than maximum of the three I_C values. We force the axis setting by the command **axis** which has the syntax **axis([x axis minimum value x axis maximum value y axis minimum value y axis maximum value])** i.e. a four element row matrix. Certainly the x axis variation is from 1 to 2. Discovering the y axis value from the graph from 0 to 0.012, we carry out the following at the command prompt:

```
>>axis([1 2 0 0.012]) ↵
```

Figure 4.3(d) is the outcome from the last command line. In the figure the vertical or y axis indicates I_C and the horizontal or x indicates V_{CE}. Should you add the demarcation like example 3, carry out the following:

```
>>legend('for Vbe=0.7','for Vbe=0.75','for Vbe=0.8') ↵
```

Peripheral changes on a drawn graph:

A better perceived graph needs some peripheral changes. Figure 4.3(d) shown title statement **5e-16*exp(0.7/26e-3)** does not look nice. We wish to get rid of it. Click the Edit plot icon to activate the mouse pointer, leftclick on the title statement, rightclick the mouse to see delete in a popup, and click delete. The vertical axis is I_C. We may exercise **ylabel('I_C')** to include that. Again the horizontal axis is V_{CE} which can be added by **xlabel('V_C_E')**. Note that the underscore or _ is for the subscript. Every character requires one underscore. Sometimes dotted line is also used to identify the curve. For

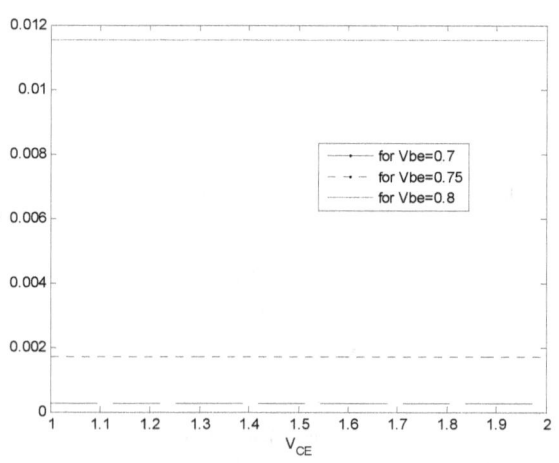

Figure 4.3(e) Graph of figure 4.3(d) using dotted lines and proper labeling

this Click the Edit plot icon to activate the mouse pointer, bring the mouse pointer on any line/curve, rightclick the mouse to see **Linestyle** in a popup, choose any style you like e.g. **Solid, Dot, Dash**, etc, and get the curve as well as legend updated. We did so and figure 4.3(e) presents the change.

4.4 Resistive divider biasing circuits

Figure 4.4(b) depicts a resistive divider biasing circuit. In previous sections we concentrated on transistor modeling considering a controlled current source but it has another model (so called large signal model) which applies a diode in conjunction with the controlled current source. The schematic circuit of figure 4.4(a) models the NPN transistor of figure 4.4(b). Consistency of the terminals you also find in the figure e.g. B for base of the transistor. In section 3.1 detailed discussion on diode modeling is found. The **Controlled Current Source** of section 4.1 we connect in accordance with the figure

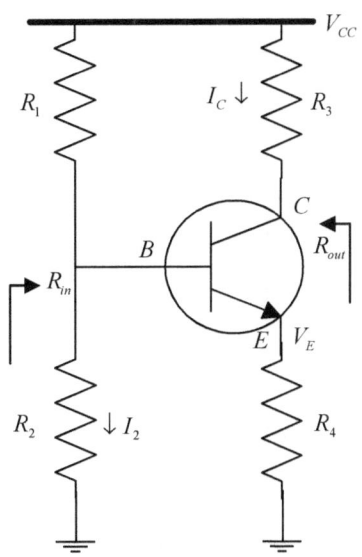

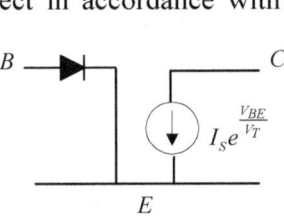

Figure 4.4(a) Transistor model using a diode and a controlled current source

Figure 4.4(b) Resistive divider biasing circuit of an NPN transistor

4.4(c) which simulates the NPN transistor of figure 4.4(b). The **Voltage Measurement** senses the diode voltage, output of which goes to the **s** terminal of **Controlled Current Source** via an **Fcn** block. The **Fcn** works considering the input signal as **u**, which basically holds the code of $I_S e^{\frac{V_{BE}}{V_T}}$. In figure 4.4(c) you see the code for $I_S = 5 \times 10^{-17} A$ and $V_T = 26\,mV$. Also note that the **Controlled Current Source** is flipped from its default appearance. Terminal consistency is indicated in the figure too e.g. **C** for collector.

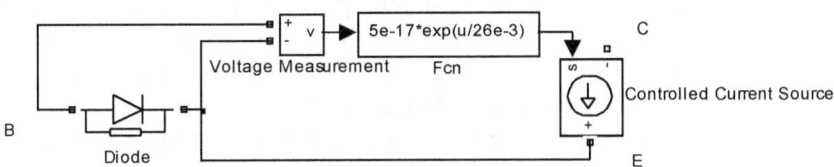

Figure 4.4(c) Modeling NPN transistor of figure 4.4(a) using a diode and a controlled current source

Let us take a specific example. The V_E of figure 4.4(b) should be 0.1029 V subject to $V_{CC} = 2.5 V$, $R_1 = 16 K\Omega$, $R_2 = 9 K\Omega$, $R_3 = 1 K\Omega$, $R_4 = 100 \Omega$, $I_S = 5 \times 10^{-17} A$, $V_T = 26 mV$, and $V_\gamma = 0.8 V$ (for diode model of the transistor) which we intend to verify through SIMULINK modeling.

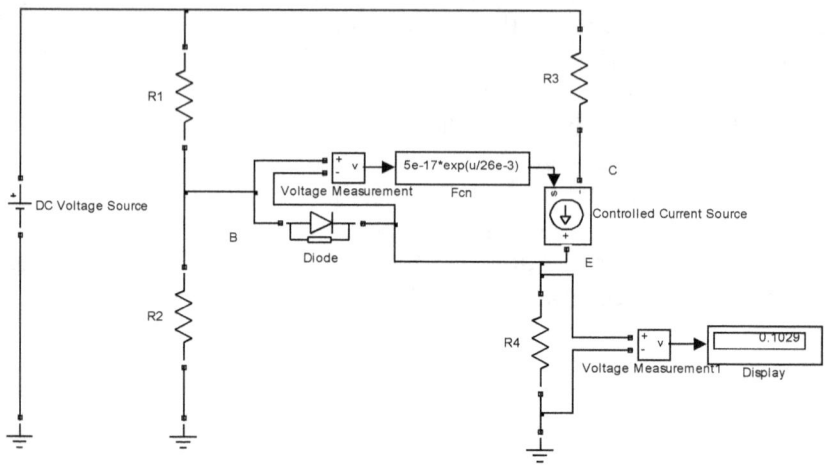

Figure 4.4(d) Modeling resistive divider biasing circuit of figure 4.4(b)

First of all let us build the model as shown in figure 4.4(d) by the following steps:

> ⇒ open a new SIMULINK model file (chapter 1),
> ⇒ get one **DC Voltage Source** block in the model (appendix B), doubleclick the **DC Voltage Source**, and enter the value of **Amplitude (V)** as **2.5** (for V_{CC}),
> ⇒ get one **Series RLC Branch** block in the model, rename it as **R1**, doubleclick **R1**, select **R** from the popup, enter its **Resistance** as **16e3**, rightclick on the block, and click **Rotate Block** under the **Format** popup to turn the resistor vertical,
> ⇒ copy the **R1** in clipboard, paste it three times to see the other three resistors, doubleclick each, and enter the **Resistance** as **9e3**, **1e3**, and **100** for R2, R3, and R4 respectively,
> ⇒ get one **Diode**, two **Voltage Measurement**, three **Ground**, one **Fcn**, one **Display**, and one **Controlled Current Source** blocks in the model,
> ⇒ doubleclick the **Diode**, make sure the value of **Forward voltage (Vf) V** is **0.8**, and leave the other parameters as default,

⇒ rightclick on the **Controlled Current Source** and click **Flip Block** under the **Format** popup to make sense with collector-emitter vertical position,

⇒ doubleclick the **Fcn**, enter its code as **5e-17*exp(u/26e-3)**, and enlarge the block to see its contents,

⇒ place the blocks relatively and connect them like figure 4.4(d),

⇒ include annotations **E**, **C**, or **B** to make sense for emitter, collector, or base identification respectively, and

⇒ click the menu **Simulation** down the **Configuration parameters** at the menu bar and select **ode23s** (section 3.2) as the model **Solver**.

Finally click simulation icon ▶ at the icon bar to run the model and find the V_E at **Display**.

4.5 Transistor modeling by common emitter current gain

So far we have been addressing a BJT as controlled current source subject to I_S, V_T, and V_{BE} variations. Common emitter current gain β models a transistor too which we present in this section.

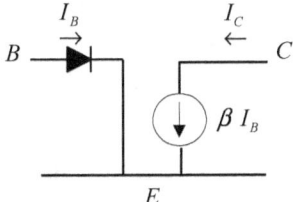

Figure 4.5(a) Transistor model using common emitter current gain

Referring to figure 4.5(a) the NPN transistor of figure 4.1(a) is modeled by using the common emitter current gain β.

SIMULINK counterpart of the model in figure 4.5(a) is shown in figure 4.5(b). How is this model different from the one in figure 4.4(c)? The **Current Measurement** block (section 2.5) carries as well as returns (at the i port) the diode current or I_B of figure 4.5(a). In figure 4.5(b) we chose β as 100. Unlike the ongoing model we feed the β by a **Gain** block just as a digression. If the modeling were conducted by an **Fcn**, code of the **Fcn** would have been **100*u**. For any circuit modeling just follow the lead of collector-emitter-base or **C-E-B** annotations of figure

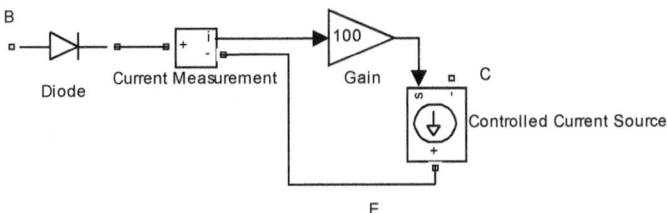

Figure 4.5(b) SIMULINK model of NPN transistor using common emitter current gain

4.5(b). Since **Diode** parameters introduce some impedance to the electronic circuit, modeling of the **Diode** had better be ideal. In order to do so, enter values of the **Snubber resistance Rs (Ohms)** and **Snubber capacitance Cs (F)** as infinity (has MATLAB code inf) and 0 respectively (section 3.1) in the parameter window. Also enter the **Resistance Ron (Ohms)** as minimum as possible for instance 10^{-9} or **1e-9** (appendix A).

In order to construct the model of figure 4.5(b), open a new SIMULINK model file (chapter 1), get one **Diode**, one **Current Measurement**, one **Controlled Current Source**, and one **Gain** blocks in the model (appendix B), doubleclick the **Gain**, enter **Gain** as 100 for β in the parameter window, doubleclick the **Diode**, enter the **Snubber resistance Rs**, **Snubber capacitance Cs**, and **Resistance Ron** as inf, 0, and **1e-9** respectively in the parameter window, rightclick on the **Controlled Current Source**, click the **Flip Block** under the **Format** popup to fit with the collector-emitter vertical position, connect the blocks like figure 4.5(b), and finally include the annotations **B-C-E**.

Similar technique is applied to a PNP transistor too. Figures 4.5(c), 4.5(d), and 4.5(e) present PNP transistor symbol with various currents, model using common emitter current gain, and SIMULINK implementation respectively. How is the model in figure 4.5(e) different from the one in figure 4.5(b)? The **Diode** and **Current Measurement** blocks both need to be flipped and we do not flip the **Controlled Current Source** during model building.

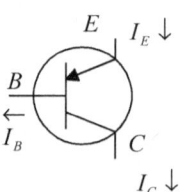

Figure 4.5(c) A PNP bipolar junction transistor with various currents

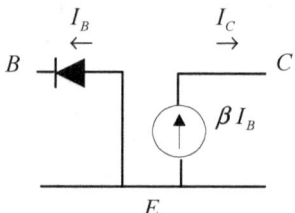

Figure 4.5(d) PNP transistor model using a diode and a controlled current source

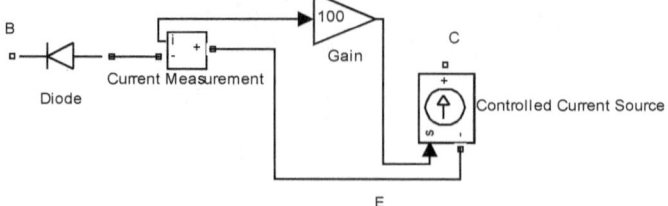

Figure 4.5(e) SIMULINK model of PNP transistor using common emitter current gain

4.6 Transistor modeling by transconductance

Small signal model of a transistor employs a resistor and a voltage dependent current source. Figure 4.6(a) shows model of the NPN transistor in figure 4.1(a) using transconductance g_m. The g_m and r_i are obtained from

$g_m = \dfrac{I_C}{V_T}$ and $r_i = \beta \dfrac{V_T}{I_C}$ respectively that means collector current I_C and

common emitter current gain β should be available for the electronic circuit we wish to model. The V_T is usually a standard value.

Depicted in figure 4.6(b) is the SIMULINK model of figure 4.6(a) circuit. To implement the model of figure 4.6(b), let us open a new SIMULINK model file (chapter 1), get one **Voltage Measurement**, one **Controlled Current Source**, one **Gain**, and one **Series RLC Branch** blocks in the model (appendix B), rightclick on the **Controlled Current Source**, click **Flip Block** under the **Format** popup to fit with the collector-emitter vertical position, rename the **Series RLC Branch** as **R**, doubleclick **R**, select **R** from the popup in the parameter window, rightclick on the **R**, click **Rotate Block** under the **Format** popup to turn the resistor vertical, connect the blocks like figure 4.6(b), and finally include the annotations **B-C-E**. The default value of the **Gain** block is 1 indicating $g_m = 1\,S$. For other

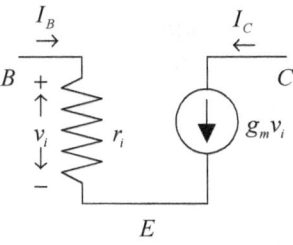

Figure 4.6(a) Small signal model of an NPN transistor using transconductance

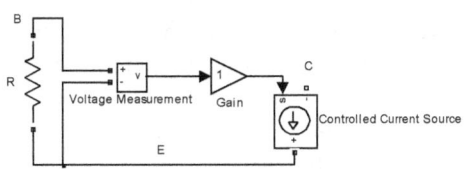

Figure 4.6(b) SIMULINK model of an NPN transistor using transconductance

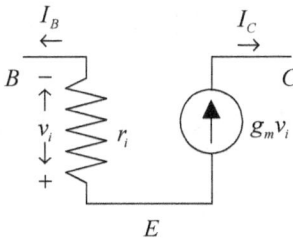

Figure 4.6(c) Small signal model of a PNP transistor using transconductance

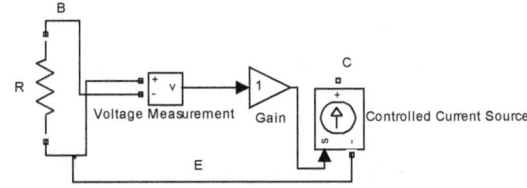

Figure 4.6(d) SIMULINK model of a PNP transistor using transconductance

g_m you may enter doubleclicking the block. The r_i we enter as **Resistance** of the **R** block.

How is the PNP counterpart going to be modeled? Shown in figures 4.6(c) and 4.6(d) are the PNP transistor model using transconductance and its SIMULINK counterpart respectively. The positive and negative ports of the **Voltage Measurement** in figure 4.6(d) get connected to the bottom and up nodes of the resistor **R** respectively. Since overlapping connection lines do not mean short circuit, this is okay and do not flip the **Controlled Current Source** during model building. In any electronic circuit we just follow the lead of collector-base-emitter or **C-B-E**.

4.7 Voltage gain after modeling a transistor

Voltage gain in an electronic circuit is basically the ration of two voltages. The two voltages are user or circuit defined. If the input and output voltages are V_i and V_o respectively, the voltage gain A_V is simply $A_V = \dfrac{V_o}{V_i}$. The V_i can be base, emitter, external, or other voltages. The V_o can be collector, emitter, collector-emitter, or any other voltage.

We collect each of V_i and V_o by a **Voltage Measurement** block (section 2.4) and view the A_V in a **Display**. The division is performed by a **Divide** block where $\div$ and $\times$ ports of the **Divide** receive V_i and V_o respectively. This is the fundamental but circuit condition may require extra tactic anyhow let us see the following examples in this regard.

✦ Example 1

The voltage gain of circuit in figure 4.1(d) is defined as $A_V = \dfrac{V_{CE}}{V_{BE}}$ which is 1.654. We intend to obtain the gain.

We start from the model of circuit which you find in figure 4.1(e).

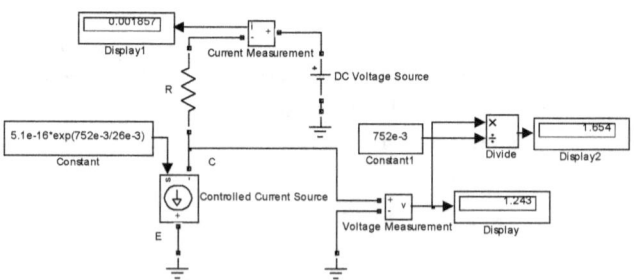

Figure 4.7(a) SIMULINK model for voltage gain of circuit in figure 4.1(d)

Figure 4.7(a) shows the SIMULINK model for this example. The V_{BE} is a constant value that we generate by **Constant1** block. The V_{CE} is already available at the v port of **Voltage Measurement**. Bring one **Constant**, one **Divide**, and one **Display** blocks in the model of figure 4.1(e), and connect the later blocks like figure 4.7(a). Doubleclick the **Constant1**, enter **752e-3** for V_{BE} in the parameter window, and enlarge the block to see its contents. Also note that overlapping lines (entry to **Divide**) do not mean short circuit. Finally run the model and find the A_V at **Display2**.

✦ Example 2

In example 1 we dealt with single A_V. Suppose the V_{BE} changes over $0.6V \le V_{BE} \le 0.8V$. How will the voltage gain A_V change?

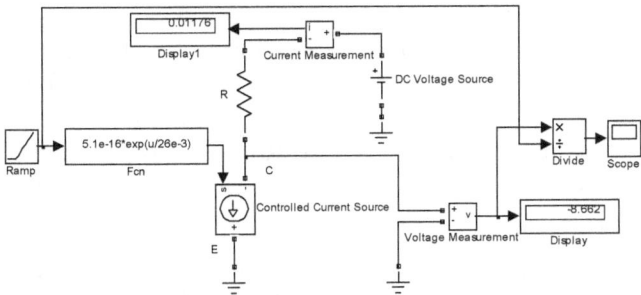

Figure 4.7(b) SIMULINK model for range voltage
gain of circuit in figure 4.1(d)

In the model of figure 4.7(a) the single value of V_{BE} is fed by a **Constant** block which is to be replaced by a **Ramp** (by default which generates t values). If we set the **Start** and **Stop** times of the solver (section 1.2.6) as 0.6 and 0.8 (lower and upper bounds of V_{BE}) respectively, our objective will be served. The **Display** is for a single value, any variation as a graph is seen in **Scope**.

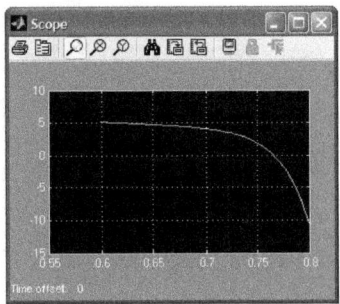

Figure 4.7(c) **Scope** output
from model in figure 4.7(b)

Figure 4.7(b) depicts the model with modification on the model of figure 4.7(a). Get one **Fcn**, one **Ramp**, one **Divide**, and one **Scope** blocks in the model of figure 4.7(a). Delete both **Constant** blocks, enter the code of **Fcn** as **5.1e-16*exp(u/26e-3)** on doubleclicking it, enlarge the **Fcn** to see its contents, connect the later blocks like figure 4.7(b), enter the **Start** and **Stop** times of the solver as **0.6** and **0.8** respectively, run the model, doubleclick the **Scope**, click the autoscale icon (figure 3.2(e)), and find the A_V variation like figure 4.7(c). In figure 4.7(c) the horizontal and vertical axes refer to V_{BE} and A_V respectively. This way you see only the waveshape, accessing to the values of A_V is not possible.

✦ **Example 3**

In example 2 we wish to determine
the bounds of A_V.

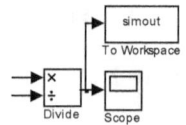

To get hand on exact value of A_V,
we need to export the data from SIMULINK
to MATLAB which occurs by a **To
Workspace** block. In the model of figure
4.7(b) all we do is get a **To Workspace**

Figure 4.7(d) **Scope** data
transporting to MATLAB

block and connect it in line with the **Scope** like figure 4.7(d), the rest portion
of the model is not shown for space reason. Since **To Workspace** holds
many data options, array or matrix type of which is suitable for us,
doubleclick the block and select **Array** from the **Save format**. Run the
model, the A_V will be available to default variable **simout** of **To
Workspace**. Because of numerical computing, there will be many A_V values
from which first and last are needed and which we pick by **simout(1)** and
simout(end) respectively. At the MATLAB prompt carry out the following
to see the two values side by side:

>>[simout(1) simout(end)] ↵

ans =
5.1577 -10.8274

As the return says voltage gain bounds we read as $5.1577 \geq A_V \geq -10.8274$.

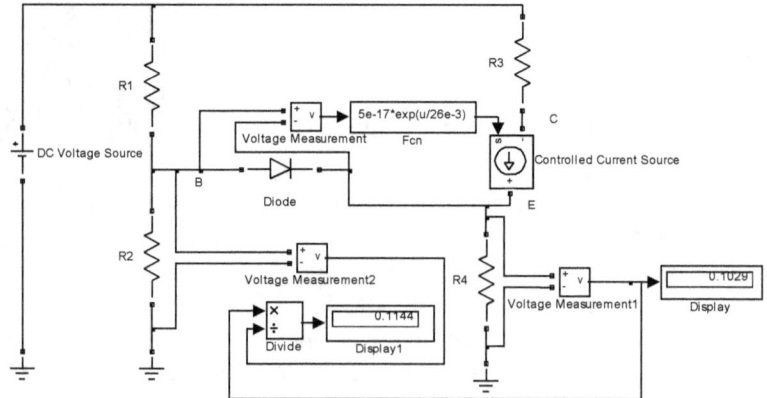

Figure 4.7(e) SIMULINK model for voltage gain of circuit in figure
4.4(b)

✦ **Example 4**

As another example, think about the resistive divider biasing circuit
of figure 4.4(b). Suppose the voltage gain A_V is defined as $A_V = \dfrac{V_o}{V_i}$ where V_i

and V_o are base to ground and emitter to ground or voltage across R_2 and R_4 respectively. We wish to determine the A_V.

Shown in figure 4.7(e) is the necessary model starting from the model in figure 4.4(d). Needless to mention, you have to get one **Divide**, one additional **Display**, and one additional **Voltage Measurement** blocks. Get them and connect like figure 4.7(e). Run the model and find the A_V =0.1144 at the **Display1**. Note that we turn the **Diode** as ideal to avoid transistor property shift (section 3.1).

4.8 Current gain after modeling a transistor

First of all current flows in a loop, gain from model without loop can not be implemented. Current gain in an electronic circuit is the ration of two currents. The two currents of coarse are user or circuit defined. If the input and output currents are I_i and I_o respectively, the current gain A_I is $A_I = \dfrac{I_o}{I_i}$.

The I_i or I_o can be base, emitter, external, or other branch currents. The two currents are collected by a **Current Measurement** block (section 2.5). The division is performed by a **Divide** block where ÷ and × ports of the **Divide** receive I_i and I_o respectively and we view the A_I in a **Display** followed by **Divide**. Next two examples demonstrate some current gain simulations.

◆ **Example 1**

Electronic circuit of figure 4.1(d) is modeled in figure 4.1(e). Suppose the current gain A_I is defined as $A_I = \dfrac{I_C}{I_B}$. In figure 4.1(e) there is no base current loop therefore A_I can not be simulated.

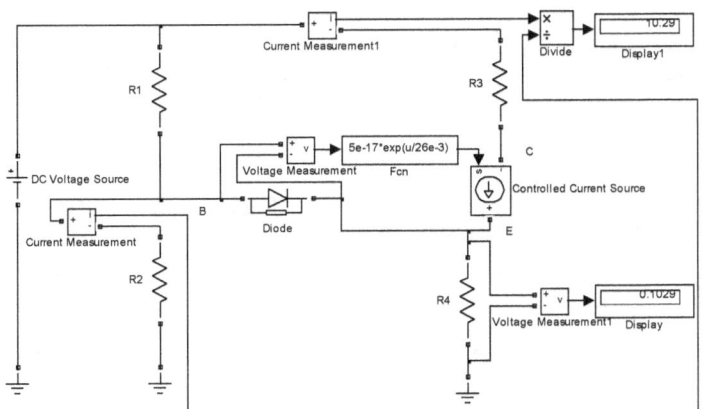

Figure 4.8(a) SIMULINK model for current gain of circuit in figure 4.4(b)

✤ Example 2

Current gain A_I of the resistive divider biasing circuit in figure 4.4(b) is defined as $A_I = \dfrac{I_C}{I_2}$. We intend to obtain the A_I through SIMULINK modeling.

Model of the circuit in figure 4.4(b) you find in figure 4.4(d) which is where we start from. Get two **Current Measurement**, one **Divide**, and one **Display** blocks in the model. Delete the upper connecting line between R1 and R3. Delete the connecting lines originating from junction B. Insert the later blocks and connect like figure 4.8(a) which is the model of this example. After running the model we find $A_I = 10.29$ at **Display1**.

4.9 Transistor modeling with early voltage

When early voltage is included in the small signal model of an NPN transistor, the model in figure 4.6(a) is modified as figure 4.8(b) which employs an additional resistor r_A across collector and emitter. The r_A is obtained from $r_A = \dfrac{V_A}{I_C}$ where V_A is the user-supplied early voltage. The SIMULINK counterpart you see in figure 4.8(c). Say we have $V_A = 10\,V$ and $I_C = 0.01\,A$, copy-paste the **R** to get **R1** in the model of figure 4.6(b), rename the **R1** as **RA**, enter the **Resistance** of RA as **10/0.01** for r_A in the parameter window, and connect like figure 4.8(c). In a similar fashion we model a PNP transistor too starting from the model of figure 4.6(d).

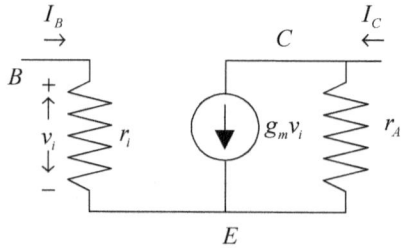

Figure 4.8(b) Small signal model of an NPN transistor subject to early effect

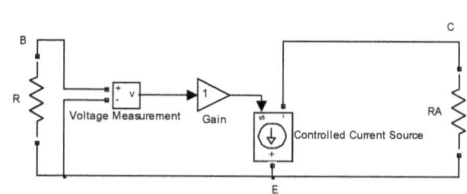

Figure 4.8(c) SIMULINK model of an NPN transistor on inclusion of early effect

4.10 Special purpose block for an electronic circuit

If we have to employ same block for different electronic circuits, we may design the block on our own which facilitates the model building to a great extent. Here in this section we illustrate two examples on that. The first one is for equivalent resistance finding and the second one is for small signal transistor model. In fact you can design any specific block this way. Once designed, save the block in a SIMULINK file, remember the path or folder name of the file, and invoke the file during model building.

✦ Designing an equivalent resistance finding block

Although th-ere is specific block for impedance, uses of those tools for academic circuit become very clumsy. We suggest the technique of section 2.6 for finding the input and output impedances of transistors. Since we apply same block both for input and output impedances, implementing a single block is desirable.

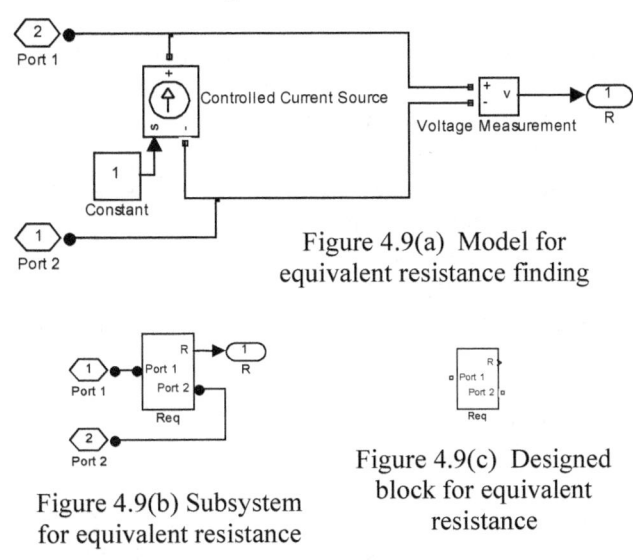

Figure 4.9(a) Model for equivalent resistance finding

Figure 4.9(b) Subsystem for equivalent resistance

Figure 4.9(c) Designed block for equivalent resistance

Let us put the block elements in determining the equivalent resistance of section 2.6 in a new SIMULINK model file like figure 4.9(a). Get two **Connection Port** and one **Out1** blocks in the model (appendix B), rename (chapter 1) them as **Port 1**, **Port 2**, and **R** respectively, and connect like figure 4.9(a), just to make sense with the equivalent resistance. Click **Edit** menu in the model followed by **Select all** or press **Crtl + A** key to select all elements in the model, bring mouse pointer on the selection, rightclick on the mouse to see a popup, select **Create Subsystem** in the popup, and be prompted with the figure 4.9(b). We renamed the **Subsystem** as **Req**. Delete the **Port 1**, **Port 2**, **R**, and connecting lines to get figure 4.9(c) shown block. What did we obtain? We

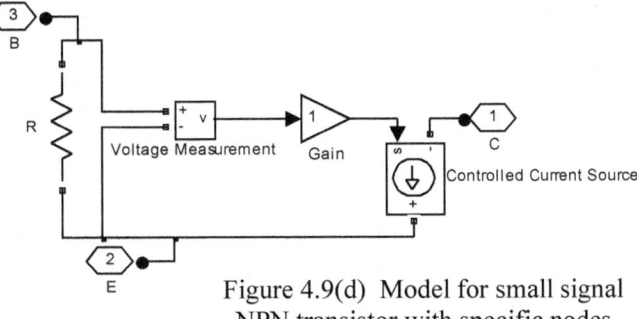

Figure 4.9(d) Model for small signal NPN transistor with specific nodes

designed the block to compute the equivalent resistance between any two nodes of an electrical circuit. Save the SIMULINK file by name **Req** and connect it wherever necessary. You also need to remember the path or folder where the **Req** is saved in. In other words the **Req** imitates the action of an ohmmeter.

✢ Designing a small signal NPN or PNP transistor

In section 4.6 we addressed how the small signal model can be implemented by using transconductance which you find in figure 4.6(b). Get three **Connection Port** blocks in the model of figure 4.6(b), delete the annotations **B-C-E**, rename the three **Port** blocks as **B, C,** and **E** to make sense with base-collector-emitter, and connect them like figure 4.9(d). You may need to flip some **Port**. Anyhow click the **Edit** menu in the model followed by **Select all** or press **Crtl + A** key to select all elements in the model, bring mouse pointer on the selection, rightclick on the mouse to see a popup, select **Create Subsystem** in the popup, and be prompted with the figure 4.9(e). We renamed the **Subsystem** as **NPN**. Delete the **B, C, E,** and connecting lines to get

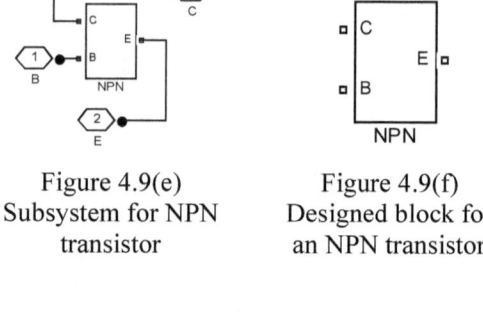

Figure 4.9(e)
Subsystem for NPN
transistor

Figure 4.9(f)
Designed block for
an NPN transistor

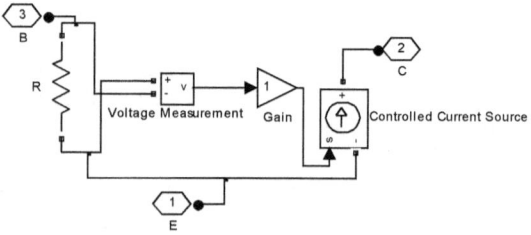

Figure 4.9(g) Model for small signal PNP
transistor with specific nodes

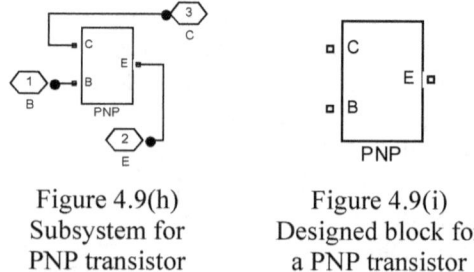

Figure 4.9(h)
Subsystem for
PNP transistor

Figure 4.9(i)
Designed block for
a PNP transistor

figure 4.9(f) shown block which is equivalent to the symbol of figure 4.1(a). Finally save the SIMULINK file by the name **NPN** for later use.

For the PNP counterpart you design the block like figure 4.9(g) starting from the model in figure 4.6(d). The **Subsysetm** and iconic appearance are depicted in figures 4.9(h) and 4.9(i) respectively. Whether

NPN or PNP, all you need is follow the lead of **B-E-C** or base-emitter-collector.

4.11 Cascaded transistor modeling

Cascaded transistor is no different from the models we have introduced in previous sections. A two stage cascade means we have to model two transistors applying aforementioned techniques. Regarding the transistor connection, it can be side by side, on top of another, back to back, face to face, etc. Here we bring some examples along with the modeling description without voltage or current finding, just to make the text less voluminous.

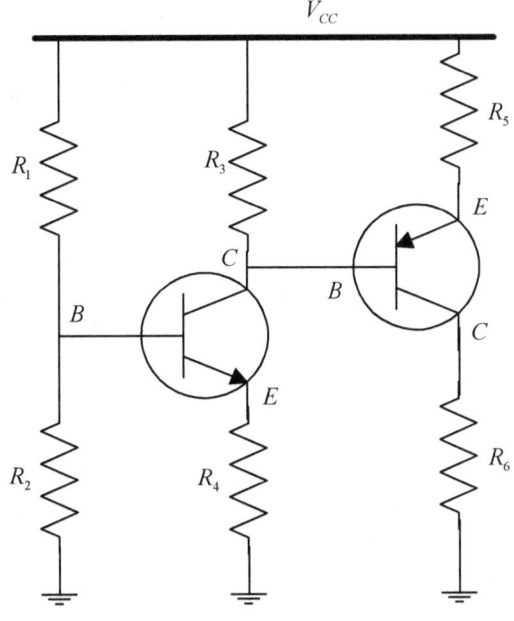

Figure 4.10(a) Two stage cascade of NPN and PNP transistors

✦ **Example 1**

Figure 4.10(a) presents a two stage cascade of NPN and PNP transistors. We wish to model the two stage using common emitter current gain, each being 100.

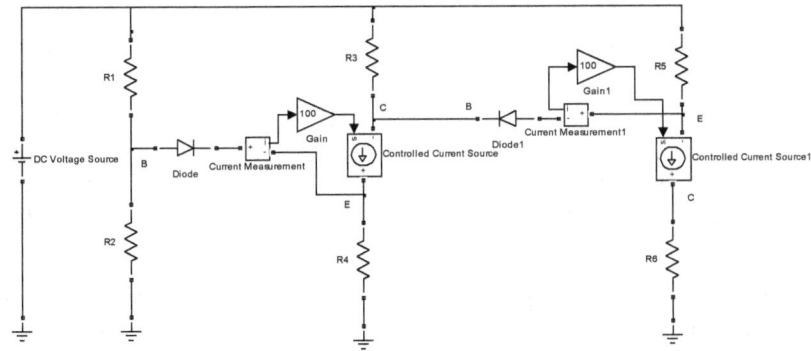

Figure 4.10(b) SIMULINK model of two stage cascade in figure 4.10(a)

Since the NPN transistor is connected as resistive divider biasing, the best way to start is from the model in figure 4.4(d). It is of utmost importance that in multistage connection the diode (section 3.1) should be ideal otherwise there will be property shift of each transistor. Anyhow in figure 4.4(d) delete both **Voltage Measurement** blocks, get two **Gain** and two **Current Measurement** blocks, turn the **Diode** as ideal by entering appropriate parameters, copy-paste the **Controlled Current Source**, copy-paste the **Diode**, flip the **Diode1** and **Current Measurement1**, delete the **Fcn** and **Display**, copy-paste the **R4** two times, copy-paste the **Ground**, enter both **Gains** as 100, add the annotations for example B for base, and finally connect the blocks according to figure 4.10(b).

◆ **Example 2**

Figure 4.10(d) depicts a vertical cascade of two transistors. In order to model the cascade, you may start from the model in figure 4.10(b). The model in its entirety you find in figure 4.10(c). Delete unnecessary blocks or elements. The input, output, and base

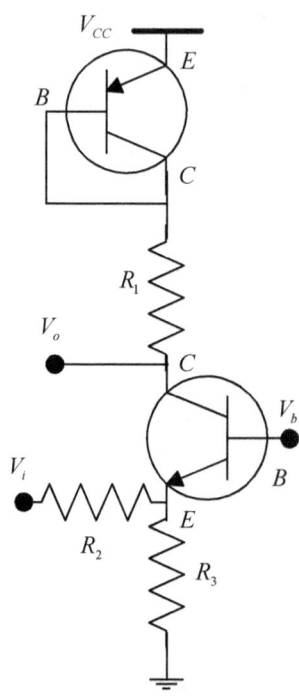

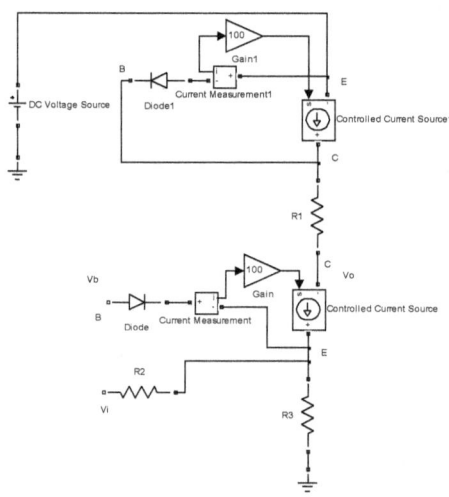

Figure 4.10(c) SIMULINK model of two stage cascade in figure 4.10(d)

Figure 4.10(d) Two stage vertical cascade of NPN and PNP transistors

applied nodes are labeled by annotations Vi, Vo, and Vb respectively. Suppose the input V_i or base voltage V_b is some constant voltage which you can get from another **DC Voltage Source**. If the output V_o value or waveshape is needed, you can connect **Display** or **Scope** as we did before.

◆ **Example 3**

Can we implement example 2 using designed block of transistors?

Certainly, we can. Figure 4.10(e) is the answer. The two stage cascade of figure 4.10(d) is now modeled by using the NPN and PNP subsystems of section 4.10. Obviously this approach is better since we do not have to go through the details of a transistor.

✦ Example 4

As another example of cascaded transistors, what if we have the circuit in figure 4.10(f)?

Having gone through example 3, one would not find it difficult to implement. Figure 4.10(g) is the SIMULINK model using our predesigned NPN block. Unlike previous examples now we have two

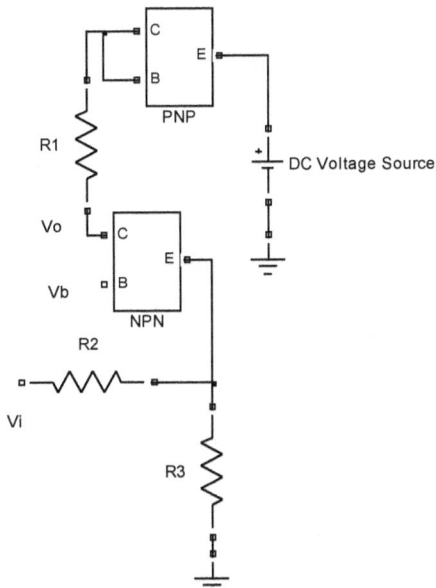

Figure 4.10(e) Two stage vertical cascade using designed NPN and PNP blocks

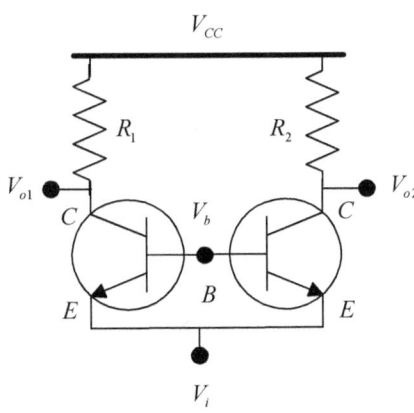

Figure 4.10(f) Two back to back NPN transistors

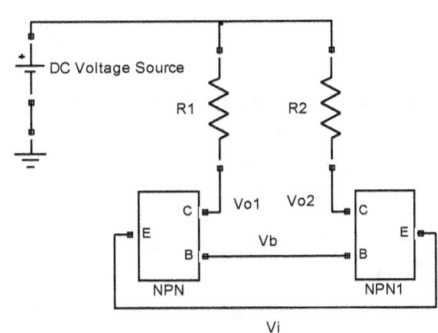

Figure 4.10(g) Model of back to back stage in figure 4.10(f)

outputs. In conjunction with one **Ground** block we can apply V_i or collect outputs if it is necessary.

4.12 Input and output impedances after transistor modeling

Although there is specific block for impedance, use of the tool for academic circuit becomes very clumsy. We suggest the technique of section 4.10 for finding the input and output impedances of transistors. Figure 4.9(c) shows the icon appearance of **Req** which we exercise for input-output impedance finding. All we do is connect the **Port 1** and **Port 2** at the required nodes and view the equivalent by connecting a **Display** at R port of the block. We know that first operating voltages and currents are found from direct current analysis and then small signal model is derived to find input or output impedance. Let us see the following two examples in this regard.

✦ Example 1

It is given that the input resistance R_{in} for the resistive divider biasing circuit in figure 4.4(b) is 3955.5 Ω which we intend to obtain through SIMULINK modeling.

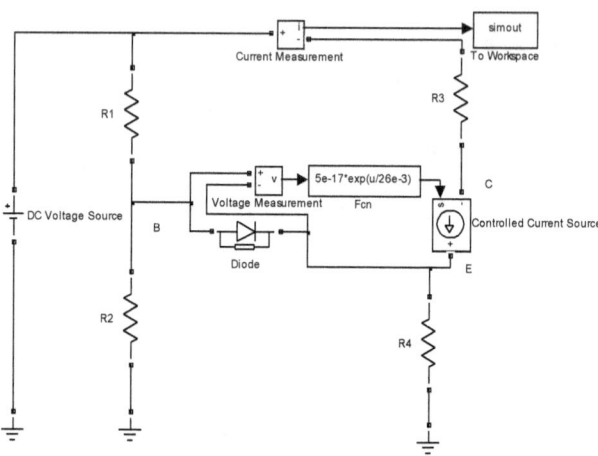

The circuit is modeled in figure 4.4(d) from which collector current I_C is needed for r_i and g_m so we get them

Figure 4.11(a) Model for small signal parameters of circuit in figure 4.4(b)

from the model in figure 4.11(a) after modification. Inserted is a **Current Measurement** block in series with R_3 to find the I_C and sent is the I_C to MATLAB workspace through **To Workspace** block. We must set the **Save format** of **To Workspace** as **Array** because it has other options which are not helpful for this problem. Also did we delete unnecessary blocks. Run the model and values of I_C are available to MATLAB as default variable **simout**. Many values are due to the time adaptive strategy of SIMULINK from which we require only the last one so get it by:

>>Ic=simout(end); ↵

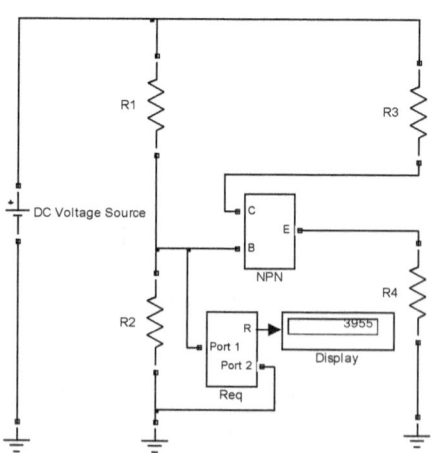

Figure 4.11(b) Model of figure 4.11(a) for input resistance finding

The g_m and r_i of figure 4.6(a) or 4.6(b) are obtained from $g_m = \dfrac{I_C}{V_T}$ and $r_i = \beta \dfrac{V_T}{I_C}$ respectively which are calculated considering $\beta = 100$ as:

```
>>gm=Ic/26e-3; ri=100*26e-3/Ic; ┘
```

Small signal parameters are at our hand so we need to remodel the circuit of figure 4.11(a). Figure 4.11(b) shows the small signal model in which we replace the transistor by NPN of section 4.10. Also do we connect **Req** of the same section at the required input resistance ports. In order to enter the small signal parameters, enter the **Amplitude (V)** of the **DC Voltage Source** as 0, doubleclick the **NPN** of figure 4.11(b), enter **Gain** as **gm** on doubleclicking it, enter **Resistance** as ri on doubleclicking the **R**, run the model, and find the R_{in} at the **Display**.

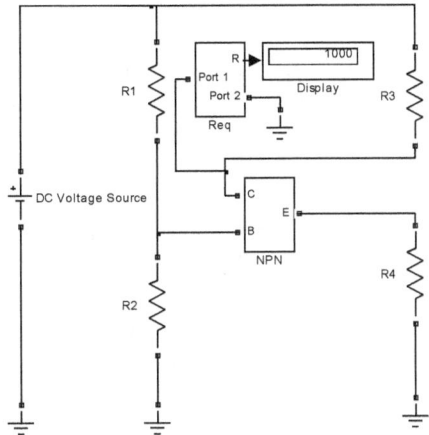

Figure 4.11(c) Model of figure 4.11(a) for output resistance finding

♦ Example 2

The output resistance (collector to ground as in figure 4.4(b)) of example 1 circuit is $1000\,\Omega$ that we wish to verify.

Start from the model in figure 4.11(b) and connect the **Req** block at the required ports. Figure 4.11(c) is the model for output resistance finding, run the model and find the R_{out} at the **Display**.

** Note that drawing another model is not appealing. The best approach is open a new model file, choose **Select all** in working model (e.g. figure 4.11(a)) from the **Edit** menu (or use **Ctrl+A** key), paste the selection in new model, and modify whatever is necessary. As another option, copy-paste the whole file in folder managing window and work on the copied one.

4.13 Connection topology - CB, CE, and CC

Having gone through previous sections, one would not find it difficult to model any topology – common base (CB), common collector (CC), or common emitter (CE). We know that out of the three branches, where we connect input and output decides the type of topology in a transistor. Figure 4.1(d) depicts input at the base and output at the collector so it is common emitter topology and the model is in figure 4.1(e). In figure 4.10(f) the output is from collector and the input is towards emitter so each transistor is in CB topology and the model you find in figure 4.10(g). We hope following these two leads one would easily be able to model the third topology which is CC.

4.14 Transistor operating mode - active and saturation

MATLAB or SIMULINK whichever platform we work in, the computing is numerical and discrete. The mode of operation of a transistor if we have to decide, that also takes place through some numerical tactic. Now we illustrate how to decide active or saturation mode of a transistor numerically.

♦ Active mode finding numerically

We know that for the transistor in figure 4.1(d), the operating mode will be active if the condition $V_{CE} > V_{BE}$ is satisfied or $V_{CE} - V_{BE} > 0$ or some positive number. We collect both V_{CE} and V_{BE} by **Voltage Measurement** or other block and determine

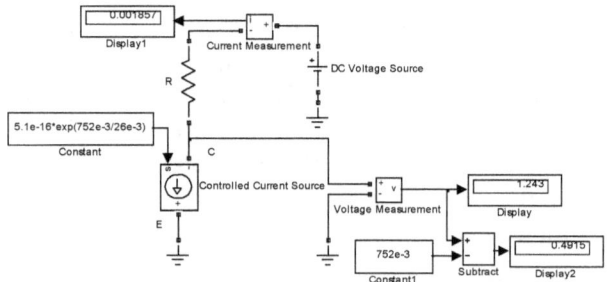

Figure 4.12(a) Model for determining active mode of transistor in figure 4.1(d)

$V_{CE} - V_{BE}$. The difference we view in a **Display**. Figure 4.1(e) is the model of circuit in figure 4.1(d) from which we start and derive the model of figure 4.12(a) for determining active mode. Since the V_{BE} is inside the **Constant**, we generate it by **Constant1**. Copy-paste the **Constant** in the model and enter **752e-3** as **Value** on doubleclicking it. Copy-paste **Display**, get one **Subtract** block, connect later blocks like figure 4.12(a), run the model, and find 0.4915 at **Display2** which is a positive quantity indicating active mode.

♦ Saturation mode finding numerically

In just addressed transistor if the condition $V_{CE} < V_{BE}$ is satisfied or $V_{CE} - V_{BE} < 0$ or some negative number, the transistor is in saturation mode. In the model of figure 4.12(a) the R is $1\,K\Omega$. What if we make it $5\,K\Omega$? Doubleclick the **R**, enter its **Resistance** as **5e3**, run the model, and find $V_{CE} - V_{BE} = -6.935$ at the **Display2** indicating saturation mode.

◆ Finding the transition between active and saturation modes numerically

From last two simulations it is evident that if R changes from $1\ K\Omega$ to $5\ K\Omega$, the transistor of model 4.12(a) undergoes from active to saturation mode. How do we determine the transitional value of R? In this sort of problem MATLAB-SIMULINK combined strategy is required. In SIMULINK we make the R as variable and control the R from MATLAB. The **To Workspace** transports SIMULINK data i.e. $V_{CE} - V_{BE}$ or output of **Subtract** (figure 4.12(a)) into MATLAB (through default variable **simout**) for every R in the model. The **To Workspace** needs **Save Format** setting as **Array**.

Now move on to MATLAB. User has to select some resolution of R and generate a row/column vector. Data accumulation of appendix D.3 will be applied to hold the $V_{CE} - V_{BE}$ for different R. We generate different R as a row vector by **R1=[1:0.1:5]*1e3**; where **R1** is user-chosen and resolution chosen is $0.1\ K\Omega$. SIMULINK solver adaptively selects step size by default from which the last value of **simout** refers

to $V_{CE} - V_{BE}$ we are interested in i.e. only **simout(end)** is required. We select every R of **R1** by a for-loop (appendix D.4). Doubleclick the **R** of model in figure 4.12(a), enter its **Resistance** as **R**, get a **To Workspace**, connect the **To Workspace** like figure 4.12(b), and save the file by some name **test**. Staying in MATLAB you can run the **test** by the command **sim**.

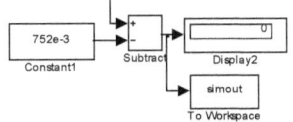

Figure 4.12(b) Transporting $V_{CE} - V_{BE}$ into MATLAB (rest portion of the model is not shown)

Anyhow the complete command line is the following:

```
>>R1=[1:0.1:5]*1e3; D=[ ]; for R=R1,sim('test.mdl'), D=[D simout(end)];end ⏎
```

Run the above line in MATLAB. Ignore MATLAB displayed warnings. The $V_{CE} - V_{BE}$ difference values are available in **D** where **D** is a user-chosen variable. Both **R1** and **D** are a row matrix, side by side viewing first tuning to a column matrix (chapter 1) occurs by:

```
>>[R1' D'] ⏎
```

```
ans =
   1.0e+003 *
            1.0000    0.0005
            1.1000    0.0003
            1.2000    0.0001
            1.3000   -0.0001
            1.4000   -0.0003
               ⋮
```

As the return indicates, the transition is between 1.2 and 1.3 $K\Omega$ or 1.25 $K\Omega$ (average). The **1.0e+003 *** is basically 10^3 or $1\ K\Omega$ (appendix A).

With this simulation we bring an end to the chapter.

Exercises

1. An NPN transistor Q has the parameters $I_S = 8.3 \times 10^{-15} A$, $V_{BE} = 777 \, mV$, and $V_T = 25 \, mV$ where the symbols have their usual meanings. Model the transistor in SIMULINK.

2. Referring to figure 4.1(d), the electronic circuit has $R = 1.25 \, K\Omega$ and $V_{CC} = 10 V$ along with transistor parameters $I_S = 7.3 \times 10^{-16} A$, $V_{BE} = 750 \, mV$, and $V_T = 25.5 \, mV$. Determine the current I_C and voltage V_C of the transistor using SIMULINK modeling.

3. The PNP transistor of figure 4.1(f) has the parameters $I_S = 0.8 \times 10^{-15} A$ and $V_T = 25 \, mV$ in conjunction with bias voltages $V_{BB} = 2.1 \, V$ and $V_{EE} = 2.9 \, V$ where the symbols have their usual meanings. The V_{audio} changes from 0 to $50 \, mV$ and the collector circuit resistance is 50Ω. Determine the current I_C and voltage V_R for both voltage bounds of V_{audio} by SIMULINK modeling.

4. Obtain the load line characteristic of the transistor circuit in figure 4.2(a) where $V_{CC} = 10 V$ and R_L changes from 0 to $800 \, \Omega$. The transistor parameters are $I_S = 8.3 \times 10^{-16} A$, $V_{BE} = 777 \, mV$, and $V_T = 25 \, mV$ where the symbols have their usual meanings. What inference do you get from the load line?

5. In question (4) determine the value of R_L such that $V_C \geq 0$.

6. Graph I_C versus V_{BE} of the NPN transistor in figure 4.3(a) for $I_S = 9 \times 10^{-15} A$ and $V_T = 25.5 \, mV$ over $400 mV \leq V_{BE} \leq 750 mV$ where the symbols have their usual meanings.

7. In question (6) graph I_B versus V_{BE} of the transistor for $\beta = 75$.

8. In question (6) graph I_C versus V_{BE} of the transistor for $I_{S1} = 9 \times 10^{-15} A$, $I_{S2} = 45 \times 10^{-15} A$, and $I_{S3} = 90 \times 10^{-15} A$.

9. Say we have $V_{BE1} = 0.65 \, V$, $V_{BE2} = 0.7 \, V$, and $V_{BE3} = 0.75 \, V$ for the transistor of question (6) with $I_S = 15 \times 10^{-16} A$ and wish to graph I_C versus V_{CE} over $1.1V \leq V_{CE} \leq 1.7V$.

10. In the resistive divider biasing circuit of figure 4.4(b), we have $V_{CC} = 3.5 \, V$, $R_1 = 10 \, K\Omega$, $R_2 = 12 \, K\Omega$, $R_3 = 3 \, K\Omega$, $R_4 = 300 \, \Omega$, $I_S = 1 \, 5 \times 10^{-16} A$, $V_T = 26 \, mV$, and $V_\gamma = 0.7 V$ (for diode model of the transistor) where the symbols have their usual meanings. Compute V_E and I_C by SIMULINK modeling.

11. In the resistive divider biasing circuit of figure 4.4(b), we have $V_{CC} = 15 V$, $R_1 = 100 \, K\Omega$, $R_2 = 50 \, K\Omega$, $R_3 = 5 \, K\Omega$, $R_4 = 3 \, K\Omega$, $\beta = 100$, and $V_\gamma = 0.7 V$ (for diode model of the transistor) where the symbols have their usual meanings. Compute V_E and I_C by SIMULINK modeling.

12. Voltage gain of the transistor in figure 4.1(d) is defined as $A_V = \dfrac{V_{CE}}{V_{BE}}$ where

    $V_{BE} = 0.8\,V$, $I_S = 5.77 \times 10^{-17}\,A$, and $V_T = 26\,mV$ and the symbols have their usual meanings. Obtain the A_V through SIMULINK modeling.

13. In question (12) the V_{BE} changes over $0.55V \leq V_{BE} \leq 0.75V$. Obtain the A_V versus V_{BE} variation by SIMULINK modeling.

14. In question (13) determine the bounds of A_V through MATLAB-SIMULINK tactic.

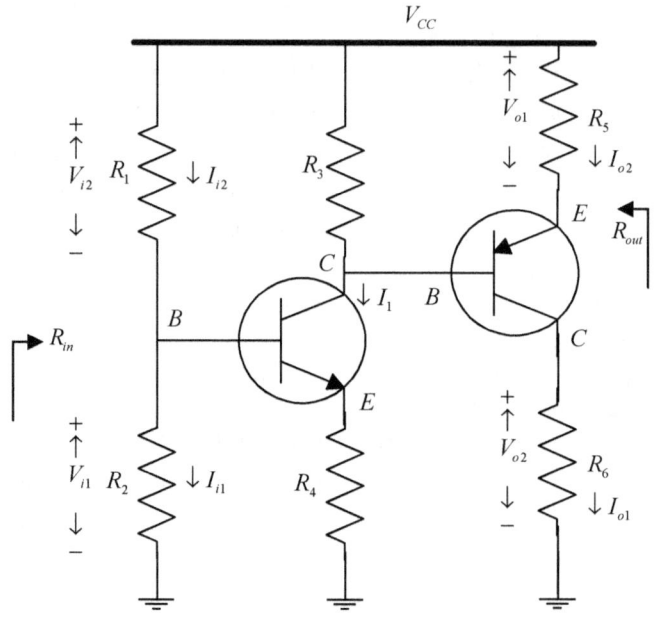

Figure E4.1(a) Two stage cascade of
NPN and PNP transistors

15. In question (10) the voltage gain A_V is defined as $A_V = \dfrac{V_o}{V_i}$ where V_i and V_o

    are base to ground and emitter to ground voltages respectively. Determine the A_V employing SIMULINK approach.

16. In question (10) the current gain A_I is defined as $A_I = \dfrac{I_C}{I_2}$. Determine the

    A_I employing SIMULINK modeling.

17. In the two stage cascade of figure E4.1(a), we have $V_{CC} = 15\,V$, $R_1 = 70\,K\Omega$, $R_2 = 50\,K\Omega$, $R_3 = 5\,K\Omega$, $R_4 = 3\,K\Omega$, $R_5 = 3.5\,K\Omega$, $R_6 = 5\,K\Omega$, $\beta = 90$ for each transistor, and $V_\gamma = 0.7\,V$ (for diode model of each transistor) where the

symbols have their usual meanings. The voltage gain A_V is defined as $A_V = \dfrac{V_{o2}}{V_{i1}}$. Determine the A_V by SIMULINK modeling.

18. In question (17) now the voltage gain A_V is defined as $A_V = \dfrac{V_{o1}}{V_{i2}}$.

19. In question (17) the current gain A_I is defined as $A_I = \dfrac{I_{o1}}{I_{i1}}$. Determine the gain employing SIMULINK modeling.

20. In question (19) the current gain A_I is defined as $A_I = \dfrac{I_{o2}}{I_{i2}}$.

21. In question (17) determine the collector currents of both transistors (i.e. I_1 and I_{o1}).

22. In question (17) determine the input and output resistances of the stage as indicated in figure E4.1(a). You need to apply the results of question (21).

23. In question (17) determine the operating mode of the NPN transistor (i.e. active or saturation). Find the same for the PNP transistor.

24. In question (17) suppose we make $R_1 = 50\ K\Omega$. Do the same as in question (23). If the NPN transistor is in saturation, determine the transitional (i.e. from active to saturation) value of R_1.

25. In question (11) consider an early voltage of $15\,V$ and determine the input and output resistances of the circuit as indicated in figure 4.4(b).

Mohammad Nuruzzaman

Answers:

(1) Hint: section 4.1

(2) $I_C = 0.004332\,A$ and $V_C = 4.585\,V$ Hint: section 4.1

(3) $I_C = 0.06317\,A$ and $V_R = 3.159\,V$ for $0\,mV$ and $I_C = 0.008549\,A$ and $V_R =$ $0.4275\,V$ for $50\,mV$ Hint: section 4.1

(4) Figure A4.1

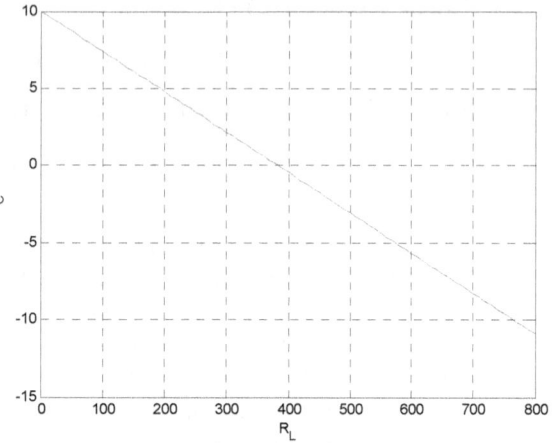

For some R_L, the V_C becomes negative.

Hint: section 4.2

(5) The R_L and V_C are both stored as a row matrix in RL and VC respectively. At the command prompt, form a two column matrix using [RL' VC'] in order to see the two values side by side (appendix D.3). From MATLAB command window,

Figure A4.1   Load line of the transistor in question 4

you find $R_L = 380\,\Omega$ subject to $10\,\Omega$ step size for R_L.

(6) Command: ezplot('9e-15*exp(Vbe/25.5e-3)',[0.4 0.75]). Graph is not shown for space reason.

Hint: section 4.3

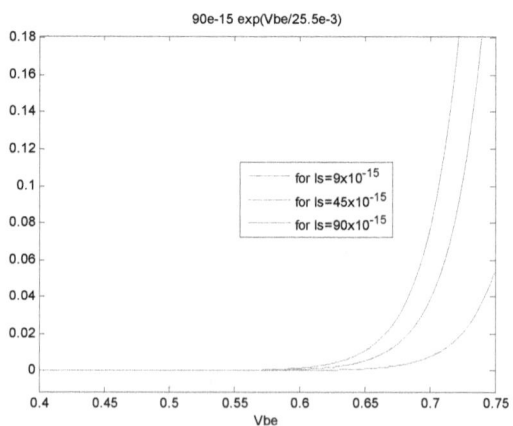

Figure A4.2 I_C versus V_{BE} for different reverse saturation currents

(7) Command: ezplot('9e-15*exp(Vbe/25.5e-3)/75',[0.4 0.75]). Graph is not shown for space reason.

Hint: section 4.3

(8) Command: ezplot('9e-15*exp(Vbe/25.5e-3)',[0.4 0.75])
hold
ezplot('45e-15*exp(Vbe/25.5e-3)',[0.4 0.75])
ezplot('90e-15*exp(Vbe/25.5e-3)',[0.4 0.75])
legend('for Is=9x10^-^1^5','for Is=45x10^-^1^5','for Is=90x10^-^1^5')

Figure A4.2.   Hint: section 4.3

(9) Figure A4.3.   Hint: section 4.3

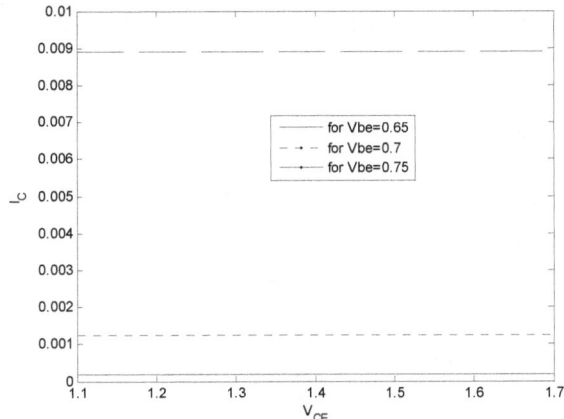

Figure A4.3 I_C versus V_{CE} for different V_{BE}

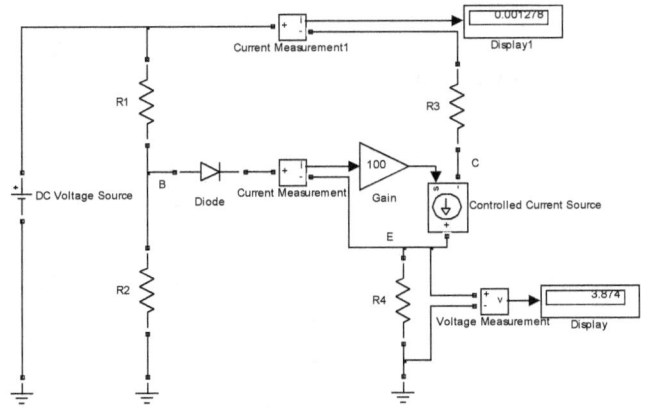

Figure A4.4   Modeling resistive divider biasing circuit using common emitter current gain

(10) $V_E = 0.2732\,V$ and $I_C = 0.000739\,A$ Hint: section 4.4

(11) Figure A4.4. $V_E = 3.874\,V$ and $I_C = 0.001278\,A$.

     Hint: sections 4.4 and 4.5

(12) $A_V = 2.212$

     Hint: section 4.7

(13) Figure A4.5.   Hint: section 4.7

(14) $5.6362 \geq A_V \geq 3.8740$

     Hint: section 4.7

(15) $A_V = 0.2807$ Hint: section 4.7

(16) $A_I = 9.112$ Hint: section 4.8

(17) $A_V = 1.842$

     Hint: sections 4.7 and 4.11

(18) $A_V = 0.8021$

(19) $A_I = 18.42$

     Hint: sections 4.8 and 4.11

(20) $A_I = 16.04$

(21) $I_1 = 0.001653\,A$ and $I_{o1} = 0.002105\,A$

(22) $R_{in} = 26.36\,K\,\Omega$ and $R_{out} = 65.9\Omega$ Hint: sections 4.10–4.12

(23) NPN in active indicated by $V_{CE} - V_{BE} = +1.137\,V$

     PNP in saturation indicated by $V_{EC} - V_{EB} = -3.672\,V$ Hint: section 4.14

(24) NPN in saturation indicated by $V_{CE} - V_{BE} = -2.05\,V$

     PNP in saturation indicated by $V_{EC} - V_{EB} = -8.431\,V$

     Active to saturation transition occurs when $R_1 = 61.95\,K\Omega$

(25) Figure A4.6 is the model for input resistance. $R_{in} = 28.2\,K\,\Omega$ and $R_{out} = 4775\Omega$.
Hint: sections 4.6, 4.9, 4.10, and 4.12

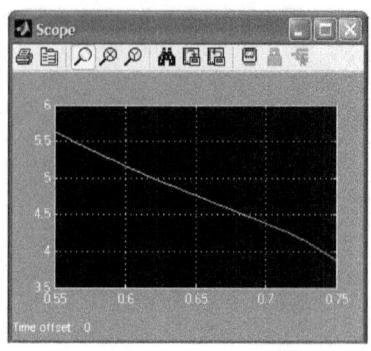

Figure A4.5   Voltage gain variation

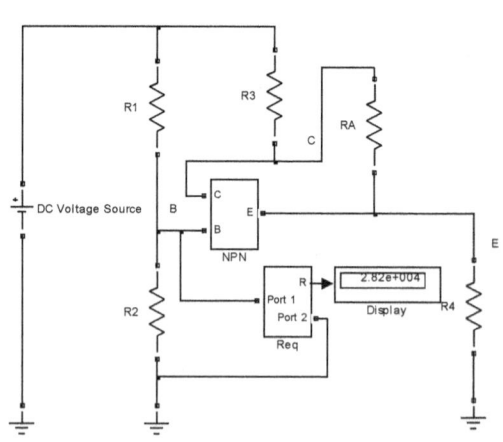

Figure A4.6   Input resistance with early voltage

Chapter 5

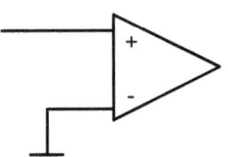

Modeling Operational Amplifiers

Most mathematical operations conducted in electronics circuitry take place by operational amplifiers (OP AMPs), examples of which can be addition, subtraction, integration, etc. An OP AMP is originally consisted of specially designed amplifiers unfortunately there is no particular block in SIMULINK for OP AMP. But simple programming tactic and subsystem formation facilitate the operation which we demonstrate by implementing the following in this short chapter:

 ❖ ❖   Operational amplifier subsystem design in SIMULINK
 ❖ ❖   Implementation on gain, adder, integrator, etc circuits
 ❖ ❖   Some simulation on OP AMP based applications

5.1 How to model an operational amplifier?

There is no dedicated block only for an operational amplifier (abbreviated as OP AMP) for this reason the modeling of an OP AMP needs us to concentrate on its ideal characteristics. Figure 5.1(a) shows the schematic circuit of an ideal OP AMP. Some characteristics of the ideal OP AMP are the following:

(a) the input impedance is infinity between the inverting and non inverting terminals of OP AMP as regards to figure 5.1(a),

(b) the output impedance is zero, and

(c) voltage gain i.e. $A_V = \dfrac{V_0}{V_i}$ is negative infinity.

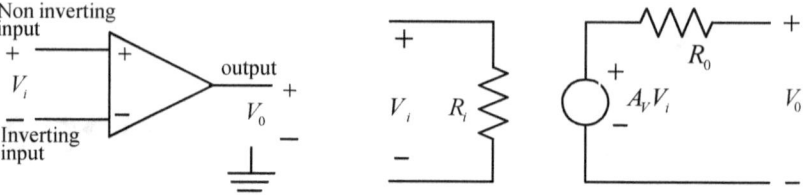

Figure 5.1(a) An ideal OP AMP

Figure 5.1(b) Schematic representation of a practical OP AMP

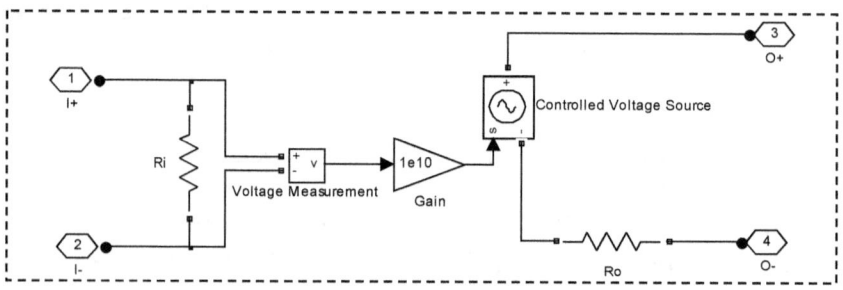

Figure 5.1(c) Modeling a practical OP AMP

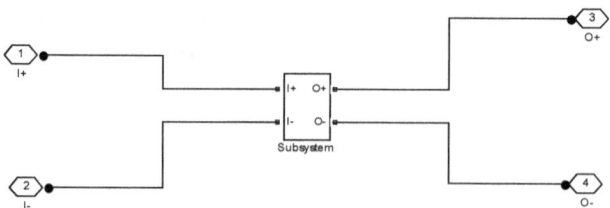

Figure 5.1(d) Forming a subsystem of circuit in figure 5.1(b)

It is absurd to have such type of OP AMP model because a computer never simulates infinity numerically which we need for the input impedance or voltage gain. Figure 5.1(b) presents the circuit model of a practical OP AMP in which R_i, R_0, and A_V indicate input resistance, output resistance, and voltage gain respectively. Concerning the figure, values of the R_i, R_0, and A_V should be ∞, 0, and $-\infty$ for an ideal OP AMP respectively. In our simulation we select some large value for ∞ and negligible value for 0. How small or large depends on the user and electrical circuit we are handling. For instance 10^{10} and 10^{-10} can be considered as ∞ and 0 accordingly codes of the two numbers are **1e10** and **1e-10** respectively (appendix A).

In order to model a practical OP AMP, we need to model the circuit in figure 5.1(b) which is shown in figure 5.1(c). The **Ri**, **Ro**, and **Gain** of figure 5.1(c) correspond to the R_i, R_o, and A_V respectively. The **Voltage Measurement** of figure 5.1(c) basically determines V_i as a function of time and converts the same function as electrical voltage of shown polarity by the **Controlled Voltage Source** block. However we go through the following steps to construct the model:

$\Rightarrow$ open a new SIMULINK model file (subsection 1.2.2),

$\Rightarrow$ get one **Series RLC Branch** block in the model (for R_0) following the link mentioned in appendix B, rename it as **Ro**, doubleclick the **Ro**, select R from the popup (section 2.1), and enter its **Resistance** value as **1e-10** in the parameter window (assuming that R_0 has negligible value of $10^{-10} \Omega$),

$\Rightarrow$ copy just modeled **Ro** in the clipboard, paste it to see **Ro1**, rename it as **Ri** (for R_i), doubleclick the **Ri**, enter its **Resistance** as **1e10** in the parameter window (assuming that R_i has large value $10^{10} \Omega$), rightclick on the **Ri**, and click the **Rotate Block** under the **Format** popup to model a vertical resistor,

$\Rightarrow$ get one **Voltage Measurement**, one **Gain**, one **Controlled Voltage Source**, and one **Connection Port** blocks in the model, and

$\Rightarrow$ doubleclick the **Gain** and enter its **Gain** value as **-1e10** in parameter window assuming that A_V has the large value -10^{10}.

There are four nodes in figure 5.1(b) – two inputs and two outputs, each of which is modeled by one **Connection Port** block. Actually we wish to construct a single system block that holds the whole circuit information of figure 5.1(b) for which the rest procedure is the following:

$\Rightarrow$ rename the **Connection Port** as **I+** (our chosen name for non-inverting node), copy just modeled **I+** in the clipboard, paste it to see **I+1**, rename it as **I-** (our chosen name for inverting node), doubleclick each of the **I+** and **I-**, and make sure the port orientation is like figure 5.1(c) (can be checked by doubleclicking the port and choosing the button box e.g. **Left** in the parameter window), again copy the **I+** in the clipboard, paste it to see **I+1**, rename it as **O+** (our chosen name for positive output node), rightclick on the **O+**, click the **Flip Block** under the **Format** popup to model it as a left facing node, again copy the **O+** in the clipboard, paste it to see O+1, and rename it as O- (our chosen name for negative output node),

⇒ place the blocks relatively like figure 5.1(c) and connect them as shown in the figure,

⇒ use the mouse to select all elements in the model as seen within the dotted target area of figure 5.1(c), and

⇒ rightclick the mouse on having the target area selected, find the option **Create Subsystem** in the popup menu, click the **Create Subsystem**, and find the subsystem as shown in figure 5.1(d).

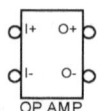

Figure 5.1(e) User-designed icon for practical OP AMP of figure 5.1(b)

Figure 5.1(d) seen **Subsystem** is the block in which the complete circuit of practical OP AMP is hidden. Now the reader can rename the **Subsystem** as **OP AMP** and delete the external nodes of the **OP AMP** so that the block icon of figure 5.1(e) is obtained. Definitely just designed block of figure 5.1(e) has the correspondence as follows: **I+** with non inverting terminal of figure 5.1(a), **I-** with inverting terminal of figure 5.1(a), **O+** with positive node of output V_0, and **O-** with negative node of output V_0.

The reader can save the SIMULINK model by any chosen name and use the **OP AMP** in subsequent modeling.

5.2 How to model a gain circuit?

We use an OP AMP to design a gain circuit, example of which you see in figure 5.2(a). Electronic circuit theory says that $\dfrac{V_o}{V_i} = -\dfrac{R}{R_1}$ and the gain $\dfrac{R}{R_1}$ is implemented according to user-chosen resistors say $R = 10\Omega$ and $R_1 = 1\Omega$. When we have $V_i = 5\,V$, the V_o should be

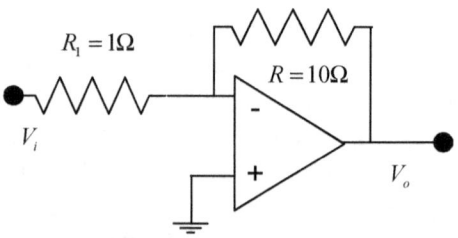

Figure 5.2(a) A gain circuit using OP AMP

$-50V$. We wish to simulate this gain circuit property.

In last section we elaborately mentioned OP AMP design which will be used in this gain circuit. The complete model you see in figure 5.2(b) that needs us to carry out the following procedure:

⇒ open a new SIMULINK model file (subsection 1.2.2),

⇒ get the **OP AMP** subsystem of figure 5.1(e) from the saved SIMULINK to untitled model through copy-paste (the OP AMP parameters R_i, R_0, and A_V are chosen as 10^{10} Ω, 10^{-10} Ω, and 10^{10} respectively),

⇒ get one **Series RLC Branch** block in the model (appendix B), rename it as **R1** for R_1, doubleclick the **R1**, select **R** from the popup (section 2.1), and enter its **Resistance** value as **1** in the parameter window,

⇒ copy just modeled **R1** in the clipboard, paste it to see the **R2**, rename it as **R** (for R), doubleclick the **R**, and enter its **Resistance** value as **10** in the parameter window,

⇒ get one **DC Voltage Source** block in the model and it simulates the $V_i=5V$; doubleclick the **DC Voltage Source** and enter value of **Amplitude (V)** as **5** in the parameter window,

⇒ get one **Voltage Measurement** and one **Display** blocks (for V_0) in the model (section 2.4), and

⇒ get one **Ground** block in the model, copy it to the clipboard, and paste it three times to see the other three blocks.

Place the blocks relatively and connect them like figure 5.2(b) and finally click the simulation icon ▶ at the icon bar to run the model and find the $V_0=-50V$ at **Display** which confirms gain circuit design employing the OP AMP.

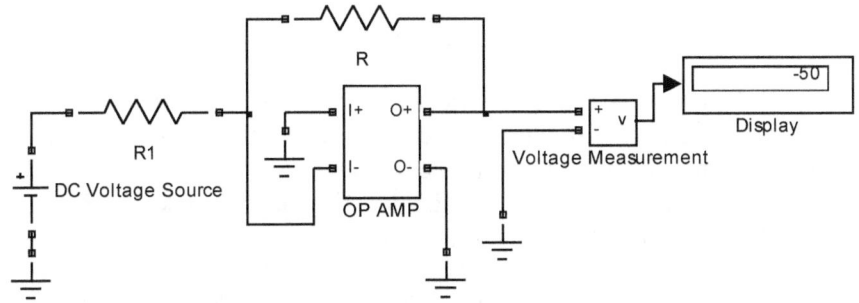

Figure 5.2(b) SIMULINK model for a gain circuit

5.3 How to model a voltage adder circuit?

An adder circuit needs multiple input resistors. Figure 5.3(a) shows a voltage adder circuit implementation using an OP AMP. The electronic circuit intakes three inputs V_1, V_2, and V_3. From the theory of OP AMP we know that output voltage V_o is given by $V_o = -\dfrac{R}{R_1}(V_1+V_2+V_3)$ with $R_1=R_2=R_3$.

Not to mention all resistor values are user-chosen. If there were two inputs, the voltage output equation would have been $V_o = -\dfrac{R}{R_1}(V_1 + V_2)$. Let us consider some example values of the resistors and voltages; $R = R_1 = R_2 = R_3 = 1\Omega$, $V_1 = 5V$, $V_2 = 10V$, and $V_3 = 15V$. What do we expect as the output voltage? Certainly the V_o should

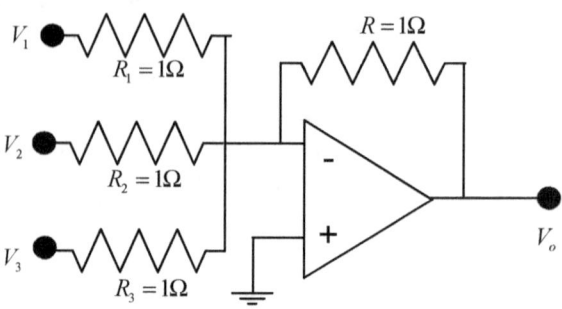

Figure 5.3(a) An adder circuit using OP AMP

be $-30V$ by applying earlier equation. We wish to simulate this circuit phenomenon.

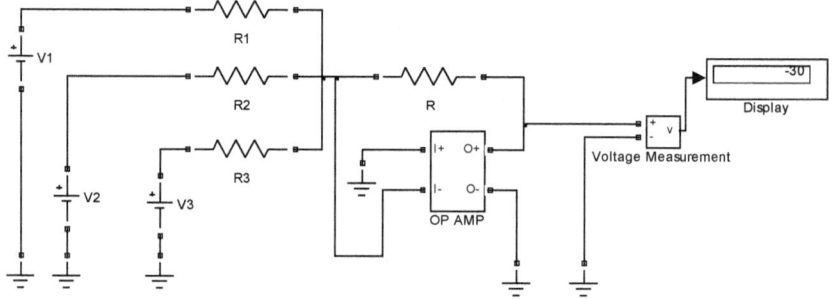

Figure 5.3(b) SIMULINK model for an adder circuit

Last two sections are prerequisite for this problem. Figure 5.3(b) is model for the adder of figure 5.3(a). You had better start from the model in figure 5.2(b) i.e. model for the gain. In order to do so, copy R1 in the clipboard, past it two times for R2 and R3, doubleclick the R and enter its Resistance as 1 in the parameter window, rename the DC Voltage Source as V1, copy V1 in the clipboard, past it two times for V2 and V3, doubleclick the V2 and V3 and enter the value of Amplitude (V) as 10 and 15 in the parameter window respectively, copy one Ground in the clipboard and paste it two times because we need two more, place the blocks relatively and connect them like figure 5.3(b), and finally click simulation icon ▶ to run the model and find $V_o = -30V$ in Display that is we have been successful in designing the adder circuit of figure 5.3(a).

Figure 5.3(b) shows the model considering three inputs in fact any number of inputs you can design in a similar fashion. For instance four inputs require V1-V2-V3-V4 and R1-R2-R3-R4 in the model and one Ground in addition.

5.4 How to model an integrator and a differentiator circuits?

An integrator circuit is also designed by the OP AMP. Figure 5.4(a) depicts an integrator circuit in which the input V_i and output V_o are related by $V_0 = -\frac{1}{RC}\int V_i dt$. The values of R and C are user-defined. Suppose we have $R=1\,\Omega$, $C=1\,F$, and $V_i =1\,V$ for $t \geq 0$. We should be having $V_o = -t$ for $t \geq 0$; we wish to simulate this circuit behavior.

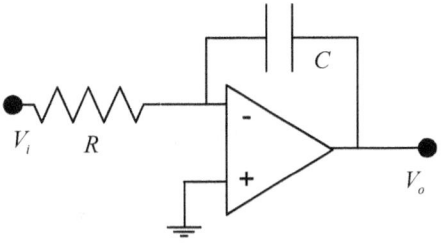

Figure 5.4(a) Integrator using an OP AMP

Figure 5.4(b) shows model of the integrator circuit. The V_i had better be modeled by **Controlled Voltage Source** which provides more flexibility. If the **DC Voltage Source** were applied to implement the integrator, only had the $V_i =1\,V$ been modeled; other waveshapes e.g. triangular had not been possible. Besides SIMULINK assumes some initial voltage on **DC Voltage Source** which deviates output from the standard result. Whatever time function enters to **s** port of **Controlled Voltage Source** appears as electrical voltage from plus and minus ports of the block.

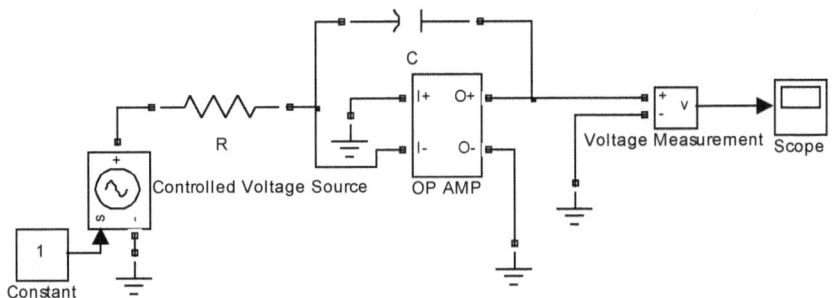

Figure 5.4(b) Modeling an integrator circuit

In order to construct the model of figure 5.4(b), we carry out the following:
  ⇒ open a new SIMULINK model file (subsection 1.2.2),
  ⇒ get the **OP AMP** subsystem of figure 5.1(e) in the model from the saved SIMULINK file (section 5.1, the OP AMP parameters R_i, R_0, and A_V are chosen as $10^{10}\,\Omega$, $10^{-10}\,\Omega$, and 10^{10} respectively),
  ⇒ get one **Series RLC Branch** block in the model (appendix B), rename it as **R** for R, doubleclick the R, select R from the popup (section 2.1), and enter its **Resistance** value as **1** in the parameter window,

⇒ copy just modeled **R** in the clipboard, paste it to see **R1**, rename it as **C** (for C), doubleclick the **C**, select **C** from the popup (section 2.12), and enter its **Capacitance** value as 1 in the parameter window,

⇒ get one **Constant** and one **Controlled Voltage Source** blocks for V_i in the model,

⇒ get one **Voltage Measurement** and one **Scope** blocks in the model for viewing V_o, and

⇒ get one **Ground** block in the model, copy it to the clipboard, and paste it three times to see the other three blocks.

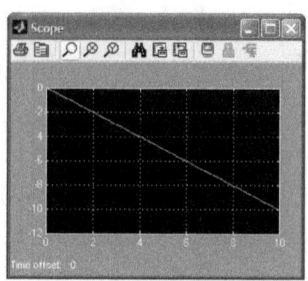

Figure 5.4(c) **Scope** output from integrator

Place the blocks relatively and connect them like figure 5.4(b) and finally click the simulation icon ▶ at the icon bar to run the model. The output $V_o = -t$ is a ramp or straight line passing through origin with slope −1. Doubleclick the **Scope** and click the autoscale icon (figure 3.2(e)), response of which is in figure 5.4(c) i.e. exactly the plot of $V_o = -t$ over default interval $0 \le t \le 10 \sec s$.

Truly speaking the mathematical operation performed is $V_o = -\dfrac{1}{RC}\int_{t=0}^{t=a} V_i dt$ where a is some user-defined t value. If the function after integration is 0 at $t = 0$, no offset or average value is found in the output V_o else some offset will be observed in the V_o which is nothing but V_o at $t = 0$.

Differentiator circuit using OP AMP:

In a differentiator circuit the resistor and capacitor basically interchange their positions which you see in figure 5.4(d). The input-output voltage relationship is given by $V_o = -RC\dfrac{dV_i}{dt}$.

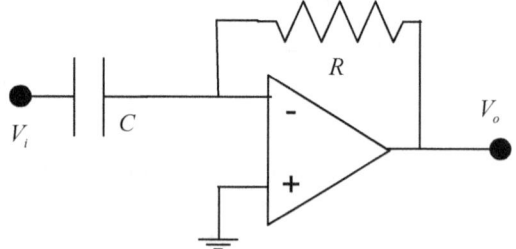

Figure 5.4(d) Differentiator using an OP AMP

There is some crucial point in a differentiator model. If V_i changes from 0 to some other value abruptly, $\dfrac{dV_i}{dt}$ becomes infinity which machine can not handle or in other words stiff differential equation solver is needed in such model.

Suppose $V_i = t$ so V_o should be -1 for $t \geq 0$ with $R = 1\,\Omega$ and $C = 1\,F$ which we intend to simulate.

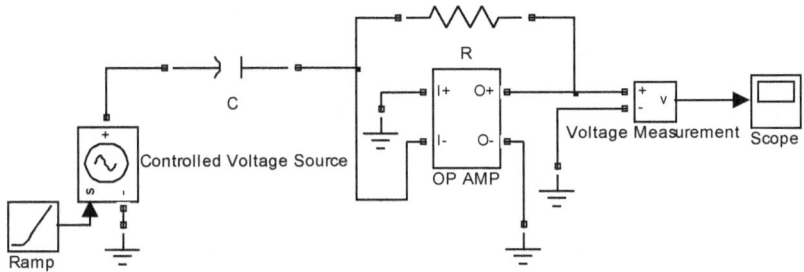

Figure 5.4(e)   Modeling a differentiator circuit

Figure 5.4(e) presents model for the differentiator which we derived from the model of figure 5.4(b) by interchanging capacitor and resistor placements. The $V_i = t$ is simulated by a **Ramp** block that is why **Ramp** is connected to **s** port of the **Controlled Voltage Source**. As far as the correct solver selection is concerned, choose any stiff solver through **Configuration parameters** window (e.g. **ode23s**, figure 1.7(g)). Finally click the

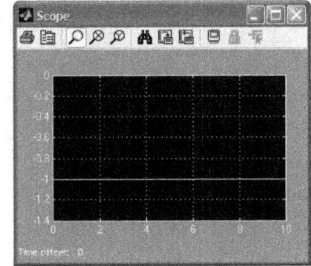

Figure 5.4(f) **Scope** output from differentiator

simulation icon ▶ at the icon bar to run the model. Doubleclick the **Scope** to view the output V_o like figure 5.4(f) with autoscale setting therefore the differentiator simulation is conducted successfully.

Since default interval setting for time is 10 secs in SIMULINK, you find the **Scope** output of figure 5.4(c) or 5.4(f) spanned so.

Note: For modeling complications it is better to avoid simulating a differentiator wherever possible. We can solve the problem by reverse integration.

5.5 How to model a noninverting amplifier?

In previous simulations one undesired fact is the output V_o is inverted i.e. a negative sign appears in the output whether the circuit is gain, adder, or other. When noninversion of the output is

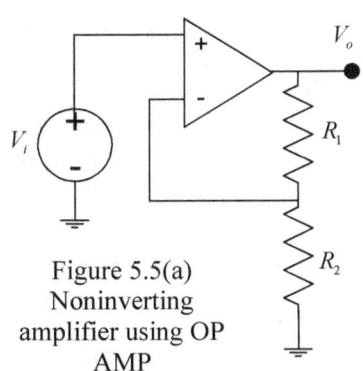

Figure 5.5(a) Noninverting amplifier using OP AMP

required, we employ the circuit of figure 5.5(a). The gain is given by $\dfrac{V_o}{V_i}$

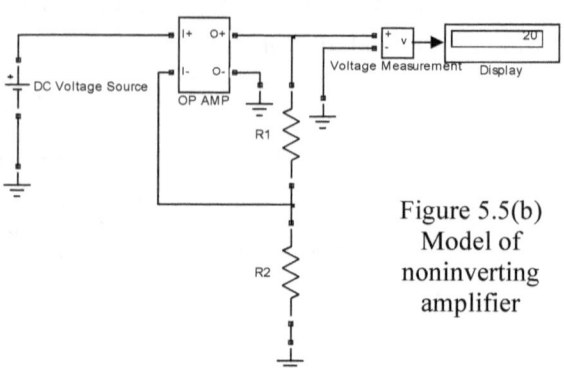

$= \dfrac{R_1 + R_2}{R_2}$. We wish to

model the noninverting amplifier subject to $R_1 = R_2 = 1\Omega$ and verify that $V_0 = 20V$ when $V_i = 10V$.

Figure 5.5(b)
Model of
noninverting
amplifier

In section 5.2 we illustrated the gain circuit (figure 5.2(a)) whose model you find in figure 5.2(b). We wish to rearrange the model of figure 5.2(b) as in figure 5.5(b) that is the model of noninverting amplifier. First delete all functional lines in model of figure 5.2(b), rename R as R2, rightclick on the R1, click the Rotate Block under the Format popup to model it as a vertical resistor, do the same for the R2, doubleclick the DC Voltage Source, enter value of Amplitude (V) as 10 in the parameter window, connect the blocks like figure 5.5(b), and finally run the model to see output V_0 at the Display.

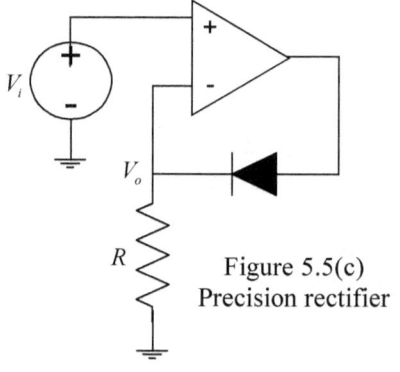

Figure 5.5(c)
Precision rectifier

5.6 How to model a precision rectifier?

We know that a rectifier converts an alternating signal to direct one by making the use of diode. The diode drop $0.7V$ causes the output to remain zero for certain time. In electronics terminology we call it dead band response. For example the swing of a sine wave is $-110V$ to $110V$, after rectification we get the swing from 0 to $109.3V$ due to diode inclusion. This problem is

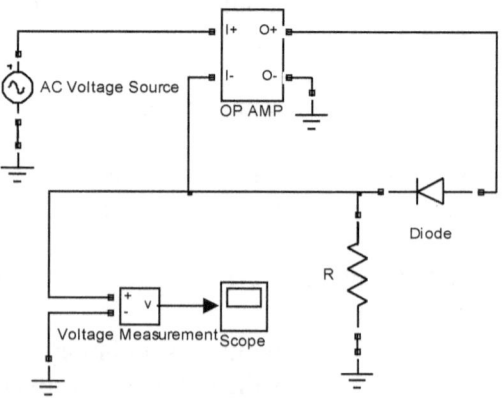

Figure 5.5(d) Model of the precision rectifier in figure 5.5(c)

overcome by a diode and an OP AMP as shown in figure 5.5(c).

As an example consider a sine wave of amplitude $\pm 5V$ and frequency $10\,Hz$. Assume $R = 1000\Omega$ and the diode drop $V_\gamma = 0.7V$. We wish to model the circuit and view output V_o over two periods i.e. $0 \leq t \leq 0.2$ secs.

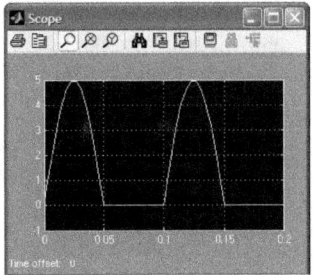

The model for precision rectifier is depicted in figure 5.5(d). In order to construct the model, open a new SIMULINK model (chapter 1), get one **AC Voltage Source** (section 2.14), one **OP AMP** (section 5.1), one **Voltage Measurement**, one **Scope**, one **Series RLC Branch**, one **Diode** (section

Figure 5.5(e) **Scope** output from precision rectifier

3.1), and four **Ground** blocks in the model. Doubleclick the **AC Voltage Source**, enter the values of **Peak amplitude (V)** and **Frequency (Hz)** as **5** and **10** in parameter window respectively, rename the **Series RLC Branch** as **R** (section 1.2.3) to make sense with given element, doubleclick the **R**, select **R** from the **Branch type** popup, enter the **Resistance** as **1000** in parameter window, rightclick on the **R**, click **Rotate Block** under the **Format** popup to model a vertical resistor, doubleclick the **Diode**, enter parameters of the **Diode** as mentioned in section 3.1, flip the **Diode**, click the menu **Simulation** (section 1.2.6) down **Configuration parameters** in model file, select **ode23s** as model **Solver**, and enter the **Stop** time as **0.2**.

Finally place the blocks relatively like figure 5.5(d), connect them as shown in the figure, run the model, and find the output waveshape in the **Scope** like figure 5.5(e) with autoscale setting (figure 3.2(e)).

As quoted, there is no dead band over first half period and the output is from 0 to $5V$.

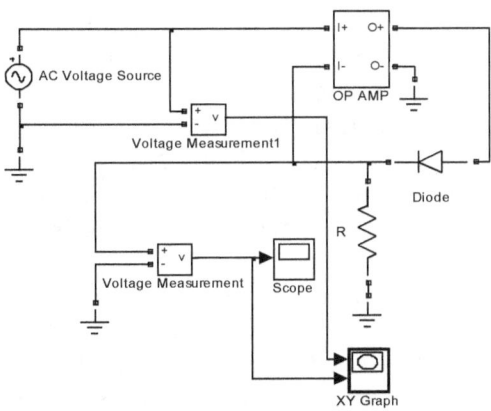

Figure 5.6(a) Model for output-input relationship of precision rectifier

5.7 Relating input and output voltages of an OP AMP

Given an OP AMP circuit, we might need to connect input and output voltages evolving from the circuit. We addressed similar problem in section 3.8 so that is a prerequisite.

Let us consider the last section mentioned input-output voltage of precision rectifier. The input swing is from $-5V$ to $+5V$, so from $0V$ to $+5V$ is output. We intend to view the output versus input voltage.

Figure 5.6(a) shown model is to be formed for the output-input relationship of the precision rectifier starting from the model in figure 5.5(d). Copy-paste the **Voltage Measurement** in the model, get one **XY Graph** block, doubleclick the **XY Graph**, enter x-min and x-max as -5 and 5 respectively for input, enter y-min and y-max as 0 and 5 respectively for output, connect the later blocks like figure 5.6(a), run the model, and doubleclick the **XY Graph** to view the V_o versus V_i like figure 5.6(b) (i.e. **Y Axis** and **X Axis**

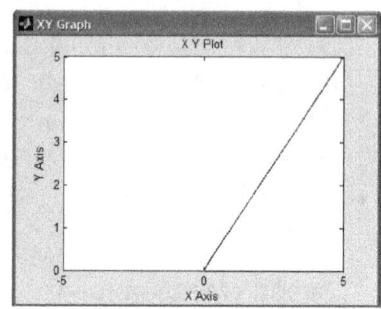

Figure 5.6(b) Output-input relationship from the precision rectifier model

respectively). Certainly the variation makes sense with the output of figure 5.5(e).

In a similar fashion you may obtain other voltage-voltage or current-voltage variation.

5.8 Handling offset in an OP AMP

Offset is just a constant or DC value that can be additive or subtractive for which we employ the **Add** or **Subtract** block respectively. We generate offset by a **DC Voltage Source** or **Constant** block depending on the electronic circuit and add/subtract the offset to/from the concerning function.

Take the example of section 5.6 quoted precision rectifier. We wish to add an offset $0.5V$ to noninverting terminal of the OP AMP in figure 5.5(c).

The electronic circuit model

Figure 5.6(c) Model for offset inclusion

is shown in figure 5.5(d). Starting from that model we have to form the model of figure 5.6(c) for the offset inclusion. Get a **DC Voltage Source** block in the model, rightclick on the **DC Voltage Source**, click **Rotate Block** under the **Format** popup to model a horizontal source, doubleclick the **DC Voltage Source**, enter value of **Amplitude (V)** as 0.5 for offset in the parameter window, delete the upper connection line of **AC Voltage Source**, connect the later block like figure 5.6(c), run the model, and find the **Scope** output as seen in figure 5.6(d) on autoscale setting (figure 3.2(e)).

Now for the same circuit we wish to subtract an offset $1V$ from the output. We need one **Constant** and one **Subtract** blocks in the model of figure 5.5(d). Connection of the later two blocks is seen in figure 5.6(e).

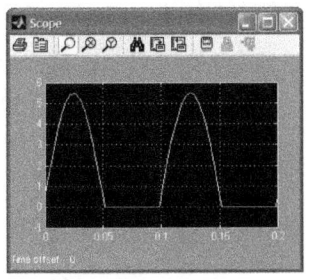

Figure 5.6(d) **Scope** output from model in figure 5.6(c)

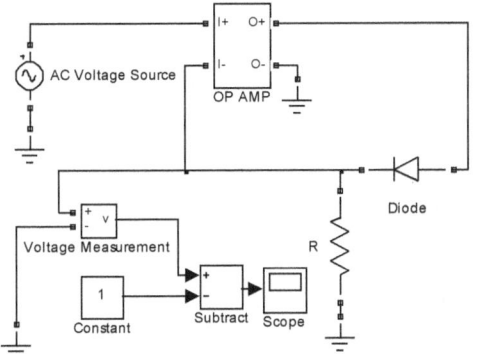

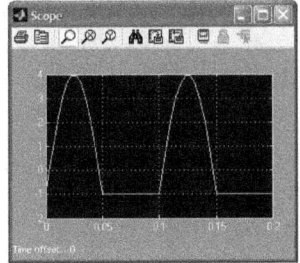

Figure 5.6(e) Model for offset inclusion at the output

Figure 5.6(f) **Scope** output from model in figure 5.6(e)

After running the model you find the **Scope** output like figure 5.6(f) on autoscale setting. How is this offset different from the beginning one? The former has the polarity, and the later does not.

The problem of offset you may encounter in integrator simulation for which later technique can

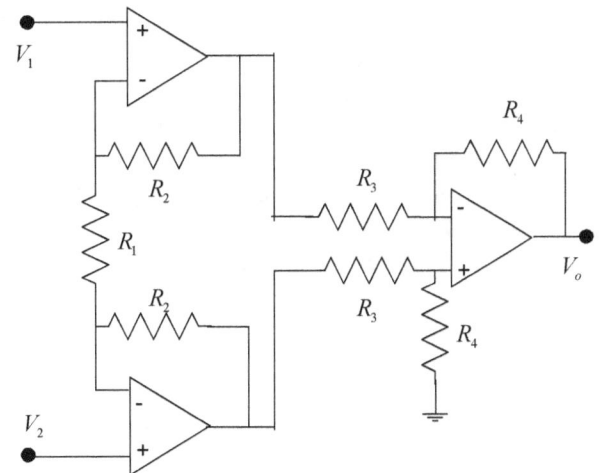

Figure 5.7(a) Instrumentation amplifier using OP AMP

be useful.

5.9 Model on some OP AMP application

Here in this section we illustrate one application simulation on OP AMP. Figure 5.7(a) depicts the electronic circuitry of an instrumentation amplifier employing three OP AMPs where the output of the instrumentation amplifier is given by $V_o = \dfrac{R_4}{R_3}\left(1 + 2\dfrac{R_2}{R_1}\right)V_d$ with $V_d = V_2 - V_1$. Suppose we have $R_1 = 1\ K\Omega$, $R_2 = 2\ K\Omega$, $R_3 = 2\ K\Omega$, $R_4 = 1\ K\Omega$, and $V_d = 5V$. The output should be $V_o = 12.5V$ which needs to be simulated.

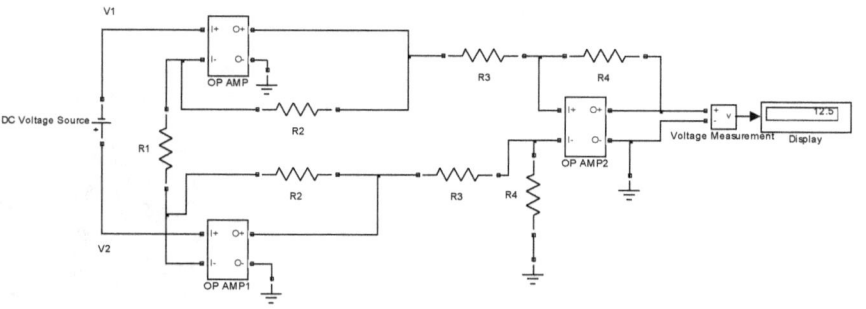

Figure 5.7(b)   Model for instrumentation amplifier

Figure 5.7(b) is the SIMULINK model of the instrumentation amplifier. You need three **OP AMP** (section 5.1) blocks. Open a new SIMULINK model file and get the three blocks. Get one **Series RLC Branch** block in the model, rename it as **R1** (subsection 1.2.3), doubleclick the **R1**, select R from the **Branch type** popup, and enter the **Resistance** as **1000** for R_1 in parameter window. Copy-paste the **R1** and do necessary manipulations (like rotation by 90 degrees) for the other 6 resistors. Get four **Ground**, one **DC Voltage Source**, one **Voltage Measurement**, and one **Display** blocks in the model, rotate twice the **DC Voltage Source** by 90 degrees, and connect the blocks like figure 5.7(b). Enter proper value to each element and run the model. Value at the **Display** confirms the modeling.

Simulation on the instrumentation amplifier brings an end to the chapter.

Exercises

1. Model a practical OP AMP in SIMULINK.
2. Model the gain circuit of figure 5.2(a) in SIMULINK based on gain 4. Choose the feedback resistor as $100\,\Omega$. What is the value of the input resistor R_1? Verify that when $V_i = 9\,V$, the circuit return is $V_0 = -36\,V$.
3. Model the voltage adder circuit of figure 5.3(a) in SIMULINK subject to $R = R_1 = R_2 = R_3 = 100\Omega$, $V_1 = 50V$, $V_2 = 70V$, and $V_3 = -35V$. Verify that the output V_0 is $-85\,V$.
4. Choose the feedback resistor as $1000\,\Omega$ while the others as 100Ω in problem (3). What should be the output V_0? Verify the output by SIMULINK modeling.
5. In question (3) another input resistor R_4 is added and $V_4 = 55V$. What should be the output V_0? Verify the output by SIMULINK modeling.
6. Modify the circuit of figure 5.3(a) such that we get a subtractor designed with $R = R_1 = R_2 = 100\Omega$. Verify the design with $V_1 = 50V$ and $V_2 = 70V$ (i.e. $V_0 = V_2 - V_1$).
7. The integrator of the circuit in figure 5.4(a) has $R = 1\,M\Omega$ and $C = 1\,\mu F$. Suppose the input V_i is $3\,V$. What should be the output V_0? Verify the output by SIMULINK modeling over the interval $0 \le t \le 1\sec$.
8. In problem (7) now V_i is $120\pi \sin 2\pi t f\;V$ where frequency f is $60\,Hz$. Verify the waveshape over two cycles.
9. The differentiator of circuit in figure 5.4(d) has $R = 2\,\Omega$ and $C = 0.5\,F$. Suppose the input V_i is $\dfrac{\sin 2\pi t f}{120\pi}\;V$ where frequency f is $60\,Hz$. What should be the output V_0? Verify the output by SIMULINK modeling over the interval $0 \le t \le \frac{2}{60}\sec s$.
10. Model the instrumentation amplifier of figure 5.7(a) in SIMULINK and verify that $V_o = 15V$ subject to $R_1 = 2\,K\Omega$, $R_2 = 3\,K\Omega$, $R_3 = 2.4\,K\Omega$, $R_4 = 3\,K\Omega$, and $V_d = 3V$.
11. Model the noninverting amplifier of figure 5.5(a) in SIMULINK based on $R_1 = 3\,K\Omega$ and $R_2 = 2\,K\Omega$ and verify that $V_o = 10V$ when $V_i = 4\,V$.
12. Model the precision rectifier of figure 5.5(c) in SIMULINK assuming $R = 750\Omega$ and diode drop $V_\gamma = 0.7\,V$. Obtain output V_o waveshape over two periods if the input is a sine wave of amplitude $\pm 20V$ and frequency $20\,Hz$.
13. Obtain the graphical link between the output and input voltages of the precision rectifier in question (12).
14. In question (7) add an offset $1\,V$ at the noninverting terminal.
15. In question (7) subtract an offset $1\,V$ from the output.

Answers:

(1) Hint: section 5.1   (2) $R_1 = 25\Omega$ Hint: section 5.2

(3) Hint: section 5.3   (4) $V_0 = -850\,V$ (5) $V_0 = -140\,V$

(6) Figure A5.1(a)   Hint: section 5.3

(7) $V_0 = -3t \quad V$.   Figure A5.1(b).

Hint: enter Stop time as 1. Section 5.4 and chapter 1.

(8) $V_0 = -\dfrac{1}{RC}\int_{t=0}^{t=a} 120\pi \sin 2\pi t\, f\, dt = -1 + \cos 2\pi t\, f \quad V$ Figure A5.1(c)

(9) $V_0 = -RC\dfrac{d}{dt}\left(\dfrac{\sin 2\pi t\, f}{120\pi}\right) = -\cos 2\pi t\, f \quad V$ Figure A5.1(d)

Hint: Section 5.4 and chapter 1.

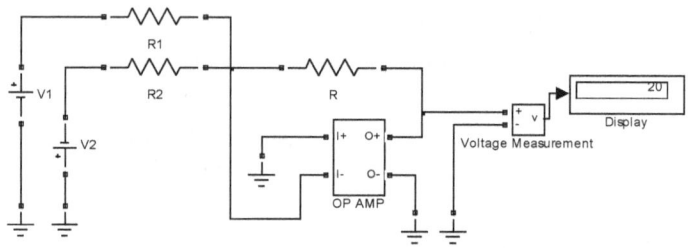

Figure A5.1(a) Modeling a subtractor

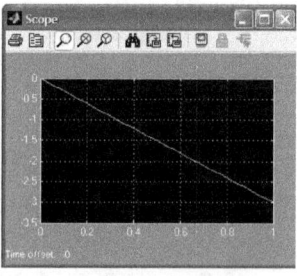

Figure A5.1(b) Scope
output of problem 7

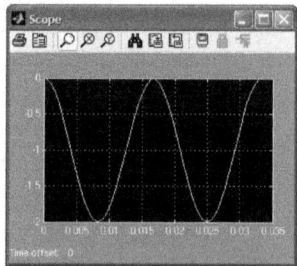

Figure A5.1(c) Scope
output of problem 8

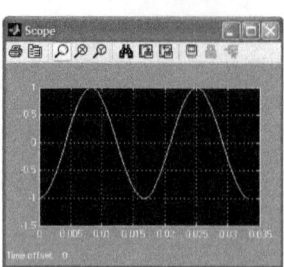

Figure A5.1(d) Scope
output of problem 9

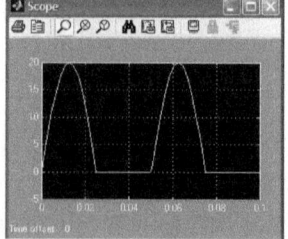

Figure A5.1(e) Scope
output of problem 12

(10) Hint: section 5.9
(11) Hint: section 5.5
(12) Figure A5.1(e)   Hint: section 5.6
(13) Figure A5.2(a)   Hint: section 5.7
(14) Figure A5.2(b)   Hint: section 5.8
(15) Figure A5.2(c)   Hint: section 5.8

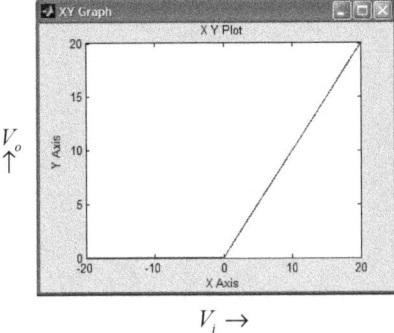

$V_o \uparrow$

$V_i \to$

Figure A5.2(a) Output-input voltage relationship of problem 13

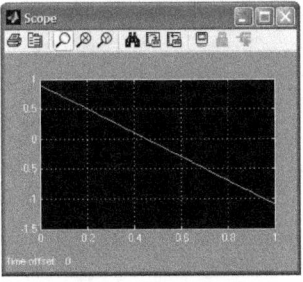

Figure A5.2(b) Output with offset at inverting terminal

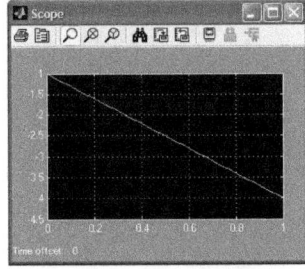

Figure A5.2(c) Output after offset subtraction

Mohammad Nuruzzaman

Chapter 6

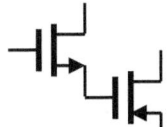 # Modeling Field Effect Transistors

This chapter chiefly focuses simulation on metal-oxide-semiconductor field effect transistor or MOSFET or MOS. Most treatments of MOS electronic circuits follow the same as conducted for BJT in other chapters for this reason very often we have drawn parallels between the two during implementation. Illustrated below outlines the conceptual sequence addressed in this chapter:

♦ ♦ MOSFET I/V characteristics as dependent current source
♦ ♦ Biasing circuit and transconductance based model
♦ ♦ Gain implementation on MOS whether voltage or current
♦ ♦ Input/output impedance and multistage topology

6.1 Implementing I/V characteristics of MOSFETs

In section 4.3 we exercised the I/V characteristics of a transistor by using **ezplot**. The same can be conducted for a metal-oxide-semiconductor field effect transistor or MOSFET. Mainly here we graph the MOS transistor I/V characteristics which are often addressed in introductory electronic circuit text book. All the while our approach has been hand-on computing and graphing on electronic circuits, this chapter is no exception. There is no specific functional tools for MOSFET circuits. We hope our plain illustration will make the study easy for which following examples are devised.

✦ Example 1

Figure 6.1(a) displayed NMOS circuit has the I_D expression which is given as $I_D = \frac{1}{2}\mu_n C_{ox}\frac{W}{L}[2(V_{GS}-V_{TH})V_{DS}-V_{DS}^2]$ subject to device aspect ration $\frac{W}{L}=0.5$, $\mu_n C_{ox}=100$ $\mu A/V^2$, $V_{GS}=0.9V$, and $V_{TH}=0.4V$ where μ_n is the mobility in $m^2/V/s$ and C_{ox} is the gate capacitance per unit area in F/m^2. We intend to graph I_D versus V_{DS} for the transistor.

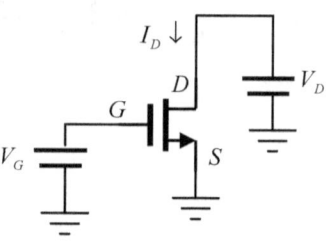

Figure 6.1(a) Basic NMOS transistor circuit

The solution is very simple, just exercise the **ezplot** (appendix E) with the given parameter as follows:

```
>>Id='0.5*100*0.5*(2*(0.9-0.4)*Vds-Vds^2)'; ↵
>>ezplot(Id,[0 1]) ↵
```

The **Id** and **Vds** are user-chosen and indicate I_D and V_{DS} respectively. The variation for V_{DS} is chosen over $0 \le V_{DS} \le 1V$ that is why the second input argument is **[0 1]**. Figure 6.1(b) depicts the I_D versus V_{DS} variation. Since the $\mu_n C_{ox}$ is in $\mu A/V^2$, the vertical axis of the graph shows I_D in μA.

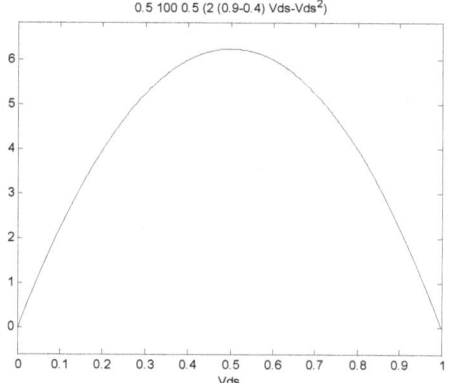

✦ Example 2

In example 1 mentioned circuit now we wish to plot I_D versus V_{DS} for three cases of V_{GS}; for $V_{GS}=0.7V$, $V_{GS}=0.8V$, and $V_{GS}=0.9V$.

Figure 6.1(b) Graph of I_D versus V_{DS} for NMOS transistor circuit

In section 4.3 we exercised similar problem. Graph the first I_D versus V_{DS}, hold the graph, and superimpose the subsequent plots by:

```
>>Id='0.5*100*0.5*(2*(0.7-0.4)*Vds-Vds^2)'; ↵
>>ezplot(Id,[0 1]) ↵
>>hold ↵
>>Id='0.5*100*0.5*(2*(0.8-0.4)*Vds-Vds^2)'; ↵
>>ezplot(Id,[0 1]) ↵
>>Id='0.5*100*0.5*(2*(0.9-0.4)*Vds-Vds^2)'; ↵
>>ezplot(Id,[0 1]) ↵
```

Identifying marks for the trajectories are included by:
>>legend('Vgs=0.7','Vgs=0.8','Vgs=0.9') ⏎
The text **Vgs=0.7** is user-chosen and stands for $V_{GS}=0.7V$, so is the others.
Figure 6.1(c) shows the outcome of the last command line execution. The
ezplot graphs all curves by the same color which makes difficult to identify

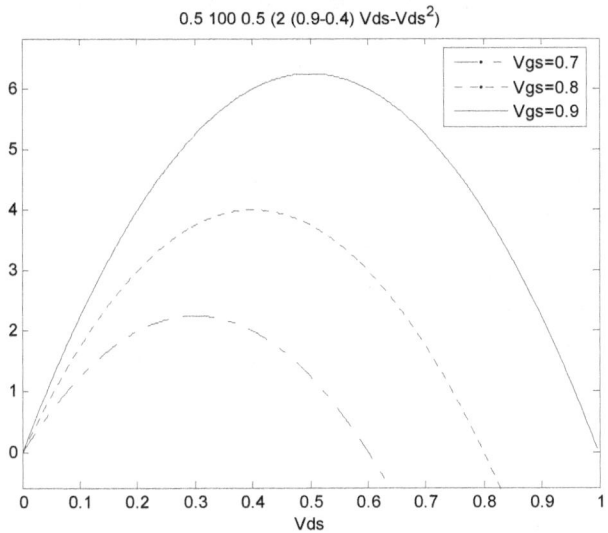

Figure 6.1(c) Graph of I_D versus V_{DS} for several V_{GS}

each curve. Click the Edit plot icon (figure 1.2(c)), select each curve on the
plot, rightclick on the curve to view **Line Style** property in a popup, and
change the **Line Style** property. You may also select the title text and get rid
of the text by deleting it.

◆ **Example 3**
The approximate linear on resistance of the NMOS in example 1 is
defined as $R_{on}=\dfrac{V_{DS}}{I_D}\approx\dfrac{1}{\mu_n C_{ox}\dfrac{W}{L}(V_{GS}-V_{TH})}$. Should the reader need R_{on} versus

V_{GS} plot, exercise **Ron='1/0.5/100/0.5/2/(Vgs-0.4)'; ezplot(Ron,[0 1])** at
the command prompt. The graph is not shown for space reason.

◆ **Example 4**
In example 1 we intend to determine the V_{DS} when $I_D|_{max}$ occurs.

In **maple** package (appendix D.12) there is a function by the name
maximize which solves the problem. The syntax of the function is
maple(maximize under quote, expression to be maximized, reserve word

location under quote). Assuming the I_D is stored in **Id**, we execute the following:

>>maple('maximize',Id,'location') ↵

ans =

6.2500000000000000000000, {[{Vds = 0.5000000000000000000000}, 6.2500000000000000000000]}

The first return is for $I_D|_{max}$. Second brace contents in above indicate that V_{DS} =0.5 V when $I_D|_{max}$=6.25 μA.

6.2 MOSFET as a voltage controlled current source

Much of the discussion of section 4.1 on BJT is applicable here. Figure 6.2(a) shows the symbol of n-type MOSFET or NMOS transistor whose voltage dependent current source model is in figure 6.2(b). Link of the model to SIMULINK block you also find in figure 6.2(c). The symbols have their usual meanings e.g. G or **G** for gate of the transistor. Let us go through the following examples.

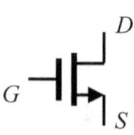

Figure 6.2(a) Symbol of NMOS field effect transistor

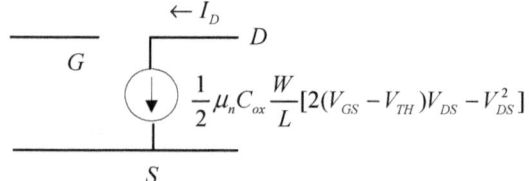

Figure 6.2(b) NMOS as a voltage controlled current source

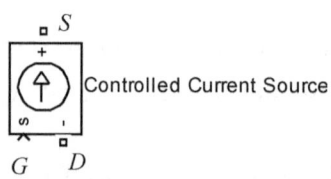

Figure 6.2(c) Linking NMOS terminology to SIMULINK block

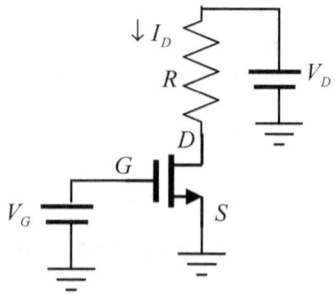

Figure 6.2(d) A basic NMOS transistor circuit

✦ **Example 1**

The NMOS transistor circuit in figure 6.2(d) carries I_D=0.0002 A and V_{DS}=0.8 V in the saturation region whose governing equation is

-144-

$I_D = \frac{1}{2}\mu_n C_{ox}\frac{W}{L}(V_{GS}-V_{TH})^2$ where aspect ration $\frac{W}{L}=\frac{2}{0.18}$, $\mu_n C_{ox}=100\,\mu A/V^2$, V_{GS} $=1\,V$, $V_{TH}=0.4\,V$, $R=5\,K\Omega$, and $V_D=1.8\,V$. We wish to verify the I_D and V_{DS} by SIMULINK modeling.

Details of the model construction (figure 4.1(e)) you find as example 1 in section 4.1. In the **Constant** block we need to feed the code for $\frac{1}{2}\mu_n C_{ox}\frac{W}{L}(V_{GS}-V_{TH})^2$ which is **0.5*100e-6*2/0.18*(1-0.4)^2**, doubleclick the block and do so. Also enter the **R** and **DC Voltage Source** as 5e3 and 1.8 by doubleclicking each for $R=5\,K\Omega$ and $V_D=1.8\,V$ respectively. Include the annotations by deleting previous ones e.g. **G** for G. Figure 6.2(e) is the model we need. Run the model and find the I_D and V_{DS} at **Display1** and **Display** respectively.

Figure 6.2(e)
Model of the
NMOS transistor
circuit in figure
6.2(d) – right side
figure

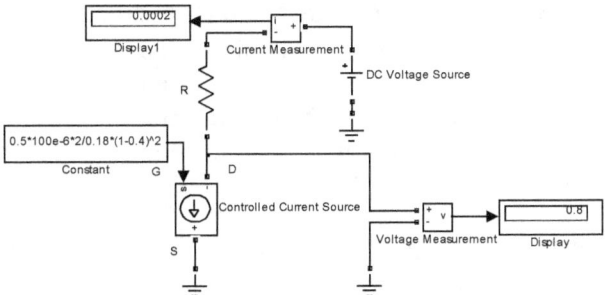

✦ Example 2

Figure 6.3(a) shows a PMOS transistor circuit which is subject to $\frac{W}{L}=\frac{2}{0.18}$, $\mu_n C_{ox}=110$ $\mu A/V^2$, $V_{TH}=0.4\,V$, $R=1\,K\Omega$, $V_G=$ $1\,V$, and $V_S=2.5\,V$ and operating in the saturation region. We wish to determine the I_D and V_R by SIMULINK modeling.

Analogous modeling to this problem you find as example 2 in section 4.1. Figure 6.3(b) shows SIMULINK model for this problem which we derived from

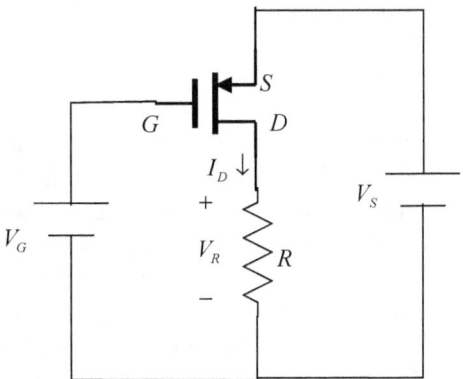

Figure 6.3(a) A PMOS transistor circuit

the model in figure 4.1(g). The modification you need is delete the **DC Voltage Source 2**, enter **R, DC Voltage Source, DC Voltage Source 1**, and **Fcn** related parameters as 1e3, 2.5, 1, and -0.5*110e-6*2/0.18*(u-0.4)^2 respectively, delete previous annotations, include new annotations for example S for source, and run the model after connecting the line which was disrupted due to deletion. The Display and Display1 show the V_R and I_D as $0.7394\,V$ and $0.0007394\,A$ respectively.

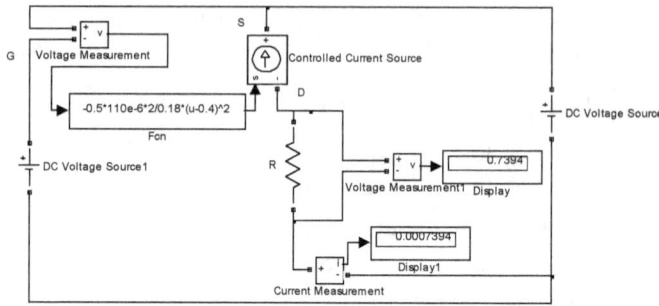

Figure 6.3(b)
Model of the
PMOS transistor
circuit in figure
6.3(a) – right side
figure

6.3 Resistive divider biasing circuits of MOSFETs

Figure 6.3(c) depicts the resistive divider biasing circuit of an NMOS transistor which has the following circuit parameters: $\frac{W}{L}=\frac{5}{0.18}$, $\mu_n C_{ox}=100\,\mu A/V^2$, $V_{TH}=$ $0.5\,V$, $V_{DD}=2\,V$, $R_1=7\,K\Omega$, $R_2=$ $10\,K\Omega$, $R_3=1\,K\Omega$, and $R_4=600\,\Omega$. Determine the I_D and V_{DS} by SIMULINK modeling provided that the transistor operates in the saturation mode.

The circuit's BJT counterpart including modeling process is addressed in section 4.4. The complete model for this problem is seen in figure 6.3(d) which we derived from figure 4.4(d). Alter the model by the following: delete the Diode, enter R1, R2, R3, R4, DC Voltage Source, and Fcn related

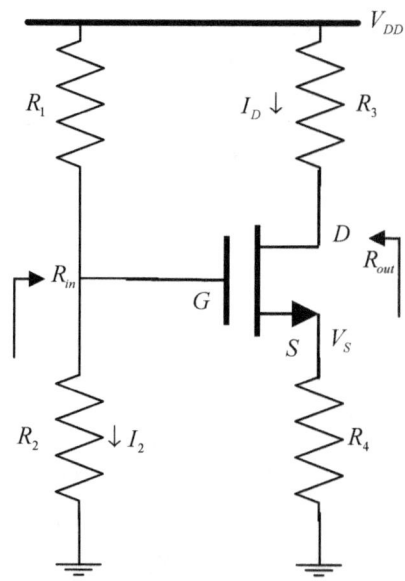

Figure 6.3(c) Resistive
divider biasing circuit of an
NMOS transistor

parameters as 7e3, 10e3, 1e3, 600, 2, and 0.5*100e-6*5/0.18*(u-0.5)^2 respectively, delete previous annotations, include new annotations for example G for gate, delete the line between R3 and Controlled Current Source, get one Current Measurement and one Display blocks for I_D, delete Voltage Measurement connection lines, reconnect the model like figure 6.3(d), run the model after reconnecting, and find the I_D and V_{DS} at Display1 and Display as $0.0003233\ A$ and $1.483\ V$ respectively.

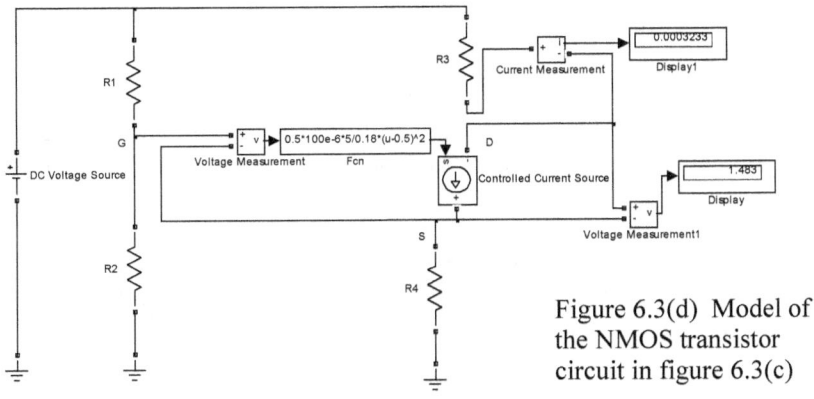

Figure 6.3(d) Model of the NMOS transistor circuit in figure 6.3(c)

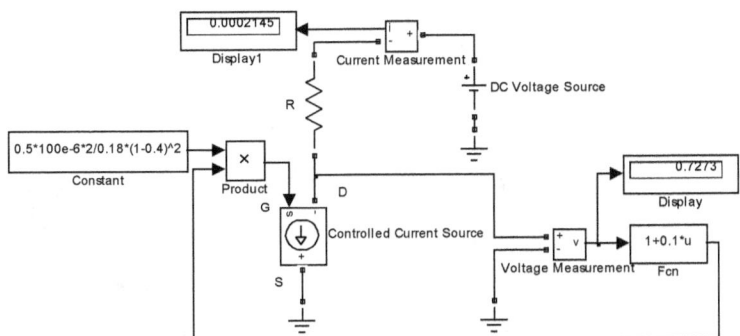

Figure 6.4(a) Model of the NMOS transistor circuit in figure 6.2(d) subject to channel length modulation

6.4 MOS modeling due to channel length modulation

NMOS/PMOS addressed in previous sections is without the effect of channel length modulation. When the channel length modulation is considered, the drain current of the transistor in saturation region is given by $I_D = \frac{1}{2}\mu_n C_{ox}\frac{W}{L}(V_{GS} - V_{TH})^2(1 + \lambda V_{DS})$ where λ is the channel length modulation in V^{-1}. Other symbols have earlier quoted meanings. In section 6.2 suppose the I_D and V_{DS} are to be found with $\lambda = 0.1\ V^{-1}$.

The model of figure 6.2(e) is modified to incorporate the channel length modulation which is shown in figure 6.4(a). In the figure the **Voltage Measurement** output is V_{DS} that is passed through an **Fcn** block whose **Expression** code is **1+0.1*u** in the parameter window (doubleclick the **Fcn**

-147-

to see) or $1+\lambda V_{DS}$. For another λ e.g. $\lambda=0.2\,V^{-1}$, the **Expression of Fcn** would be **1+0.2*u**. The **Fcn** assumes its input signal to be **u**. The $1+\lambda V_{DS}$ is multiplied with $\frac{1}{2}\mu_n C_{ox}\frac{W}{L}(V_{GS}-V_{TH})^2$ (generated by the **Constant** block) by a **Product** block to form $I_D=\frac{1}{2}\mu_n C_{ox}\frac{W}{L}(V_{GS}-V_{TH})^2(1+\lambda V_{DS})$.

Reconnect the later blocks, run the model of figure 6.4(a), and find $I_D=0.0002145\,A$ and $V_{DS}=0.7273\,V$ at **Display1** and **Display** respectively.

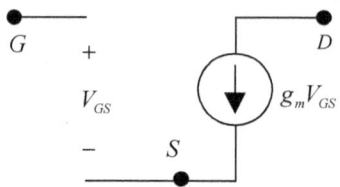

Figure 6.4(b) NMOS transistor
using transconductance

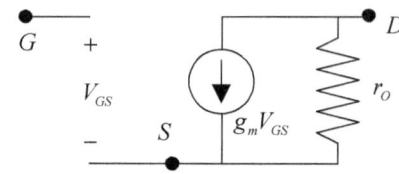

Figure 6.4(c) NMOS transistor
using transconductance and
channel length modulation

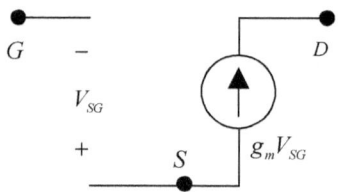

Figure 6.4(d) PMOS transistor
using transconductance

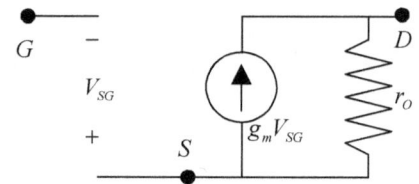

Figure 6.4(e) PMOS transistor
using transconductance and
channel length modulation

6.5 MOS modeling by transconductance

Similar to section 4.6 a MOS transistor has transconductance based model too. We first determine I_D by using previous sections mentioned techniques then apply $g_m=\sqrt{2\mu_n C_{ox}\frac{W}{L}I_D}$ for the model. Figures 6.4(b), 6.4(c), 6.4(d), and 6.4(e) show the transconductance based transistor circuits of NMOS without λ (channel length modulation), NMOS with λ, PMOS without λ, and PMOS with λ respectively. Their SIMULINK counterparts you see in figures 6.5(a), 6.5(b), 6.5(c), and 6.5(d) respectively where the symbols have their usual meanings for example G or **G** for gate.

Suppose we have the following circuit parameters: $\frac{W}{L} = \frac{5}{0.18}$, $\mu_n C_{ox} =$ $100\,\mu A/V^2$, and $I_D = 0.005\,A$ then g_m in standard unit (assigned to **gm**) is computed at MATLAB prompt by:

```
>>gm=sqrt(2*100e-6*5/0.18*0.005); ↵
```

The r_o in figure 6.4(c) or 6.4(e) is due to λ which we calculate by

$$\frac{1}{\frac{1}{2}\mu_n C_{ox}\frac{W}{L}(V_{GS}-V_{TH})^2\lambda}$$ in saturation region. Suppose for the transistor we have

$V_{GS} = 1\,V$, $V_{TH} = 0.4\,V$, and $\lambda = 0.1\,V^{-1}$ then calculate r_o (assigned to **ro**) at the command prompt by:

```
>>ro=1/0.5/100e-6/(5/0.18)/(1-0.4)^2/0.1; ↵
```

After that enter **gm** as **Gain** and **ro** as **Resistance** in the parameter windows of **Gain** (in figures 6.5(a) through 6.5(d)) and vertical resistor **RO** (in figures 6.5(b) or 6.5(d)) respectively.

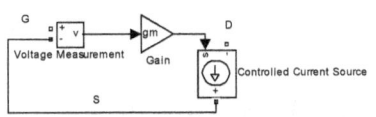

Figure 6.5(a) NMOS model
using transconductance

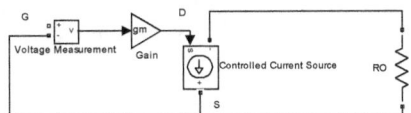

Figure 6.5(b) NMOS model
using transconductance and
channel length modulation

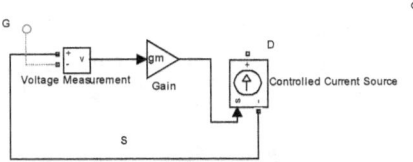

Figure 6.5(c) PMOS model
using transconductance

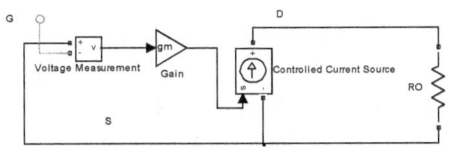

Figure 6.5(d) PMOS model
using transconductance and
channel length modulation

In section 4.10 we addressed how to form subsystem for single NPN or PNP transistor, the same can be applied in the case of MOS. Figures 6.6(a), 6.6(b), 6.6(c), and 6.6(d) present SIMULINK subsystem of NMOS, designed NMOS block, SIMULINK subsystem of PMOS, and designed PMOS block respectively. Design the blocks of NMOS or PMOS and keep them in your folder for later simulation.

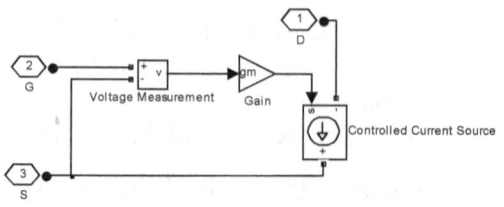
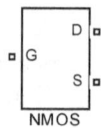

| Figure 6.6(a) SIMULINK subsystem of NMOS | Figure 6.6(b) Designed NMOS block |

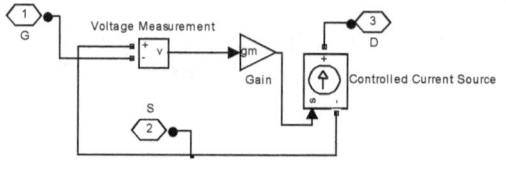
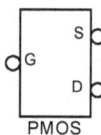

| Figure 6.6(c) SIMULINK subsystem of PMOS | Figure 6.6(d) Designed PMOS block |

6.6 Voltage gain after modeling a MOS transistor

Voltage gain for BJT is addressed in section 4.7. Similar gain finding we exercise elucidating following examples.

✦ Example 1

The voltage gain of circuit in figure 6.2(d) is defined as $A_V = \dfrac{V_{DS}}{V_G}$ which should be 0.8 (example 1 of section 6.2). We intend to obtain the gain.

The model is similar to figure 4.7(a) in which **Constant, Constant1, R,** and **DC Voltage Source** parameters should be **0.5*100e-6*2/0.18*(1-0.4)^2, 1, 5e3,** and **1.8** for the NMOS and you find the A_V at **Display** after running the model (see example 1 of section 4.7).

✦ Example 2

The V_G now changes over $0.5V \leq V_G \leq 0.9V$ in example 1. Determine the voltage gain variation as a waveshape.

The model is similar to figure 4.7(b) with example 1 mentioned parameters but the **Fcn** code should be **0.5*100e-6*2/0.18*(u-0.4)^2**. The **Start**

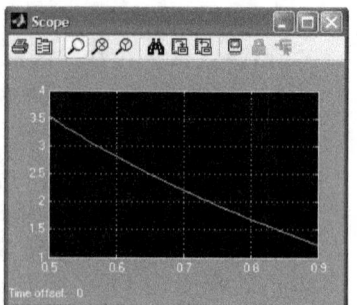

Figure 6.6(e) Voltage gain variation of NMOS

and **Stop** times of the solver are 0.5 and 0.9 respectively because of $0.5V \leq V_G \leq 0.9V$. Run the model and find the **Scope** output as in figure 6.6(e) for A_V (see example 2 of section 4.7).

◆ **Example 3**

Determine the A_V bounds of the last example.

With the help of **To Workspace** block and exercising [simout(1) simout(end)] akin to example 3 of section 4.7, you find $1.2284 \leq A_V \leq 3.5444$.

◆ **Example 4**

The resistive divider biasing circuit of section 6.3 has the voltage gain $A_V = \dfrac{V_o}{V_i} = 0.1649$ where V_i and V_o are gate to ground and source to ground or voltages across R_2 and R_4 respectively.

Verify the gain by the model of figure 6.6(f), see example 4 of section 4.7 for the similarity.

Figure 6.6(f)
Voltage gain
modeling for the
NMOS circuit of
figure 6.3(c) –
right side figure

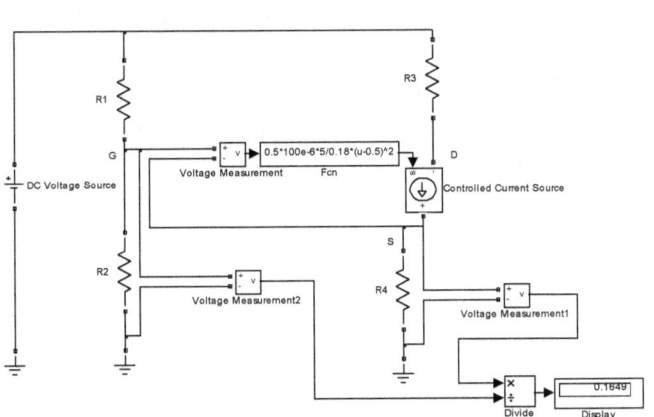

6.7 Current gain after modeling a MOS transistor

Analogous current gain for BJT you find in section 4.8. Similar modeling is applicable here too. Following example is incorporated to demonstrate the current gain of MOS.

◆ **Example**

Current gain of NMOS circuit in section 6.3 is $A_I = \dfrac{I_D}{I_2} = 2.748$ which we intend to obtain by SIMULINK modeling.

Example 2 of section 4.8 is the BJT counterpart of this current gain. Model of figure 6.7(a) depicts the current gain finding in which two **Current Measurement**s are used to get I_D and I_2. The model is derived from the model in figure 6.3(d). The altering you need is delete R2 resistor line, copy-paste **Current Measurement**, delete the **Voltage Measurement** and **Display1**, get one **Divide** block (appendix B), connect the later blocks like figure 6.7(a), run the model, and find the A_I at **Display**.

Figure 6.7(a)
Current gain
modeling for the
NMOS circuit of
figure 6.3(c) –
right side figure

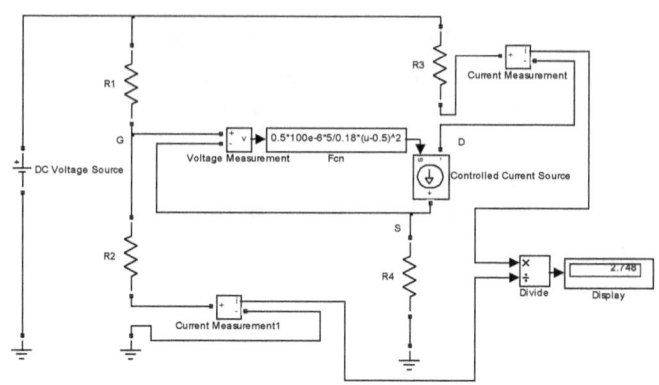

6.8 Input and output impedances after MOS modeling

Section 4.12 is the prerequisite for this section. We first determine the I_D which in turn provides g_m for the customized model of NMOS or PMOS of section 6.5. Application of customized **Req** easily provides equivalent impedance whether input or output that is illustrated in the following example.

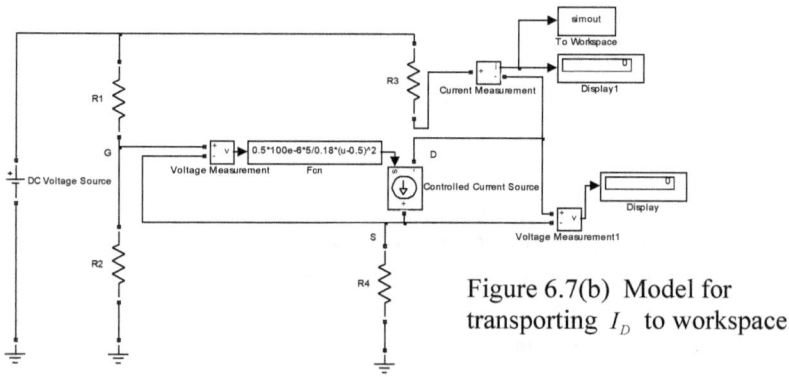

Figure 6.7(b) Model for
transporting I_D to workspace

✦ **Example**

We wish to determine R_{in} and R_{out} of the NMOS in figure 6.3(c).

From the model of figure 6.3(d) we need to transport the I_D into MATLAB whose model you see in figure 6.7(b). Make sure you set **Save format** of **To Workspace** as **Array**. Get hold on I_D by **Id** and assign computed g_m to **gm** by:

```
>>Id=simout(end); ↵
>>gm=sqrt(2*100e-6*5/0.18*Id); ↵
```

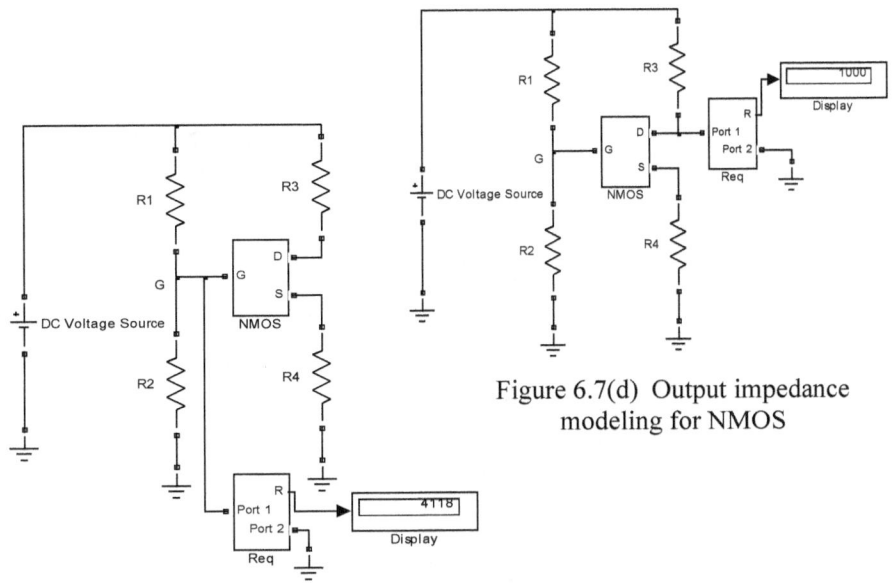

Figure 6.7(d) Output impedance
modeling for NMOS

Figure 6.7(c) Input impedance
modeling for NMOS

Then set **DC Voltage Source** at 0, remodel the circuit as in figures 6.7(c) and 6.7(d) using customized **NMOS** and **Req** for R_{in} and R_{out}, and find their values at each **Display** as **4118** and **1000** certainly in Ohm respectively.

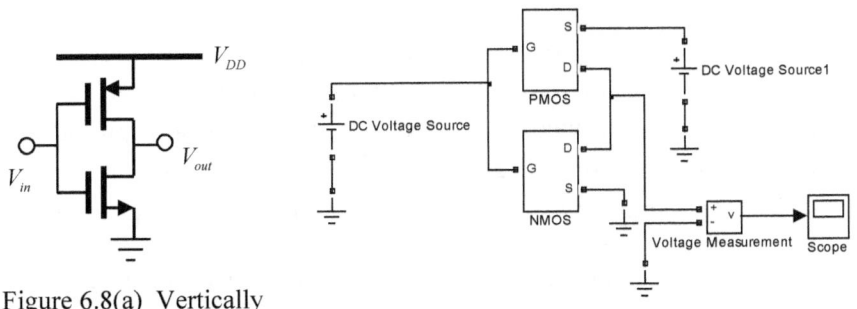

Figure 6.8(a) Vertically
cascaded MOS

Figure 6.8(b) Cascaded MOS modeling

6.9 Cascaded MOS modeling

Having customized the NMOS or PMOS block, we may model multistage or cascaded MOS transistor like section 4.11. All you need is follow the leads of gate, source, and drain of each MOS in the cascade.

Just as an example figure 6.8(a) shows two MOS transistors connected vertically. With the customized NMOS and PMOS of section 6.5, we model the circuit like figure 6.8(b) in which DC Voltage Source, DC Voltage Source1, and Voltage Measurement refer to V_{in}, V_{DD}, and V_{out} respectively.

6.10 MOS operating mode - triode and saturation

In section 4.14 we illustrated BJT operating mode checks. Similar check can be conducted on a MOS too. Theoretically the characteristic is tested by the following:

for NMOS   triode region: $V_{GS} > V_{TH}$ and $V_{DS} - V_{GS} + V_{TH} < 0$,

saturation region: $V_{GS} > V_{TH}$ and $V_{DS} - V_{GS} + V_{TH} > 0$ and

for PMOS   triode region: $V_{SG} > V_{TH}$ and $V_{SD} - V_{SG} + V_{TH} < 0$,

saturation region: $V_{SG} > V_{TH}$ and $V_{SD} - V_{SG} + V_{TH} > 0$.

Initially we may not know the operating mode. First assume triode or saturation mode, then carry out the modeling or computing, and after that verify the conditionality.

As an example consider the example 1 NMOS of section 6.2. We did

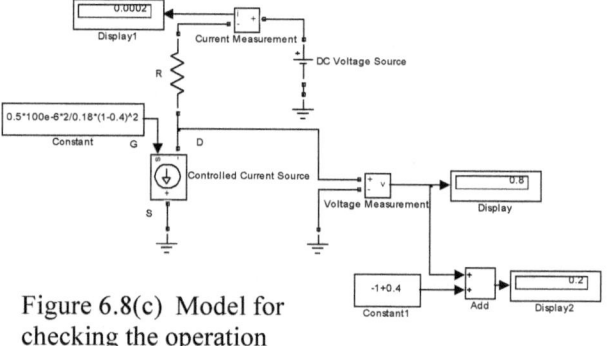

Figure 6.8(c) Model for checking the operation mode of an NMOS

the simulation assuming saturation zone of NMOS. Let us verify that.

Model of the figure 6.8(c) demonstrates the testing. The $V_{GS} > V_{TH}$ is satisfied from given values. The **Voltage Measurement** provides V_{DS}. The $-V_{GS} + V_{TH}$ is fed to the **Constant1**. The two quantities are added by the **Add** block whose output is flown to **Display2** indicating some positive value after running the model i.e. our assumption was correct.

The other modeling of section 4.14 can easily be extended for MOS transistors.

With this example we bring an end to the chapter.

Exercises

---

1. Graph I_D versus V_{DS} of the NMOS transistor in figure 6.1(a) subject to device aspect ration $\dfrac{W}{L} = 0.6$, $\mu_n C_{ox} = 110\ \mu A/V^2$, $V_{GS} = 0.82\ V$, and $V_{TH} = 0.45\ V$ over $10 mV \le V_{DS} \le 900 mV$ where the symbols have their usual meanings. Assume that the transistor is in triode region.

2. In question (1) plot I_D versus V_{DS} for three cases of V_{GS}; for $V_{GS} = 0.6\ V$, $V_{GS} = 0.7\ V$, and $V_{GS} = 0.95\ V$.

3. Graph approximate R_{on} versus V_{GS} of question (1) mentioned transistor over $0 \le V_{GS} \le 1V$ and comment on the graph.

4. Determine the $I_D|_{\max}$ and V_{DS} when the $I_D|_{\max}$ occurs in question (2) mentioned transistor.

5. Determine the I_D and V_{DS} of the NMOS transistor circuit in figure 6.2(d) by SIMULINK modeling if it functions in saturation region where aspect ration $\dfrac{W}{L} = \dfrac{4}{0.3}$, $\mu_n C_{ox} = 120\ \mu A/V^2$, $V_{GS} = 0.8\ V$, $V_{TH} = 0.4\ V$, $R = 3.5\ K\Omega$, and $V_D = 2.8\ V$.

6. Determine the I_D and V_R of the PMOS transistor circuit in figure 6.3(a) by SIMULINK modeling if it functions in saturation region where aspect ration $\dfrac{W}{L} = \dfrac{4}{0.3}$, $\mu_n C_{ox} = 120\ \mu A/V^2$, $V_G = 2\ V$, $V_{TH} = 0.4\ V$, $R = 3.5\ K\Omega$, and $V_D = 2.8\ V$.

7. Resistive divider biasing circuit of the NMOS transistor in figure 6.3(c) has the following parameters: $\dfrac{W}{L} = \dfrac{6}{0.28}$, $\mu_n C_{ox} = 110\ \mu A/V^2$, $V_{TH} = 0.45\ V$, $V_{DD} = 3\ V$, $R_1 = 8\ K\Omega$, $R_2 = 10\ K\Omega$, $R_3 = 3\ K\Omega$, and $R_4 = 5\ K\Omega$. Determine the I_D and V_{DS} by SIMULINK modeling provided that the transistor operates in the saturation mode.

8. Consider the effect of channel length modulation in NMOS of problem (5) where $\lambda = 0.1\ V^{-1}$ and determine the I_D and V_{DS} by SIMULINK modeling.

9. In problem (7) obtain the SIMULINK model using transconductance and designed NMOS block with channel length modulation $\lambda = 0.15\ V^{-1}$.

10. In question (5) determine the voltage gain $A_V = \dfrac{V_{DS}}{V_{GS}}$. Obtain the A_V variation over $0.4V \le V_{GS} \le 1V$. What are the bounds of A_V?

11. In question (6) determine the voltage gain $A_V = \dfrac{V_{SD}}{V_{SG}}$.

12. In question (7) determine the voltage gain $A_V = \dfrac{V_{DS}}{V_{GS}}$.

13. In question (7) determine the current gain $A_I = \dfrac{I_D}{I_2}$.

14. In question (9) determine R_{in} and R_{out}.

15. Figure E6.1(a) shows a self bias NMOS transistor with $\dfrac{W}{L} = \dfrac{5}{0.18}$, $\mu_n C_{ox} = 105\ \mu A/V^2$, $V_{TH} = 0.4\,V$, $R_1 = 5\ K\Omega$, $R_D = 500\ \Omega$, $R_S = 300\ \Omega$, and $V_{DD} = 3\,V$. Determine the I_D and V_{DS} of the transistor by SIMULINK modeling if it functions in the saturation region.

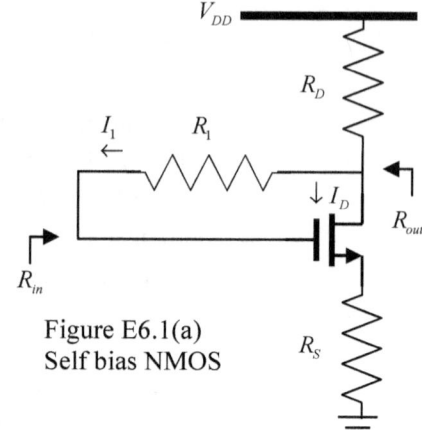

Figure E6.1(a)
Self bias NMOS

16. In question (15) determine the current gain $A_I = \dfrac{I_D}{I_1}$, voltage gain $A_V = \dfrac{V_{DS}}{V_{GS}}$, R_{in}, and R_{out}.

17. Develop the transconductance based model of question (15) mentioned self bias MOS transistor with channel length modulation $\lambda = 0.15\,V^{-1}$.

18. Develop the model of question (15) such that we are able to verify the operating mode of the transistor.

Answers:

(1) Commands: Id='0.5*110*0.6*(2*(0.82-0.45)*Vds-Vds^2)';
               ezplot(Id,[10 900]*1e-3)

Hint: section 6.1, the I_D is in μA, and graph is not shown for space reason.

(2) Figure A6.1(a)   Hint: section 6.1

Figure A6.1(a)
Graph of I_D versus
V_{DS} for several V_{GS}
– right side figure

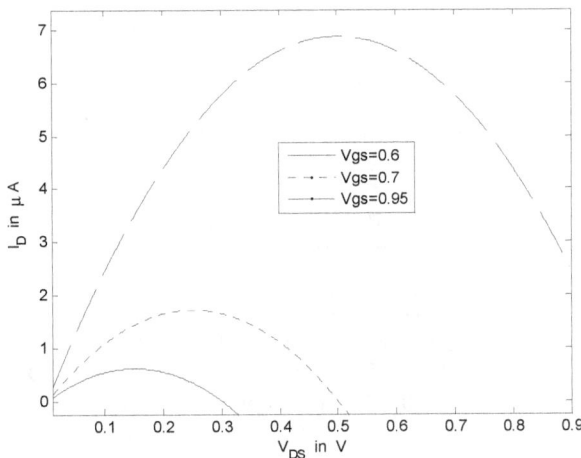

Figure A6.1(b) Graph
of R_{on} versus V_{GS} –
right side figure

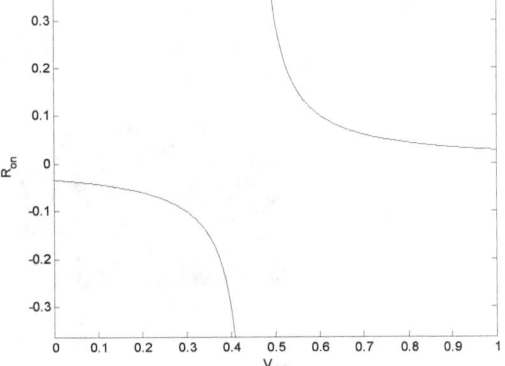

(3) Figure A6.1(b) with R_{on} in $M\Omega$. Ignore the negative R_{on} portion of the graph.   Hint: section 6.1

(4) $I_D|_{max}=0.7425\,\mu A$ when $V_{DS}=0.15\,V$, $I_D|_{max}=2.0625\,\mu A$ when $V_{DS}=0.25\,V$, and $I_D|_{max}=8.25\,\mu A$ when $V_{DS}=0.5\,V$.

Hint: section 6.1

(5) $I_D=0.000128\,A$ and $V_{DS}=2.352\,V$.   Hint: section 6.2.

(6) $I_D=0.000128\,A$ and $V_R=0.448\,V$.   Hint: section 6.2.

(7) $I_D=0.0003528\,A$ and $V_{DS}=0.178\,V$.   Hint: section 6.3.

(8) $I_D=0.0001568\,A$ and $V_{DS}=2.251\,V$. See figure 6.4(a).

Hint: section 6.4.

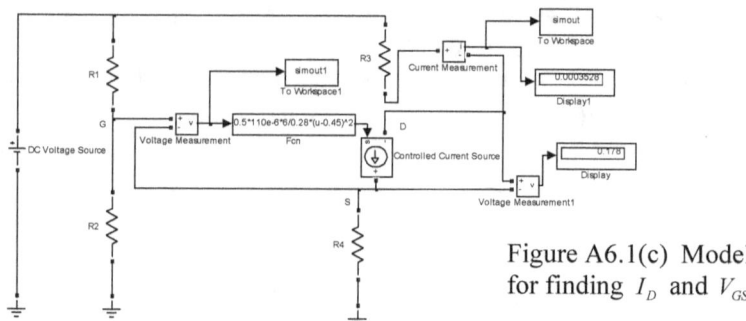

Figure A6.1(c) Model for finding I_D and V_{GS}

(9) First run the model of figure A6.1(c) then determine I_D, V_{GS}, g_m, and r_O respectively at the command prompt by:
Id=simout(end);   Vgs=simout1(end);
gm=sqrt(2*110e-6*6/0.28*Id);
ro=1/0.5/110e-6/(6/0.28)/(Vgs-0.45)^2/0.15;
Figure A6.1(d) depicts the model reckoning transconductance and channel length modulation. Set the NMOS and RO's Gain and **Resistance** as gm and ro respectively. Hint: sections 6.3-6.5.

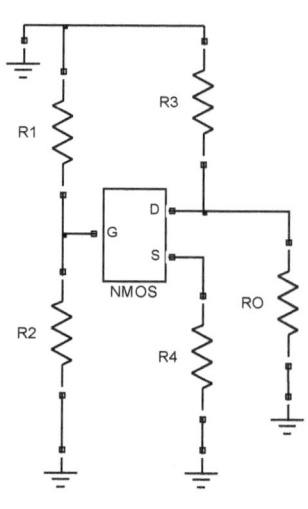

Figure A6.1(d) Model considering channel length modulation

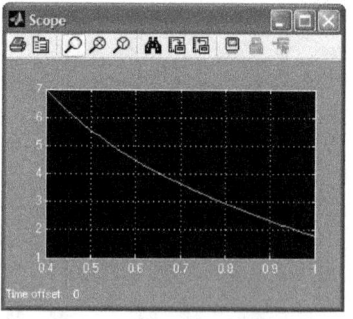

Figure A6.2(a) Voltage gain variation

(10) A_V =2.94. Figure A6.2(a). $1.792 \leq A_V \leq 7$.   Hint: section 6.6.

(11) A_V =2.94.   Hint: section 6.6.

(12) A_V =−1.833.   Hint: section 6.6.

(13) A_I =2.117.   Hint: section 6.7.

(14) R_{in} =4444Ω and R_{out} =2589Ω   Hint: section 6.8.

(15) I_D =0.001844 A and V_{DS} =1.525 V . See figure A6.2(b).   Hint: section 6.2.

(16) $A_I = \infty$, A_V =1, R_{in} =2105Ω, and R_{out} =200.4Ω.
Hint: sections 6.6, 6.7, and 6.8.

Figure A6.2(b) Modeling
a self bias MOS

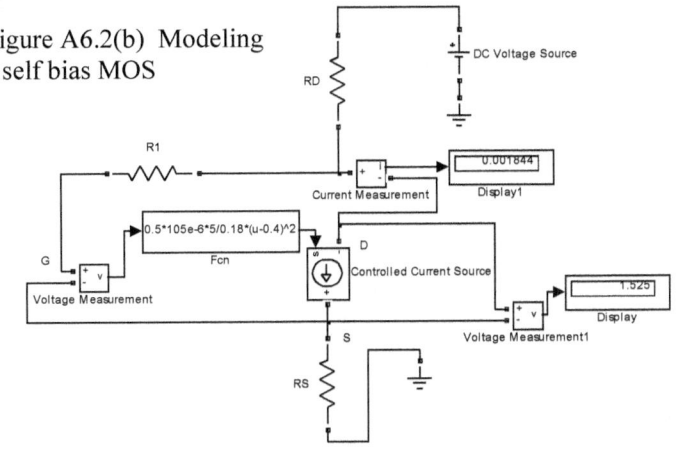

(17) Transport I_D and V_{GS} into workspace by:
Id=simout(end); Vgs=simout1(end);
Compute g_m and r_o by:
gm=sqrt(2*105e-6*5/0.18*Id);
ro=1/0.5/105e-6/(5/0.18)/(Vgs-0.4)^2/0.15;
Figure A6.2(c).
Hint: sections 6.4 and 6.5.

(18) Figure A6.2(d). The block Compare to
Constant checks $V_{GS} > V_{TH}$, parameter
window of which is set for $V_{TH} = 0.4$. The
block Add computes $V_{DS} - V_{GS} + V_{TH}$,
parameter window of which needs +-+ for
three inputs. Again Compare to

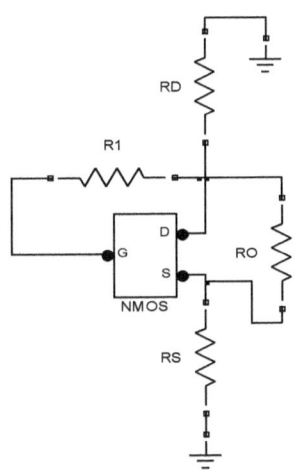

Figure A6.2(c) Self bias
MOS with customized
block

Figure A6.2(d)
Operating mode testing
of a self bias MOS

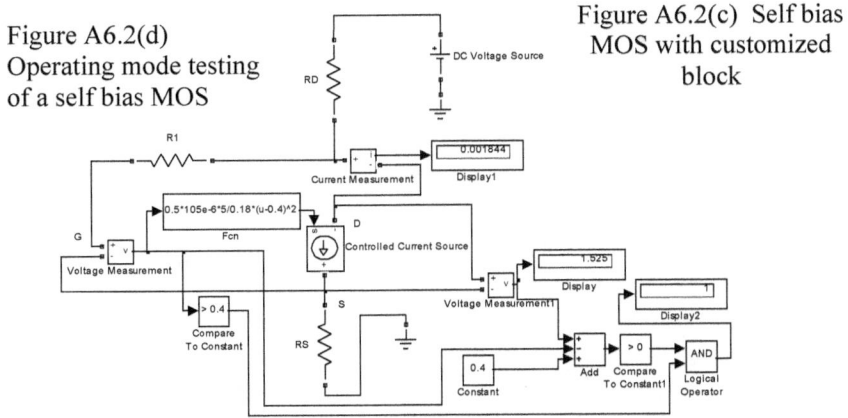

Constant1 checks $V_{DS} - V_{GS} + V_{TH} > 0$. If both conditions are true, the mode should be saturation which is checked by the **Logical Operator** in conjunction with AND setting. If the **Display2** shows 1, the mode is saturation conversely 0 means no.

Hint: section 6.10.

Chapter 7

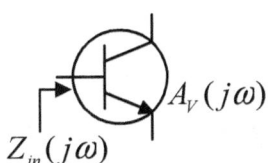

Electronic Circuits in Frequency Domain

Frequency domain analysis is a vital study on electronic circuits because every electronic circuit functions properly over a certain frequency band. The notion of frequency analysis is conducted on input-output of the electronic circuit which in turn employs transfer function. Having obtained the transfer function from MATLAB or SIMULINK, embedded tools facilitate the frequency domain analysis that is what we attempt to show by addressing:

♦♦ Way to obtain the transfer function from a circuit model

♦♦ Handling a transfer function in symbolic form

♦♦ Computing on an electronic circuit frequency response

♦♦ Graphing on an electronic circuit frequency response

7.1 How to model a passive circuit?

In an electronic circuit there may be passive elements that are resistor-inductor-capacitor; acquaintance of the design is compulsory for an electronic circuit because most electronic circuits are partly composed of these elements. We addressed the modeling in chapter 2. Exactly the same blocks we apply here i.e. model the circuit by the **Series RLC Branch** block. Unless the circuit elements are lumped, we model separately every resistor, inductor, or capacitor in the circuit. In frequency domain analysis it is extremely important that we know the input and output voltage nodes. If some source element (e.g. **AC Voltage Source** or **AC Current Source**) is connected to any nodes of a circuit in SIMULINK model, it is understood by the solver that those nodes are inputs. Again if some sink (chapter 1) element (e.g. **Voltage Measurement** or **Current Measurement** block) is connected to any nodes, it is understood by the solver that those nodes are outputs. All these instances are handled in previous chapters. It is of utmost importance that we know the transfer function related to the electronic circuit input-output because frequency response is chiefly connected with transfer function which we address in next section.

7.2 Transfer function from a passive circuit model

MATLAB and SIMULINK help us get handily the transfer function of a passive circuit without any calculation. In this process, we carry out the following steps:

(a) model the passive circuit by using the **Series RLC Branch** block in SIMULINK,

(b) connect the **AC Voltage Source** or **AC Current Source** according to the requirement to the given input nodes,

(c) connect the **Voltage Measurement** or **Current Measurement** block according to the requirement to the given output nodes,

(d) save the SIMULINK model file by your chosen name for instance **test** (extension will be .mdl automatically),

(e) use the built-in function **power_analyze** with the syntax **power_analyze**(SIMULINK file name,'**ss**') at the MATLAB command prompt for instance **power_analyze('test','ss')** where the second input argument (i.e. '**ss**') indicates the state space model of the circuit,

(f) both the file name and **ss** must be put under quote because of string type,

(g) assign the return from **power_analyze** to some variable **H** which keeps the whole circuit model information for example H=**power_analyze('test','ss')**;, and

Figure 7.1(a) An R - C filter circuit

(h) use the built-in function **tf** (abbreviation for the transfer function) on **H** for displaying transfer function of the passive circuit.

We demonstrate the transfer function finding with the following examples.

✦ **Example 1**

Figure 7.1(a) shows an R-C highpass filter which has the transfer function $H(s) = \frac{V_0}{V_i} = \frac{s}{s+1}$ with $R = 1\,\Omega$ and

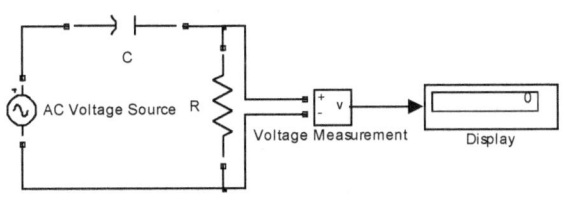

Figure 7.1(b) Modeling the filter circuit of figure 7.1(a)

$C = 1\,F$. Our objective is to obtain this transfer function starting from the given circuit whose implementation requires the following:

⇒ open a new SIMULINK model file (subsection 1.2.2), bring a **Series RLC Branch** block in the model (appendix B), rename the block as C, doubleclick the C, select C from the **Branch type** popup (sections 2.1 and 2.12), and enter the **Capacitance** as 1 in the parameter window,

⇒ bring another **Series RLC Branch** in the model, rename the block as R, doubleclick the R, select R from the **Branch type** popup, enter its **Resistance** as 1 in the parameter window, rightclick on the R, and click the **Rotate Block** under the **Format** popup to model the block vertically (because of its orientation in the circuit),

⇒ bring an **AC Voltage Source** and a **Voltage Measurement-Display** set blocks in the model (sections 2.4 and 2.14) to model the V_i and V_0 respectively, and

⇒ place the blocks relatively and connect them like figure 7.1(b) and save the model by the name **test** (can be any user-chosen name).

Now move on to MATLAB and carry out the following:

```
>>H=power_analyze('test','ss'); ⏎ ← H is any user-chosen name, ignore the warnings
>>tf(H) ⏎                          ← Calling the tf for displaying the transfer function
```

Transfer function from input "U_AC Voltage Source" to output "U_Voltage Measurement":

```
    s
  -------
  s + 1
```

Referring to the return, the **U_AC Voltage Source** indicates the AC **Voltage Source** of figure 7.1(b) or V_i of figure 7.1(a). Similar explanation follows for the **Voltage Measurement** or V_0. The return to H is an object rather than a variable. The object is in the form of state space model which involves four matrices (not shown) to describe any passive electrical circuit. Note that the **AC Voltage Source** is for input and **Voltage Measurement** is for output but explicit link of either one is not in the transfer function.

✦ Example 2

Figure 7.1(c) displayed R-L-C network implements a passband filter circuit which has the transfer function $H(s) = \dfrac{V_0}{V_i} = \dfrac{1250\,s + 8729}{s^2 + 1257\,s + 1.746 \times 10^6}$ from V_i to V_0 and which we wish to obtain by SIMULINK circuit simulation.

Now the given circuit specification is in terms of reactance at a frequency $f = 2KHz$. We assume that the parallel R-L branch is lumped in the circuit and exercise the following steps to view the transfer function:

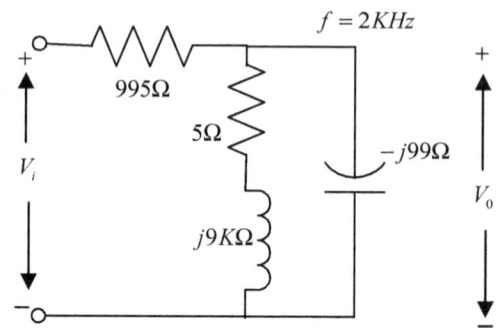

Figure 7.1(c) A passband filter circuit

⇒ open a new SIMULINK model file, bring a **Series RLC Branch** block in the model, rename the block as **R Branch**, doubleclick the **R Branch**, select **R** from the **Branch type** popup, enter its

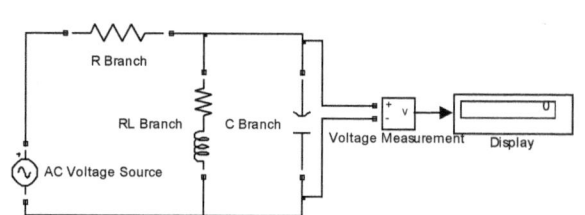

Figure 7.1(d) Modeling the passband filter circuit of figure 7.1(c) for transfer function

Resistance as 995 in the parameter window to model the series resistor,

⇒ copy and paste the **R Branch** through the clipboard to see **R Branch1**, rename the **R Branch1** as **RL Branch**, doubleclick the **RL Branch**, select **RL** from the **Branch type** popup, enter its **Resistance** and **Inductance** as 5 and 9e3/2/pi/2e3 respectively, rightclick on the **RL Branch**, and click the **Rotate Block** under the **Format** popup to model the block vertically,

⇒ copy and paste the **RL Branch** through the clipboard to see **RL Branch1**, rename the **RL Branch1** as **C Branch**, doubleclick the **C Branch**, select **C** from the **Branch type** popup, enter its **Capacitance** as 1/99/2/pi/2e3 for modeling the parallel capacitor,

⇒ get an **AC Voltage Source** and a **Voltage Measurement-Display** blocks to model the V_i and V_0 respectively, and

⇒ place the blocks relatively and connect them like figure 7.1(d) and save the file by the name **test**.

Moving on to MATLAB, we exercise the following:

>>H=power_analyze('test','ss'); ↵ ← H is any user-chosen name, ignore the warnings
during the command execution in MATLAB

>>tf(H) ↵   ← Calling the tf for displaying the transfer function

Transfer function from input "U_AC Voltage Source" to output "U_Voltage Measurement":

```
      1250 s + 8729
----------------------------------
s^2 + 1257 s + 1.746e006
```

In the denominator of transfer function from MATLAB, you find **1.746e006** which is equivalent to 1.746×10^6 and in which the **e006** means 10 to the power 6 (appendix A).

✦ How to make the transfer function coefficients available?

In last two examples we just displayed the transfer function from a passive circuit which we assigned to some workspace variable H. For the calculation reason we may need coefficients of the numerator and denominator polynomials which happens through the use of two output arguments and with the built-in function **tfdata** (abbreviation for transfer function data). The function employs the syntax [N,D]=tfdata(transfer function name,'v') where output arguments of **tfdata** are indicated by [N,D] and the N,D are user-chosen variables. The second input argument 'v' is reserved for coefficient return. Let us find the coefficients for example 2. The numerator and denominator polynomial coefficients of the transfer function $\dfrac{1250s + 8729}{s^2 + 1257s + 1.746 \times 10^6}$ should be [1250   8729] and [1   1257 1.746×10^6] respectively which we determine as follows:

>>[N,D]=tfdata(T,'v') ↵

```
N =
  1.0e+003 *
            0    1.2503    8.7289
D =
  1.0e+006 *
      0.0000    0.0013    1.7458
```

The **N** or **D** is a row matrix in which the elements are the polynomial coefficients in descending power of s. The term **1.0e+003 *** means every element in the **N** matrix is multiplied by 10^3. As you see, the coefficient data of transfer function is kept as exponential form at the workspace.

✦ What about the transfer function other than voltage to voltage?

The two examples we exercised are from voltage to voltage transfer function. Transfer function exists from voltage to current, current to current, or other. For the current source input and branch current output we employ the **AC Current Source** and **Current Measurement** blocks (section 2.5) respectively. Circuit condition and requirement manifest what the type of transfer function would be.

7.3 Defining an electronic system in MATLAB

For frequency based analysis input-output relationship of an electronic circuit is entered as a transfer function provided that its expression is known to us. A single input single output electronic system may be defined in two ways; polynomial and pole-zero-gain forms. Each of the forms has dedicated built-in function to define it in MATLAB, which we will explain first in the following. How these prototype defining functions are useful in implementing an electronic circuit is addressed later in this section.

✦ **Defining a transfer function by polynomial form**

The electronic system $H(s)$ is supposed to be expressible in numerator and denominator polynomial forms. The $H(s)$ can be voltage gain, impedance, current gain, etc. MATLAB built-in function tf (abbreviation for transfer function) is applied to define $H(s)$ with the syntax tf(numerator polynomial coefficients as a row matrix, denominator polynomial coefficients as a row matrix) where the polynomial coefficients must be in descending order and any missing coefficient is set to 0.

Suppose an electronic system is represented by transfer function $H(s) = \dfrac{7s^3 - 7s + 42}{s^4 - 118s^2 - 240s}$. The $H(s)$ has polynomial coefficient representation as [7 0 −7 42] and [1 0 −118 −240 0] for the numerator and denominator respectively. Having known so, we define the system as follows:

```
>>H=tf([7 0 -7 42],[1 0 -118 -240 0]) ↵

Transfer function:
 7 s^3 - 7 s + 42
----------------------------
s^4 - 118 s^2 - 240 s
```

In above execution we assigned the return from tf to workspace H (can be any user-supplied variable). If we append a semicolon at the end of the statement i.e. H=tf([7 0 -7 42],[1 0 -118 -240 0]);, the functional popup would not be displayed. However the variable H holds the $H(s)$ information as an electronic system.

✦ **Defining a transfer function by pole-zero form**

Not always is the transfer function in numerator-denominator polynomial form. When the transfer function numerator or denominator is given in factored or pole-zero-gain form, we employ MATLAB built-in function zpk with the syntax zpk(zeroes as a row matrix, poles as a row matrix, gain as a single number).

For example the $\begin{Bmatrix} \text{zeroes}: 2, 3, -1 \\ \text{poles}: 5, 0, 8, 6 \\ \text{gain}: 7 \end{Bmatrix}$ forms the electronic system $H(s)$ as

$\dfrac{7(s-2)(s-3)(s+1)}{(s-5)s(s-8)(s-6)}$ which we enter as follows:

>>z=[2 3 -1]; ↵ ← Assigning the zeroes as a row matrix to z, z is user-chosen
>>p=[5 0 8 6]; ↵ ← Assigning the poles as a row matrix to p, p is user-chosen
>>k=7; ↵ ← Assigning the gain as a scalar to k, k is user-chosen
>>H=zpk(z,p,k) ↵ ← Calling the zpk with the mentioned syntax and assigned the
return from zpk to H where H is user-chosen variable

Zero/pole/gain:
7 (s-2) (s-3) (s+1)
----
s (s-5) (s-6) (s-8)

We could have executed all commands in one line as H=zpk([2 3 -1],[5 0 8 6],7) instead of intermediate assigning. Appending a semicolon at the end of the statement i.e. H= zpk(z,p,k); does not show the functional popup. The last variable H holds the $H(s)$ information as a system from the pole-zero-gain description.

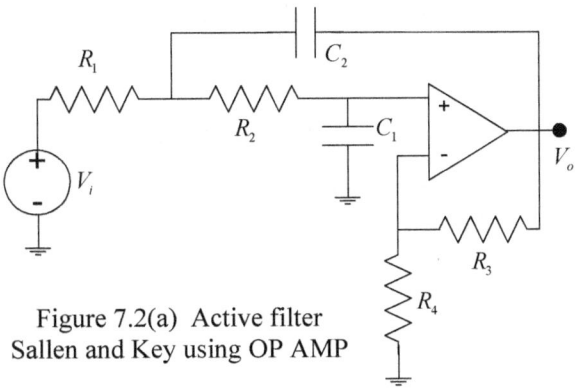

Figure 7.2(a) Active filter
Sallen and Key using OP AMP

♦ Example 1 on electronic system defining

Figure 7.2(a) indicates an active filter using OP AMP. Its output to input voltage ration is given by $\dfrac{V_o}{V_i} = \dfrac{1+\dfrac{R_3}{R_4}}{R_1 R_2 C_1 C_2 s^2 + \left(R_1 C_1 + R_2 C_1 - R_1 \dfrac{R_3}{R_4} C_2\right)s + 1}$. We

wish to define this voltage gain A_V as an electronic system in s domain based on $R_1 = 1\ K\Omega$, $R_2 = 0.7\ K\Omega$, $R_3 = 2\ K\Omega$, $R_4 = 3\ K\Omega$, $C_1 = 3\ mF$, and $C_2 = 4\ mF$. We expect the A_V to be $\dfrac{1.667}{8.4s^2 + 2.433s + 1}$.

In this sort of problem we had better use like name variable e.g. R1 for R_1 and enter all given parameters as follows:

>>R1=1e3; R2=0.7e3; R3=2e3; R4=3e3; C1=3e-3; C2=4e-3; ↵

Form the numerator and denominator coefficients and assign them to N and D respectively as follows:

>>N=1+R3/R4; D=[R1*R2*C1*C2 R1*C1+R2*C1-R1*R3/R4*C2 1]; ↵

Note that there is one space gap between two coefficients e.g. between $R_1R_2C_1C_2$ and $R_1C_1 + R_2C_1 - R_1\dfrac{R_3}{R_4}C_2$ anyhow get the voltage gain system Av as:

>>Av=tf(N,D) ↵
Transfer function:
    1.667

    8.4 s^2 + 2.433 s + 1

The N, D, and Av are user-chosen variables.

⬍ Example 2 on electronic system defining

The NMOS circuit of figure 7.2(b) has the voltage gain $A_V = \dfrac{V_i}{V_o} = \dfrac{(R_1 \| R_2)C_1s}{(R_1 \| R_2)C_1s+1} \times \dfrac{-g_m R_3(R_4C_2s+1)}{R_4C_2s + g_m R_4 +1}$. We wish to form the electronic system subject to $C_1 = 75nF$, $C_2 = 35nF$, $R_1 = 100K\Omega$, $R_2 = 70K\Omega$, $R_3 = 2.1K\Omega$, $R_4 = 4K\Omega$, and $g_m = 0.005\Omega^{-1}$.

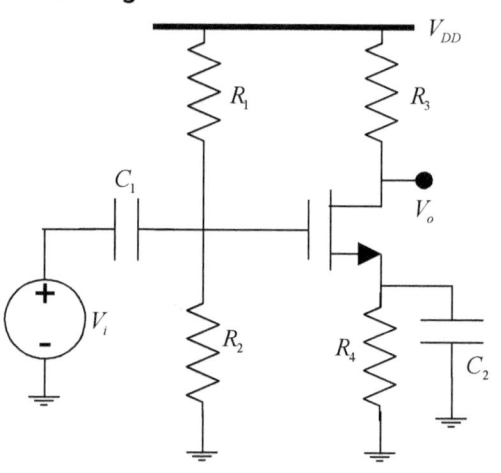

Figure 7.2(b) An NMOS circuit

Let us enter the given parameters to like name variables with proper scale factors (appendix A):

>>C1=75e-9; C2=35e-9; gm=0.005; ↵
>>R1=100e3; R2=70e3; R3=2.1e3; R4=4e3; ↵

The $R_1 \| R_2$ is parallel combination of R_1 and R_2 for which we employ Zp of appendix D.10. The A_V has two components in product form; $\dfrac{(R_1 \| R_2)C_1s}{(R_1 \| R_2)C_1s+1}$ and $\dfrac{-g_m R_3(R_4C_2s+1)}{R_4C_2s + g_m R_4 +1}$ and assign them to T1 and T2 respectively where T1 and T2 are user-chosen:

>>T1=tf([Zp(R1,R2)*C1 0],[Zp(R1,R2)*C1 1]); ↵
>>T2=tf(-gm*R3*[R4*C2 1],[R4*C2 gm*R4+1]); ↵

The * operator on component transfer functions performs the product computing hence do so by:

>>Av=T1*T2 ↵

Transfer function:

```
  -4.54e-006 s^2 - 0.03243 s
-------------------------------------
4.324e-007 s^2 + 0.06499 s + 21
```

As the return says we have $A_V = \dfrac{-4.54\times10^{-6}s^2 - 0.03243s}{4.324\times10^{-7}s^2 + 0.06499s + 21}$ which is assigned to workspace variable Av.

✦ Example 3 on electronic system defining

In example 2 we have $A_V = \dfrac{V_i}{V_o} = \dfrac{(R_1 \| R_2)C_1 s}{(R_1 \| R_2)C_1 s + 1}$ without C_2. In terms of pole-zero form we discover pole: $-1/((R_1 \| R_2)C_1)$, zero: 0, and gain: 1 because

the A_V can be rearranged as $\dfrac{s}{s + 1/((R_1 \| R_2)C_1)}$. Note that the coefficients of s are unity in this form regardless of numerator or denominator and the rearranging should follow likewise. Assuming the example 2 parameters are at MATLAB workspace, the A_V is formed as:

```
>>Av=zpk(0,-1/Zp(R1,R2)/C1,1) ↵
Zero/pole/gain:
      s
   -------------
   (s+323.8)
```

That is we get $A_V = \dfrac{s}{s + 323.8}$.

7.4 Frequency response values of an electronic system

When we have source frequency which varies from $-\infty$ to $+\infty$, what output response of an electronic system should be is termed as the frequency response. Electronic system transfer function is made available either from a model or from an expression which we addressed in previous sections. Built-in **freqresp** (abbreviation for <u>freq</u>uency <u>resp</u>onse) helps us compute the transfer function as will be explained in the following. The transfer function $H(s)$ is in s domain but frequency response is in ω domain i.e. replacing s by $j\omega$ is understood.

✦ Value of $H(j\omega)$ at a particular angular frequency

Consider the electronic system $H(s) = \dfrac{s}{s+1}$. For example at $\omega = 4\,rad/\sec$, the $H(j\omega)$ (i.e. $\dfrac{j\omega}{j\omega+1}$) becomes $0.9412 + j\,0.2353$ which we wish to compute.

The **freqresp** has a syntax **freqresp**(system, required angular frequency) which we call as follows:

```
>>H=tf([1 0],[1 1]); ↵
```
← Defining $H(s)$ as in section 7.3, H⇔$H(s)$

>>R=freqresp(H,4) ↵   ← Workspace R is any user-chosen variable

R =
      0.9412 + 0.2353i   ← R holds the $H(j4)$

✦ **Values of the $H(j\omega)$ at a set of angular frequencies**

We wish to compute above $H(j\omega)$ for $\omega=0$, 10, and 100 $rad/\sec$ which should be 0, $0.9901+ j\,0.099$, and $0.9999+ j\,0.01$ respectively.

The freqresp also keeps option for returning the frequency response at multiple angular frequencies. In the second input argument of freqresp, now we insert the required frequencies as a row matrix as follows:

>>R=freqresp(H,[0 10 100]); ↵ ← Second argument holds frequencies

At this point in MATLAB context, the return to R (user-chosen variable) for the set of frequencies is a three dimensional array with the first two dimensions empty. Discussion of the three dimensional array is beyond the scope of text (see reference 32). To remove the first two singleton dimensions from the R, we employ the command **squeeze** on the R as follows:

>>V=squeeze(R) ↵   ← The V is any user-chosen variable

V =
      0
      0.9901 + 0.0990i
      0.9999 + 0.0100i

As the return says, computed values of $H(j\omega)$ are available as a column matrix in the workspace variable V for the three frequencies respectively. The V(1), V(2), and V(3) return the three frequencies separately respectively.

✦ **Range of $H(j\omega)$ values for a range of angular frequencies**

Very often for graphing or analysis reason we need to have $H(j\omega)$ values for a range of frequencies. For instance we intend to obtain ongoing $H(j\omega)$ over $-10 \leq \omega \leq 10$ $rad/\sec$ with a ω step 0.1 $rad/\sec$ (i.e. $\Delta\omega=0.1\,rad/\sec$). Under this circumstance we generate the ω vector as a row matrix (chapter 1) using the colon operator by executing first w=-10:0.1:10; and then R=freqresp(H, w); V=squeeze(R); at the command prompt. Though the second input argument of freqresp is a row matrix, the return to V is a column matrix.

✦ **Separating various components of** $H(j\omega)$

Once we have the $H(j\omega)$ calculated, further requirements can be the separation of real, imaginary, magnitude, and phase angle components of $H(j\omega)$ i.e. $\text{Re}\{H(j\omega)\}$, $\text{Im}\{H(j\omega)\}$, $|H(j\omega)|$, and $\angle H(j\omega)$ which require the uses of built-in functions real, imag, abs, and angle respectively. For instance values stored in just mentioned V can be separated by using the commands real(V), imag(V), abs(V), and angle(V) respectively – in single or multiple frequency case. The return from the angle is by default in radian. If you wish to see the return in degrees, use the command rad2deg.

✦ **Decibel (dB) values of** $H(j\omega)$ **magnitudes**

Sometimes it is desirable to have decibel (dB) values of the magnitude $|H(j\omega)|$ which happens through the command 20*log10(abs(V)) on earlier V where $\log_{10} x$ has the code log10(x) and the decibel is defined as $20\log_{10}|H(j\omega)|$.

✦ **Frequencies in terms of Hz instead of rad/sec**

We might be interested in Hertz (Hz) frequency instead of angular one ($rad/\sec$) for what reason we employ f=w/2/pi where f is user-chosen and holds the Hertz frequency as column matrix (because $f = \dfrac{\omega}{2\pi}$).

✦ **Bypassing** $\log_{10} 0 = -\infty$ **point in dB values**

In the decibel plot it is very common that $|H(j\omega)|=0$ for some frequency. Since $\log_{10} 0 = -\infty$, the computer prints some error message. Under this type of situation, we add a negligible positive quantity epsilon to the $|H(j\omega)|$ values, which has the MATLAB code eps. For the ongoing system function, we should use 20*log10(eps+abs(V)).

✦ **Suppressing high dB values of** $H(j\omega)$ **magnitudes**

Even though $\log_{10} 0 = -\infty$ is overcome by using the eps, the values of 20*log10(eps) become too much negative like −300 dB or so which has no practical importance. In most electronic analysis, we restrict the lowest dB as −50 or −60 dB. If any dB value is less than −50 or −60, we force that to be −50 or −60. This sort of dB axes manipulation needs some programming technique. Before we do that, let us assign the calculated dB values of ongoing V to some variable D (any user-chosen name) as follows:

>>D=20*log10(eps+abs(V)); ↵ ← D holds dB values as column matrix

Let us say any dB value stored in D less than −50 will be set to −50. For this purpose the function find (appendix D.6) becomes useful as follows:

>>r=find(D<=-50); D(r)=-50; ↵

Concerning above implementation, the r (any user-chosen variable) holds the integer position index of column matrix D where $20\log_{10}|H(j\omega)| \leq -50$ (conducted by the command r=find(D<=-50);). Only $20\log_{10}|H(j\omega)| \leq -50$ elements in the D are set to −50 by writing the command D(r)=-50;. Thus the last D holds the expected $20\log_{10}|H(j\omega)|$ values as a column matrix.

✦ Normalization of $H(j\omega)$ magnitudes with respect to maximum

Some electronic system may not have $|H(j\omega)|$ values ranging 0 to 1. If dB variation is needed from 0 to some value, the $|H(j\omega)|$ must be between 0 and 1. For example the electronic system $H(s)=\dfrac{3s}{s+1}$ shows non 0-1 variation. Let us form the system as follows:

>>H=tf([3 0],[1 1]); ↵ ← Defining $H(s)$, H⇔$H(s)$

We wish to find $|H(j\omega)|_{max}$ over the interval $-10 \leq \omega \leq 10$ rad/sec with a ω step 0.1 rad/sec. First we determine the $|H(j\omega)|$ values by applying earlier functions at indicated ω points as follows:

>>w=-10:0.1:10; R=freqresp(H,w); V=squeeze(R); ↵

We know that the last V is holding complex $H(j\omega)$ values as a column matrix. Appendix D.5 cited max finds the maximum of $|H(j\omega)|$ as follows:

>>M=max(abs(V)) ↵   ← M holds the $|H(j\omega)|_{max}$ value

M =
        2.9851

Normalization means finding the $\dfrac{H(j\omega)}{|H(j\omega)|_{max}}$ values so we just divide the V by the M as follows:

>>S=V/M; ↵ ← S holds the normalized complex $H(j\omega)$ values where S
                                    is user-chosen variable

Obviously the S is a column matrix as well.

7.5 Graphing frequency spectrum of an electronic system

Last section demonstrates frequency response of an electronic system for the most part on calculation context. This section is all about graphing the frequency response of the electronic system.

Frequency response of the electronic system $H(j\omega)$ in graphical form has four components namely $\text{Re}\{H(j\omega)\}$, $\text{Im}\{H(j\omega)\}$, $|H(j\omega)|$, and $\angle H(j\omega)$ versus ω over certain interval of ω. To graph the frequency response, there can be two options – either get the data and plot on your own or use the readymade function like **bode**, both of which are addressed in the following.

◆ Getting data first and plotting the response afterwards

In last section we have addressed computations all along on $H(s) = \frac{s}{s+1}$ or $H(j\omega) = \frac{j\omega}{j\omega+1}$. We wish to plot the $|H(j\omega)|$ versus ω over $0 \le \omega \le 10 rad/\sec$ by choosing a ω step $0.01\ rad/\sec$.

First we follow the techniques of last section for obtaining frequency spectrum data as follows:

```
>>H=tf([1 0],[1 1]);  ↵    ← Defining the H(s), H⇔H(s)
>>w=[0:0.01:10]';  ↵       ← The w holds the ω values as a column matrix
>>R=freqresp(H,w);  ↵      ← freqresp calculates H(jω) and assigns to R
>>V=squeeze(R);  ↵         ← Removing singleton dimensions, V holds H(jω)
                              complex values as a column matrix
```

After that we employ appendix E cited **plot** as follows:
```
>>plot(w,abs(V))  ↵    ← Graphing the |H(jω)| versus ω like figure 7.3(a)
```

The **plot** has two input arguments, the first and second of which are the ω and $|H(j\omega)|$ data as a row or column matrix respectively. In a similar fashion the commands **plot(w,real(V))**, **plot(w,imag(V))**, and **plot(w,angle(V))** graph the spectra $\text{Re}\{H(j\omega)\}$ versus ω, $\text{Im}\{H(j\omega)\}$ versus ω, and $\angle H(j\omega)$ versus ω respectively (graphs are not shown for space reason). Not only that, decibel spectrum of

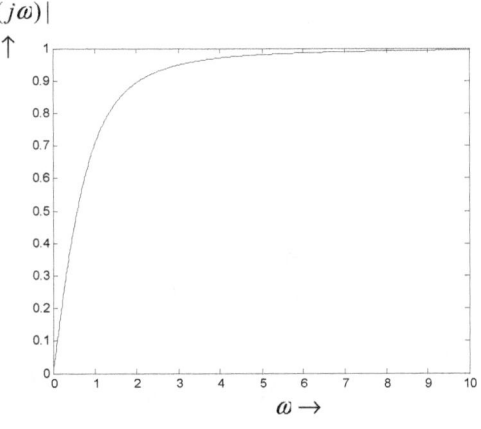

Figure 7.3(a) Plot of $|H(j\omega)|$ versus ω

section 7.4 (whose values are in **D** in the section) is graphed too by the **plot** by using **plot(w,D)** regardless of with or without dB suppression.

✦ Employing readymade bode plotter

MATLAB is so resourceful in programming and built-in function sense that we may have different options for the same type of problem. The readymade tool means we use bode plot to see the electronic system frequency response but mainly the magnitude and phase ones. The function we employ is **bode** which keeps provision for accepting different input-output arguments as will be explained in the following.

Graphing a system function:

We wish to plot the $H(j\omega)$ over ω for the electronic system

$H(s) = \dfrac{s}{s+1}$ (or $H(j\omega) = \dfrac{j\omega}{j\omega+1}$) from 0.1 to 100 $rad/\sec$ by using the

command **bode** and do so at the command prompt as follows:

>>H=tf([1 0],[1 1]); ⏎ ← Defining the $H(s)$, H⟺ $H(s)$
>>bode(H,{0.1,100}) ⏎

Now the **bode** has two input arguments, the first and second of which are the system function assignee name and the ω interval description respectively. In the ω interval description we use the second brace (not the first, nor the third brace) and input only bounds of the interval, ω step size is automatically chosen by the **bode**. Figure 7.3(b) presents the bode plot for the example $H(j\omega)$. By default the **bode** returns decibel spectrum $20\log_{10}|H(j\omega)|$ and phase spectrum $\angle H(j\omega)$ in degrees versus ω together.

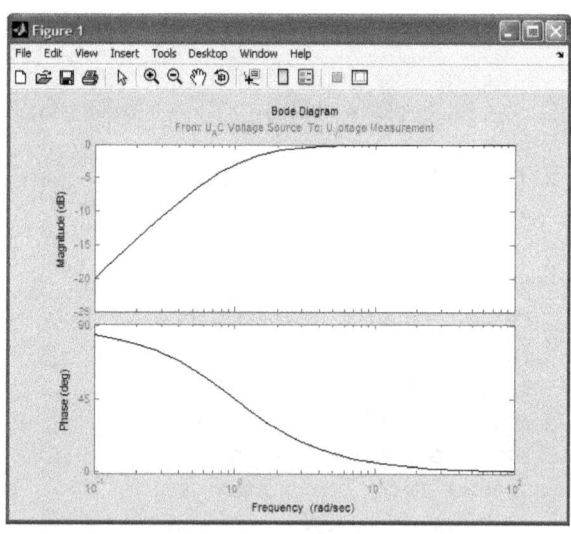

Figure 7.3(b) **Bode** plot of an electronic system

There are different options hidden in figure window of the bode plot. Rightclick on the mouse in plot area of figure 7.3(b), find the **Properties** in a popup, and click the **Properties** to see prompt window like figure 7.3(c).

To graph $|H(j\omega)|$ versus ω over the same interval, we click **Units** menu of the figure 7.3(c) and select **Absolute** in the **Magnitude** in popup.

To graph $|H(j\omega)|$ versus frequency in Hertz over the same interval, we click **Units** menu of the figure 7.3(c) and select **Hz** in the **Frequency** in popup.

If you want to see only the $20\log_{10}|H(j\omega)|$ versus ω, rightclick on the mouse in plot area of figure 7.3(b), find the **Show** in a popup, and click the **Phase** under the **Show** to see only the magnitude spectrum. Similarly you can view only the phase spectrum $\angle H(j\omega)$ by clicking the **Magnitude** under the **Show**.

The default phase in the bode plot is in degree. If you want to turn that to radian, click the **Units** menu of figure 7.3(c) and select the **radians** in the **Phase** in popup.

The default title of figure 7.3(b) is **Bode Diagram**. If you intend to change that to some other for example **System Response**, click **Labels** menu of the figure 7.3(c) and type **System Response** from keyboard deleting the **Bode**

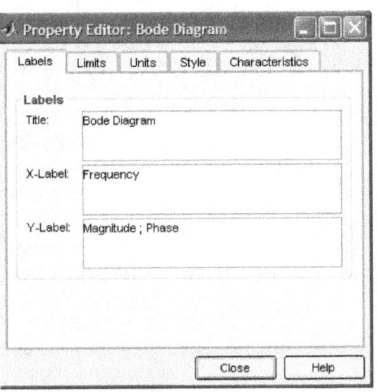

Figure 7.3(c) **Property editor of bode**

Diagram under the **Title** slot of the **Labels**. In a similar fashion you can change the horizontal and vertical axes labeling of figure 7.3(b).

Mouse based access on the graph data:

Mouse pointer helps us find frequency spectrum data. For this bring the mouse pointer at any point on the **bode** drawn curve (figure 7.3(b)) and leftclick the mouse. An indicatory box appears on top of the **bode** plot in which you find the mouse point coordinates of horizontal and vertical axes quantities.

Adding grid lines to the bode graph:

Figure 7.3(b) shows only the bode graph. If you wish to include grid lines to horizontal and vertical axes, execute the command **grid** at the command prompt.

Graphing multiple electronic systems:

We have been exercising all along keeping the $H(j\omega)$ in the H. Suppose there is another electronic system available in workspace variable **H1** (i.e. $H_1(j\omega)$) and we wish to plot the $H(j\omega)$ and $H_1(j\omega)$ over the same ω interval. The **bode** also keeps the provision for graphing multiple electronic systems. For the graphing, the command we need is **bode(H,H1,{0.1,100})** – the names of assignee holding the electronic systems and interval description all separated by a comma respectively. In case of three electronic systems, we just employ the command **bode(H,H1,H2,{0.1,100})** where **H2** indicates the third electronic system over the same interval.

Note: The **bode** does not function for negative frequencies as **freqresp** of last section does and its return is always in magnitude-phase angle form.

7.6 Electronic system bandwidth

Here in this section we address the bandwidth of an electronic system.

The bandwidth concept of an electronic system $H(s)$ depends on the magnitude spectrum behavior of $H(j\omega)$ i.e. $|H(j\omega)|$ versus ω. As you know there are four widely known types of response – lowpass, highpass, bandstop, and bandpass. The first step of bandwidth finding is view the magnitude spectrum using **bode** of section 7.5 and be confirmed

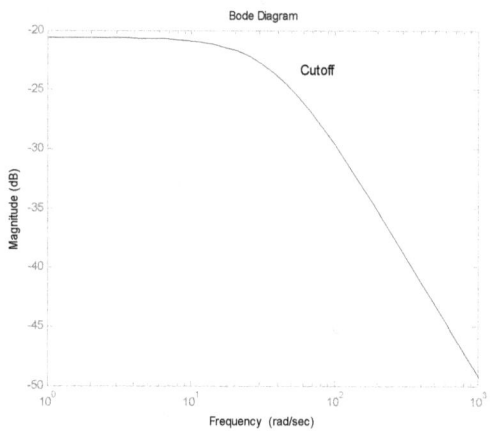

Figure 7.4(a)   Lowpass behavior of an electronic system

about the type of response. Bandwidth for each response needs slight different technique from that of the other. We address all of them in the sequel.

Lowpass:

Example of a lowpass electronic system is $H(s)=\dfrac{7}{2s+75}$. Its spectrum we view by:

```
>>H=tf(7,[2 75]); bode(H) ↵
```

Graph of figure 7.4(a) certainly shows the lowpass behavior in the magnitude spectrum after running the command. For the lowpass the cutoff frequency we determine by the function **bandwidth**:

```
>>bandwidth(H) ↵
```

ans =
            37.4108

The above return is in terms of $rad/\sec$ i.e. $\omega_c = 37.4108\,rad/\sec$. For this response, the $|H(j\omega)|_{max}$ occurs at $\omega = 0\,rad/\sec$. Cutoff frequency is the frequency when $|H(j\omega)|$ becomes $|H(j\omega)|_{max}-3\ dB$. If you need the Hertz frequency, you may execute **bandwidth(H)/2/pi**.

Highpass:

Example of a highpass electronic system is $H(s)=\dfrac{7s}{2s+150}$. View the magnitude spectrum by:

>>H=tf([7 0],[2 150]);bode(H) ↵ ← Figure 7.4(b) shows the behavior

The problem with this system is the $|H(j\omega)|_{max}$ occurs at $\omega=\infty$ or $s=\infty$ and computer never deals with infinity. The **bandwidth** does not work here instead **c_cross** is invoked (appendix D.8). In section 7.4 we explained how to get $H(j\omega)$ samples by making the use of **freqresp**. The spectrum of figure

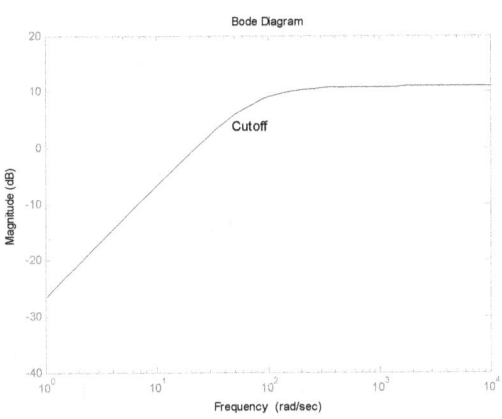

Figure 7.4(b)   Highpass behavior of an electronic system

7.4(b) indicates steady state like trajectory for large frequencies (**bode** made it from $10^0\,rad/\sec$ to $10^4\,rad/\sec$). Step selection here is not useful because of wide range. Appendix D.9 mentioned **linspace** can be used for specific sample number say 1000. With the symbol meaning of section 7.4, let us get the $H(j\omega)$ samples by:

>>w=linspace(10^0,10^4,1000); R=freqresp(H,w); V=squeeze(R); ↵

So V holds $H(j\omega)$ samples from which $|H(j\omega)|$ samples we get by:

>>A=abs(V); ↵   ← A is user-chosen, holds $|H(j\omega)|$ samples

Roughly the last value in A can be taken as $|H(j\omega)|$ for $\omega=\infty$ and get the value by:

>>L=A(end); ↵   ← L is user-chosen, holds the last $|H(j\omega)|$ sample

The L is basically approximate $|H(j\omega)|_{max}$. The cutoff occurs at $|H(j\omega)|_{max}/\sqrt{2}$ and call the **c_cross** accordingly:

```
>>c_cross(w,A,L/sqrt(2)) ↵
```

```
ans =
```
76.0676   ← i.e. ω_c =76.0676 rad / sec

Bandpass:

Bandpass electronic system is different from the lowpass or highpass owing to two cutoff frequencies. Another aspect of the spectrum behavior is the maximum is exhibited neither at $\omega=0$ nor at $\omega=\infty$. The two cutoff frequencies are before and after the maximum. Such example is $H(s)=\dfrac{10}{2s^2+0.8s+500}$. In a similar fashion view the magnitude spectrum like figure 7.4(c) by:

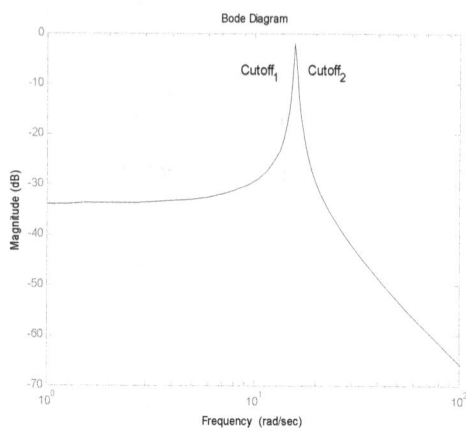

Figure 7.4(c) Bandpass behavior of an electronic system

```
>>H=tf(10,[2 0.8 500]);bode(H) ↵
```

The figure indicates relative positions of the two cutoffs too. Like the highpass counterpart let us select 1000 sample points for ω displayed by the **bode**:

```
>>w=linspace(10^0,10^2,1000); R=freqresp(H,w); V=squeeze(R); ↵
>>A=abs(V); ↵    ← A is user-chosen, holds | H(jω)| samples
```

Appendix D.5 mentioned **max** helps us determine the $| H(j\omega)|_{max}$ and its corresponding ω. This ω is termed as operating or center frequency:

```
>>[M,I]=max(A); ↵ ← M and I are user-chosen, hold | H(jω)|max and
                        integer index related to frequency respectively
>>wo=w(I) ↵           ← wo is user-chosen, holds the center frequency
```

```
wo =
```
15.7658   ← i.e. center frequency in rad / sec

The two cutoff frequencies we find by the same **c_cross**:

```
>>wc=c_cross(w,A,M/sqrt(2)) ↵      ← wc is user-chosen
```

```
wc =
```
15.6171   16.0135

In last execution the **wc** holds the two cutoff frequencies as a row matrix respectively.

Bandwidth is just the difference between the two frequencies:
>>wc(2)-wc(1) ⏎

ans =
             0.3964   ← i.e. bandwidth in *rad* / sec

Bandstop:
Bandstop is similar to the bandpass electronic system. The crucial point is whatever we computed around the maximum will be conducted now around a minimum. Another sharp contrast is cutoff is computed based on maximum. Anyhow a bandstop electronic system can be $H(s) =$

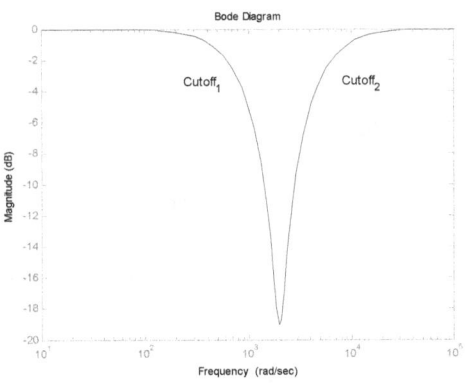

Figure 7.4(d)   Bandstop behavior of an electronic system

$\dfrac{0.2s^2 + 100s + 800000}{0.2s^2 + 900s + 800000}$. Its

related commands are the following with similar symbology:
>>H=tf([0.2 100 800000],[0.2 900 800000]);bode(H) ⏎

Figure 7.4(d) is the magnitude spectrum response over $10^1 \le \omega \le 10^5$ of coarse in *rad* / sec. The rest commands are:
>>w=linspace(10^1,10^5,1000); R=freqresp(H,w); V=squeeze(R); ⏎
>>A=abs(V); [M,I]=min(A); ⏎
>>wo=w(I) ⏎

wo =
             2.0118e+003

The min is for the minimum. The **2.0118e+003** means 2.0118×10^3 which is the center frequency in *rad* / sec (appendix A). Despite the minimum, maximum is found for the cutoff by:
>>M1=max(A); ⏎   ← M1 is user-chosen, holds $| H(j\omega) |_{max}$
>>wc=c_cross(w,A,M1/sqrt(2)) ⏎

wc =
        1.0e+003 *
             0.7607   5.1646

Having found the two cutoffs, the bandwidth we get by:
>>wc(2)-wc(1) ⏎

ans =
             4.4040e+003   ← i.e. 4.4040×10^3 *rad* / sec

7.7 Frequency based analysis using symbolic toolbox

In MATLAB a symbolic toolbox is embedded which we may apply to determine transfer function, mathematical expression derivation, voltage gain, or other. In the following we attempt to show some application of the toolbox in electronic circuit problems. The command **syms** helps us declare any variable related with the computing. For instance R_1 is a variable which we declare by writing **syms R1**; one space gap between **syms** and **R1**. Another variable **R2** we declare by **syms R1 R2** and so on. Some symbolic function may also be needed to solve our problem. Let us see the following problems.

✦ Example 1: Voltage gain finding

Voltage gain A_V of the RC network in figure 7.5(a) is given by $A_V =$

$$\frac{V_o}{V_i} = \frac{R_2(R_1 C_1 s + 1)}{R_1 R_2 s(C_1 + C_2) + R_1 + R_2}$$

which we wish to obtain.

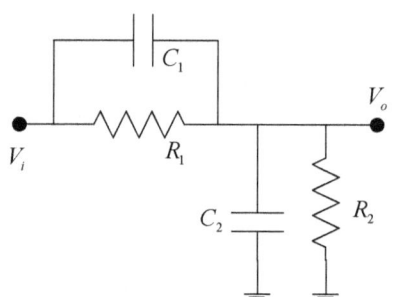

Pondering about series-parallel connection we discover that $R_1 - C_1$ combination is in series with $R_2 - C_2$ so

$$\frac{V_o}{V_i} = \frac{R_2 \| Z_{C2}}{R_1 \| Z_{C1} + R_2 \| Z_{C2}}$$ where Z_{C1} is the

impedance like contribution from C_1 and

Figure 7.5(a) Series parallel RC network

which is $\frac{1}{sC_1}$. Make the **Zp** of appendix D.10 available and carry out the following:

```
>>syms R1 C1 R2 C2 s ↵
```

We declared all related variables (e.g. **R1** for R_1) in above. The whole computing is assigned to some user-chosen variable **A** as follows:

```
>>A=Zp(R2,1/s/C2)/(Zp(R1,1/s/C1)+Zp(R2,1/s/C2)); ↵
```

The **Zp(R2,1/s/C2)** basically implements $R_2 \| Z_{C2}$. To get rid of redundant factor we may use command **simplify** and assign the result to some user-chosen variable **Av** by:

```
>>Av=simplify(A); ↵
```

Appendix D.11 quoted **pretty** helps us view the expression as follows:

```
>>pretty(Av) ↵
        (R1 s C1 + 1) R2
  ---------------------------------------------
  R1 R2 s C2 + R1 + R2 R1 s C1 + R2
```

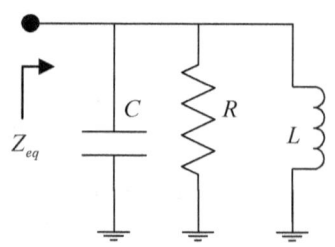

Figure 7.5(b) Parallel RLC network

Above is what we are after.

◆ Example 2: Impedance like expression finding

Figure 7.5(b) shown RLC network has s domain impedance like expression $Z_{eq} = \dfrac{LRs}{s^2 LRC + sL + R}$ which we intend to obtain.

Perform computing similar to example 1 noticing that $Z_{eq} = R \parallel \dfrac{1}{sC} \parallel sL$:

```
>>syms R L C s ↵
>>A=Zp(Zp(R,1/s/C),s*L); ↵
```

User-chosen A holds computed Z_{eq} . Assign simplification of A to Zeq by:

```
>>Zeq=simplify(A) ↵

Zeq =

L*s*R/(R+s^2*L*R*C+s*L)
```

Above is what we sought, which is held in user-chosen Zeq.

◆ Example 3: Deriving expression - OP AMP with finite gain

MATLAB or SIMULINK is not the solution of all electronic problem solving but at least the platform may ease our computing. In many analyses we wish to form an expression from some equations. Let us see how we get the expression using the method of elimination.

In chapter 5 an OP AMP is considered with extremely high or infinite gain. In the case of finite gain A we need to modify governing equations of the OP AMP. For example the differentiator of figure 5.4(d) has the OP AMP finite gain A which is depicted in figure 7.5(c) so the s domain governing equations are:

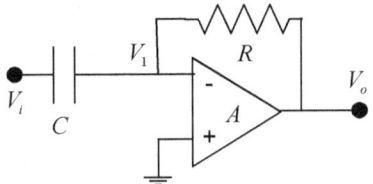

Figure 7.5(c) OP AMP with finite gain

equation 1 (current entering to and leaving from noninverting terminal): $(V_i - V_1)sC = \dfrac{V_1 - V_o}{R}$ and

equation 2 (gain of OP AMP): $V_o = -AV_1$.

Our objective is to eliminate V_1 and find a relationship for voltage gain $A_V = \dfrac{V_o}{V_i}$ which is $-\dfrac{sRCA}{RCs + A + 1}$.

First step of this sort of problem is to declare all related variables (e.g. Vo for V_o and so for the others) by syms as follows:

```
>>syms s R C V1 Vo Vi ↵
```

Rearrange the first equation so that the right hand side of the equation is zero and assign the equation without zero to some user-chosen variable e1:

>>e1=(Vi-V1)*s*C-(V1-Vo)/R; ↵

We have two equations and three variables; V_o, V_1, and V_i. All we need is eliminate V_1. The embedded **maple** package (appendix D.12) keeps a function by the name **algsubs** which we apply here for the elimination. The syntax of the function is **maple('algsubs'**, variable to be eliminated as an equation but left side of the equation must be the variable itself, expression from which the variable is to be eliminated). The return we assign to some user-chosen variable which is the resulting expression after the elimination. Although we have $V_o = -AV_1$, we write it as $V_1 = -V_o / A$ for input argumentation. Let us enter it by:

>>e2='V1=-Vo/A'; ↵

Above **e2** is a user-chosen variable which holds $V_1 = -V_o / A$. Note that once symbolic variables are declared, an expression does not need quote whereas an equation does which is evident in above executions. Anyhow call the eliminator as:

>>O1=maple('algsubs',e2,e1) ↵

O1 =

s*C*(Vi*A+Vo)/A+1/R*Vo*(1+A)/A

As you find in return (held by user-chosen **O1**), there is no **V1** and the expression is **Vi** and **Vo** related. We solve the expression for **Vo** by forming an equation that is **O1=0** and using **solve** (appendix G) hence do so by:

>>O2=solve(O1,Vo) ↵

O2 =

-s*C*Vi/(s*C*R+1+A)*R*A

The result is put to user-chosen **O2**. If we set **Vi=1** in above return, that is basically the voltage gain A_V we are looking for. Assign 1 to Vi by:

>>Vi=1; ↵

The command **subs** on **O2** substitutes the last value hence carry out the following:

>>Av=subs(O2) ↵

Figure 7.5(d) Sallen and Key filter with node voltage

Av =
-s*C/(s*C*R+1+A)*R*A

The **Av** is a user-chosen variable which holds just the code of $-\dfrac{sRCA}{RCs + A + 1}$.

✦ Example 4: Deriving expression from multiple equations

Figure 7.5(d) depicted Sallen and Key filter with labeled node voltages has the s domain governing equations as follows:

equation 1: $\dfrac{V_1 - V_2}{R_2} = sC_1 V_2$,

equation 2: $V_2 = \dfrac{R_4}{R_3 + R_4} V_o$, and

equation 3: $\dfrac{V_i - V_1}{R_1} + \dfrac{V_2 - V_1}{R_2} + sC_2(V_o - V_1) = 0$.

One would get $A_V = \dfrac{V_o}{V_i} = \dfrac{R_4 + R_3}{R_4 R_1 R_2 C_1 C_2 s^2 + \left(R_4 R_1 C_1 + R_4 R_2 C_1 - R_1 R_3 C_2\right)s + R_4}$ by solving

the above three equations. We wish to obtain the A_V using symbolic toolbox.

Example 3 shown technique and functions are exercised here with three equations. Declare all related variables in circuit of figure 7.5(d) with like names as follows:

```
>>syms s R1 C1 R2 C2 R3 R4 V1 V2 Vo Vi ↵
```

We wish to eliminate V_2 from equations 1 and 3 so we enter equation 2 as V_2 expressible and equations 1 and 3 as expression:

```
>>e1=(V1-V2)/R2-V2*s*C1; ↵
>>e2='V2=Vo*R4/(R3+R4)'; ↵
>>e3='(Vi-V1)/R1+(Vo-V1)*s*C2+(V2-V1)/R2'; ↵
```

The user-chosen e1, e2, and e3 hold the expression on equation 1, V_2 equation, and expression on equation 3 respectively. Eliminate V_2 from expressions stored in e1 and e3 and assign them to user-chosen O1 and O2 respectively as follows:

```
>>O1=maple('algsubs',e2,e1); ↵
>>O2=maple('algsubs',e2,e3); ↵
```

In O1 and O2 now we have V_1, V_i, and V_o so we need V_1 expressible equation. How do we get that? We can solve O1 (can be O2 as well) for V1 using solve and assign the result to V1. The last V1 we substitute in O2 by:

```
>>V1=solve(O1,V1); ↵
>>O3=subs(O2); ↵
```

The substitution result is assigned to user-chosen O3 which has V_i and V_o and the O3 is solved for V_o:

```
>>O4=solve(O3,Vo); ↵
```

Mathematics readable form of O4 we view by:

```
>>pretty(O4) ↵
```

$$\dfrac{Vi\,(R3 + R4)}{R4\,s\,C1\,R1 + R4 + R4\,s\,C1\,R2 - s\,C2\,R1\,R3 + s^2\,C2\,R1\,R4\,C1\,R2}$$

Above is what we sought. Only the Vi is excess, we can remove it by substituting 1:

```
>>Vi=1; Av=subs(O4); ↵
```

User-chosen Av holds the result i.e. the expression for A_V. Verify the expression by exercising pretty(Av) at the command prompt.

7.8 Frequency based analysis of an electronic circuit

We paved the way for frequency based analysis on an electronic circuit in previous sections. Problems can be related to low or high frequency response. In this section we address different electronic circuits considering following examples.

✦ Example 1: Voltage gain versus frequency from expression

The Sallen and Key filter of figure 7.2(a) has the voltage gain A_V as defined before (example 1 of section 7.3). Graph the A_V versus ω.

The A_V is stored in workspace variable Av. Exercise the command bode(Av) at the command prompt in order to view the A_V versus ω (section 7.5). The graph is not shown for space reason, for sure the magnitude pattern would be similar to that of figure 7.4(a) i.e. lowpass type. The plot is in dB i.e. $20\log_{10}|A_V(j\omega)|$ versus ω. Should you need $|A_V(j\omega)|$ versus ω, change the property of the plot as described in section 7.6. The cutoff you may find by the command bandwidth(Av) which is $0.4676\,rad/\sec$.

As another example the voltage gain response of NMOS circuit in figure 7.2(b) is to be analyzed i.e. frequency response and its related quantities.

The A_V is stored in workspace variable Av (example 2 of section 7.3). The bode(Av) shows you frequency response over $10 \le \omega \le 10^7 rad/\sec$ which is highpass type. The graph is not included for space reason. Get the $A_V(j\omega)$ samples by (with the symbology and concept of sections 7.4 and 7.6):

```
>>w=linspace(10^1,10^7,1000);R=freqresp(Av,w);V=squeeze(R); ↵
```

So above V holds $A_V(j\omega)$ samples from which $|A_V(j\omega)|$ samples we get by:

```
>>A=abs(V); ↵      ← A is user-chosen, holds |A_V(jω)| samples
```

Roughly the last value in A can be taken as $|A_V(j\omega)|$ for $\omega = \infty$ and get the value by:

```
>>L=A(end); ↵      ← L is user-chosen, holds the last |A_V(jω)| sample
```

The L is basically approximate $|A_V(j\omega)|_{max}$. The cutoff occurs at $|A_V(j\omega)|_{max}/\sqrt{2}$ and call the c_cross accordingly:

```
>>c_cross(w,A,L/sqrt(2)) ↵

        ans =

              145155           ← i.e. $\omega_c = 145155\,rad/\sec$
```

❖ Example 2: Poles and zeroes of an electronic system

The poles are roots of the denominator of an electronic system which may appear in any transfer function representing impedance, voltage gain, or other. The command pole helps us get the value of the poles whose syntax is pole(electronic system).

The NMOS circuit in example 2 of section 7.3 has poles which we determine by (assuming Av is present at the workspace):

```
>>pole(Av) ↵
```

```
ans =
        1.0e+005 *

    -1.5000
    -0.0032
```

i.e. the poles of $A_V(s)$ are located at $s = -1.5 \times 10^5$ and $s = -0.0032 \times 10^5$ in the s plane, both of which are real.

Similar command exists for zeroes which is zero. The same NMOS circuit has some zeroes for $A_V(s)$ which we determine by:

```
>>zero(Av) ↵
```

```
ans =
      1.0e+003 *
                 0
        -7.1429
```

i.e. the zeroes of $A_V(s)$ are located at $s = 0$ and $s = -7.1429 \times 10^3$ in the s plane, both of which are real.

The Sallen and Key filter of the example 1 does not have any zero, the response is seen as:

```
>>zero(Av) ↵
```

```
ans =
    Empty matrix: 0-by-1
```

❖ Example 3: Changing filtering response on circuit parameters

The RC network (example 1) of section 7.7 can function as lowpass as well as highpass depending on circuit parameters. We demonstrate the behavior in this example. From the electronic circuit theory we know that parameter conditions $R_1C_1 < R_2C_2$ and $R_1C_1 > R_2C_2$ turn the circuit as lowpass and highpass respectively.

In section 7.7 we introduced symbolic approach to find the $A_V(s)$. Some symbolic auxiliary functions are needed to work with the bode of section 7.5 because it works only with coefficient based transfer functions. In section 7.7 the Av holds $A_V(s)$ as expression. For value substitution we exercise subs and numerator-denominator separation needs numden (appendix D.13). For the sake of **bode**, we need to form transfer function again by using the **tf** of section 7.3. Let us see the following two cases:

When $R_1C_1 < R_2C_2$:

Choose $R_1 = 1 \ K\Omega$, $C_1 = 1 \ mF$, $R_2 = 5 \ K\Omega$, and $C_2 = 3 \ mF$ and execute the commands of example 1 in section 7.7 until you get **Av** then carry out the following:

```
>>R1=1e3; C1=1e-3; R2=5e3; C2=3e-3; ↵
```

Substitute the above values to **Av** and assign the outcome to user-chosen **Av1** by:

```
>>Av1=subs(Av); ↵
```

Extract the numerator and denominator of **Av1** and assign them to user-chosen **N** and **D** respectively by:

```
>>[N,D]=numden(Av1); ↵
```

Yet the **N** and **D** are expressions so polynomial coefficients we extract by **sym2poly**. Form a transfer function object and assign that to user-chosen **Av2** by:

```
>>Av2=tf(sym2poly(N),sym2poly(D)) ↵
Transfer function:
    5 s + 5
   -----------
   20 s + 6
```

Now exercise **bode** on **Av2** to view the frequency behavior:

```
>>bode(Av2) ↵
```

The graph certainly has similarity to figure 7.4(a) and is not shown for space reason. All we need is cutoff frequency (section 7.6) which we get by:

```
>>bandwidth(Av2) ↵

ans =
    0.3304
```
$\leftarrow$ i.e. $\omega_c = 0.3304 \ rad/\sec$

When $R_1C_1 > R_2C_2$:

Now choose $R_1 = 5 \ K\Omega$, $C_1 = 3 \ mF$, $R_2 = 1 \ K\Omega$, and $C_2 = 1 \ mF$ and execute similar commands:

```
>>R1=5e3; C1=3e-3; R2=1e3; C2=1e-3; ↵
>>Av1=subs(Av); ↵
>>[N,D]=numden(Av1); ↵
>>Av2=tf(sym2poly(N),sym2poly(D)) ↵
Transfer function:
    15 s + 1
   -----------
   20 s + 6
>>bode(Av2) ↵
```

This graph is similar to that in figure 7.4(b) meaning highpass and is not shown for space reason. The **bode** displayed frequency stretches from 10^{-3} to $10 \ rad/\sec$ therefore the cutoff frequency we get by:

```
>>w=linspace(10^-3,10^1,1000);R=freqresp(Av2,w); ↵
```

```
>>V=squeeze(R); A=abs(V); ↵
>>L=A(end); ↵
>>c_cross(w,A,L/sqrt(2)) ↵
```

ans =

    0.2863   ← i.e. $\omega_c = 0.2863 \, rad/\sec$

◆ Example 4: Impedance versus frequency

   Example 2 of section 7.7 illustrates an impedance transfer function in symbolic form which is stored in **Zeq**. Reexecute the commands until you get **Zeq** for Z_{eq} of circuit in figure 7.5(b). We wish to analyze the frequency behavior of Z_{eq} considering $R = 100\Omega$, $C = 1mF$, and $L = 1mH$.

   Example 3 demonstrates the exercise on symbolic manipulation until the transfer function for **bode** is formed and do so for this Z_{eq} as follows:

```
>>R=100; C=1e-3; L=1e-3; ↵
>>Z1=subs(Zeq); ↵
>>[N,D]=numden(Z1); ↵
>>Z2=tf(sym2poly(N),sym2poly(D)) ↵
Transfer function:
      1000 s
---------------------------
s^2 + 10 s + 1e006
```

The **Z1** or **Z2** is user-chosen variable. Example 3 quoted **tf** defines voltage gain but the **Z2** here is impedance. The Z_{eq} versus ω you view by the same **bode**:

```
>>bode(Z2) ↵
```

The figure is not presented for space reason but its shape is similar to the one in figure 7.4(c) which exhibits ω variation from 10^2 to $10^4 \, rad/\sec$. Implementing symbology and concept of section 7.6, let us go through the following (since **R** is used before, section 7.6 quoted **R** is written as another user-chosen variable **R1**):

```
>>w=linspace(10^2,10^4,1000); R1=freqresp(Z2,w); V=squeeze(R1); ↵
>>A=abs(V); ↵    ← A is user-chosen, holds | Z_eq (jω) | samples
>>[M,I]=max(A); ↵ ← M and I are user-chosen, hold | Z_eq (jω) |_max and
                     integer index related to frequency respectively
>>wo=w(I) ↵       ← wo is user-chosen, holds the center frequency
```

wo =

    1.0018e+003   ← i.e. center frequency $\omega_o = 1.0018 \times 10^3 \, rad/\sec$

The two cutoff frequencies we find by the same **c_cross**:

```
>>wc=c_cross(w,A,M/sqrt(2)) ↵    ← wc is user-chosen
```

wc =

    1.0e+003 *

    0.9968   1.0068

i.e. The two cutoff frequencies are $\omega_{c1} = 0.9968 \times 10^3 \; rad/\sec$ and $\omega_{c2} = 1.0068 \times 10^3 \; rad/\sec$. Bandwidth is just the difference between the two frequencies:

```
>>wc(2)-wc(1) ↵
```

```
ans =
```
    9.9099   ← i.e. bandwidth in $rad/\sec$

◆ Example 5: Quality factor of a second order circuit

Any RLC circuit inherently gets a quality factor Q which is related to second order transfer function $as^2 + bs + c$. Standard form of second order transfer function linking to Q is $s^2 + \dfrac{\omega_n}{Q}s + \omega_n{}^2$. Equating the two equations one obtains $Q = \dfrac{\sqrt{ca}}{b}$. Circuit in example 4 is a second order system whose denominator transfer function is stored in workspace D as symbolic expression, let us get the polynomial coefficients and assign those to user-chosen p by:

```
>>p=sym2poly(D) ↵
```

```
p =
```
    1   10   1000000

We need only the polynomial coefficients for Q (i.e. a, b, and c and get by p(1), p(2), and p(3) respectively) so compute the Q and assign to Q as follows:

```
>>Q=sqrt(p(3)*p(1))/p(2) ↵
```

```
Q =
```
    100

i.e. Q is 100 for the circuit in figure 7.5(b).

◆ Example 6: Multiple frequency responses of an electronic circuit

In example 4 we analyzed Z_{eq} versus ω. Suppose we wish to view the Z_{eq} versus ω for $C = 1mF$, $C = 0.1mF$, and $C = 0.01mF$ in a single plot.

The **bode** also keeps provision for graphing multiple electronic systems with syntax **bode**(system 1, system 2, and so on). For each capacitor, we need to form the Z_{eq} transfer function similar to example 4 and do so by:

```
For  C = 1mF :
syms R L C s
A=Zp(Zp(R,1/s/C),s*L);
Zeq=simplify(A);
R=100; C=1e-3; L=1e-3;
Z1=subs(Zeq);
```

```
[N,D]=numden(Z1);
Z2=tf(sym2poly(N),sym2poly(D));
```

For $C = 0.1mF$ **:**
```
C=0.1e-3;
Z1=subs(Zeq);
[N,D]=numden(Z1);
Z3=tf(sym2poly(N),sym2poly(D));
```

For $C = 0.01mF$ **:**
```
C=0.01e-3;
Z1=subs(Zeq);
[N,D]=numden(Z1);
Z4=tf(sym2poly(N),sym2poly(D));
```

User-chosen variables **Z2**, **Z3**, and **Z4** hold the Z_{eq} system for the three capacitors respectively. Having made available, input the three systems into **bode** by:

```
bode(Z2,Z3,Z4)
```

Outcome of above is the figure 7.6(a). In order to add a mark of distinction among various curves we exercise the command **legend** with the syntax **legend**(user-chosen text under quote for the first curve, user-chosen text under quote for the second curve, and so on) hence carry out the following:

```
legend('C=1mF','C=0.1mF','C=0.01mF')
```

For example for the first curve we chose **C=1mF** in above. MATLAB displayed graphics is in color, from which you can easily identify the trajectory.

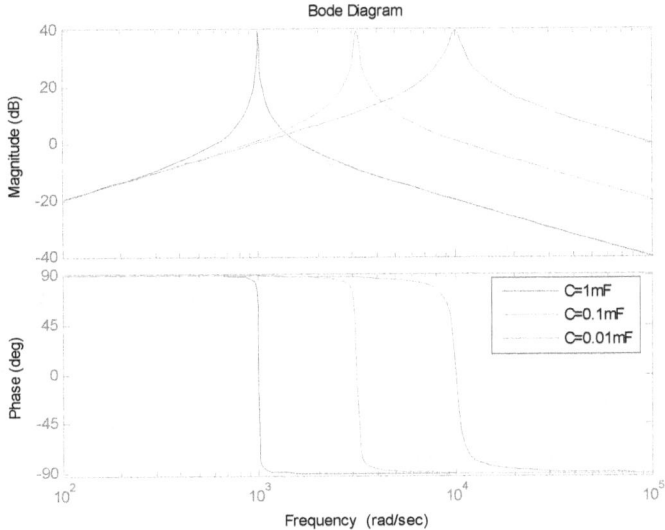

Figure 7.6(a) Impedance frequency response for
three capacitors

7.9 Model based frequency analysis of an electronic circuit

In last section most frequency related problems are implemented in MATLAB without SIMULINK modeling. In SIMULINK model based analysis we first obtain the transfer function of the concerned quantity by exercising the technique and function of section 7.2 then apply the freqresp (section 7.4) or **bode** (section 7.5) in accordance with the problem requirement. All details are addressed before, here in this section we demonstrate SIMULINK model based frequency analysis examples.

✦ **Example 1: Frequency response of RC circuit**

Analyze the $\frac{V_o}{V_i}$ frequency response of the RC circuit in figure 7.1(a) by exercising SIMULINK modeling.

The reader is referred to example 1 of section 7.2. The $\frac{V_o}{V_i}$ is stored in workspace **H** in state space form. Execute **Av=tf(H);** at the command prompt to have the $\frac{V_o}{V_i}$ in user-chosen **Av**. If we exercise **bode(Av)**, we view the graphical response similar to figure 7.4(b) which is not shown for space reason. The displayed response is highpass type spanning over 10^{-2} to $10^2 \, rad/\sec$ (from horizontal axis of the graph) hence we need to carry out the highpass commands of section 7.4 with the same symbology:

```
>>w=linspace(10^-2,10^2,1000); R=freqresp(Av,w); V=squeeze(R); ↵
>>A=abs(V); L=A(end); c_cross(w,A,L/sqrt(2)) ↵
```

```
ans =
        0.9609                    ← i.e. ωc =0.9609 rad / sec
```

✦ **Example 2: Frequency response of RLC circuit**

Analyze the $\frac{V_o}{V_i}$ frequency response of the RLC circuit in figure 7.1(c).

The model is seen as example 2 of section 7.2. Then you need:
```
>>Av=tf(H); bode(Av) ↵
```
The graph (not shown) indicates bandpass (figure 7.4(c)) behavior over $[10^{-1}, 10^5] \, rad/\sec$ so follow the section 7.6 quoted function, symbology, and tactic:
```
>>w=linspace(10^-1,10^5,1000); R=freqresp(Av,w); V=squeeze(R); ↵
>>A=abs(V); [M,I]=max(A); wo=w(I) ↵
```

```
wo =
        1.3014e+003       ← i.e. center frequency is 1.3014×10³ rad / sec
>>wc=c_cross(w,A,M/sqrt(2)) ↵
```

```
wc =
```

1.0e+003 *

0.8510   2.0521

i.e. the two cutoff frequencies are $\omega_{c1}=0.8510\times10^3\ rad/\sec$ and $\omega_{c2}=2.0521\times10^3\ rad/\sec$. Then bandwidth we get by:

>>wc(2)-wc(1) ↵

ans =

1.2012e+003   ← i.e. bandwidth is $1.2012\times10^3\ rad/\sec$

✦ Example 3: Voltage gain of a transistor at high frequency

Figure 7.6(b) shows an NPN transistor operated at high frequency. Its relevant parameters are $R_S=210\Omega$, $R_L=1.95K\Omega$, $\beta=100$, $C_\pi=105\ fF$, $C_\mu=21\ fF$, $C_{CS}=31\ fF$, $V_T=26mV$, and $I_C=1\ mA$. Obtain the $\frac{V_o}{V_i}$ function for the transistor circuit from the model.

Prerequisite for this problem is sections 4.6 and 4.10. The model for this problem is shown in figure 7.6(c). In order to construct the model, you need to open a new SIMULINK model file (subsection 1.2.2), get one saved **NPN**, one **AC Voltage Source** (for V_i), five **Series RLC Branch**, one **Voltage Measurement** (for V_o), one **Scope**, and two **Ground** blocks in the model (appendix B). Rename the **Series RLC Branch** blocks as **RS, RL, CS, Cm,** and **Cp**, doubleclick each of the last five, select **R, R, C, C,** and **C** from the **Branch type** popup of the parameter window, enter the **Resistance, Resistance, Capacitance, Capacitance,** and **Capacitance** as 210, 1.95e3, 31e-15, 21e-15, and 105e-15 for R_S,

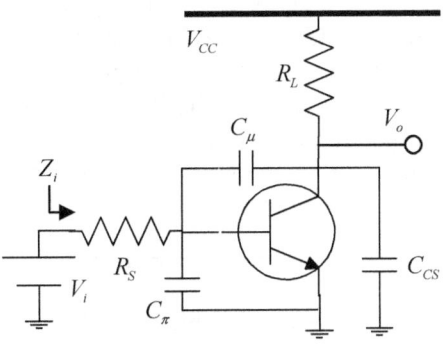

Figure 7.6(b) An NPN transistor at high frequency

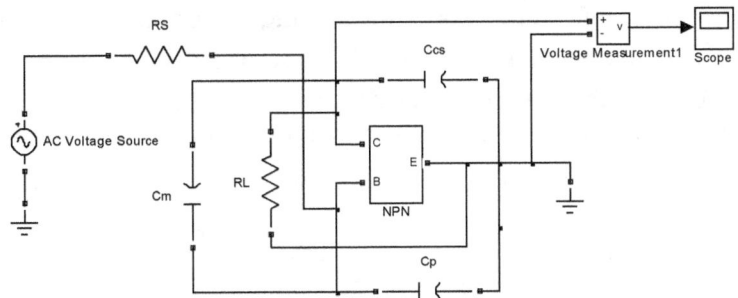

Figure 7.6(c) Model of the NPN transistor at high frequency

R_L, C_{CS}, C_μ, and C_π respectively (appendix A, f for femto or 10^{-15}). Turn the **Cm** and **RL** as vertical by clicking **Rotate Block** under the **Format** popup on doubleclicking each. Save the file by some name **test** and enter the NPN block necessary parameters to like name variable (e.g. I_c to **Ic**) as given:

>>Ic=1e-3; VT=26e-3; gm=Ic/VT; b=100; ri=b*VT/Ic; ↵

Doubleclick the **NPN** block, doubleclick **R** and **Gain** one at a time in **NPN**, and enter **ri** and **gm** as transistor model parameters respectively. Then attain the transfer function by:

>>H=power_analyze('test','ss'); ↵

Recall that the **NPN** has another **Voltage Measurement** block inside which is also interpreted as output to SIMULINK. In the circumstance of multiple inputs or outputs the **tf** determines all probable inputs-outputs based transfer functions as a matrix. How do we see that? Call the **tf** on **H** and assign the outcome to some user-chosen **Hm**:

>>Hm=tf(H) ↵

Transfer function from input "U_AC Voltage Source" to output...

$$U\_Voltage\ Measurement1:\ \frac{1.636e010\ s + 3.433e005}{s^2 + 5.437e010\ s + 4.319e020}$$

$$U\_NPN/Voltage\ Measurement:\ \frac{4.052e010\ s + 3.996e020}{s^2 + 5.437e010\ s + 4.319e020}$$

Transfer function from input "I_NPN/Controlled Current Source" to output...

$$U\_Voltage\ Measurement1:\ \frac{-2.062e013\ s - 8.422e023}{s^2 + 5.437e010\ s + 4.319e020}$$

$$U\_NPN/Voltage\ Measurement:\ \frac{-3.436e012\ s - 2.622e008}{s^2 + 5.437e010\ s + 4.319e020}$$

In the model of figure 7.6(c) the **Voltage Measurement1** refers to V_0 because of auto naming by SIMULINK, so does **AC Voltage Source** to V_i. Also note that the letter **U** is meant for voltage. As you find above, there are four transfer functions displayed, only the first of which is required as far as the $\frac{V_o}{V_i}$ is concerned. The other three we do not need for the problem. In matrix notation the above four is returned as $\begin{bmatrix} H_{11} & H_{12} \\ H_{21} & H_{22} \end{bmatrix}$ and the order is column directed and sequentially. MATLAB array indexing is likewise so our intended transfer function we get by calling **Hm(1,1)** and assign the return to some user-chosen **Av**:

>>Av=Hm(1,1) ↵
Transfer function from input "U_AC Voltage Source" to output "U_Voltage Measurement1":
   1.636e010 s + 3.433e005

---

s^2 + 5.437e010 s + 4.319e020

i.e. the Av retains $\dfrac{V_o}{V_i}$ which we read out as $\dfrac{V_o}{V_i} = \dfrac{1.636 \times 10^{10} s + 3.433 \times 10^5}{s^2 + 5.437 \times 10^{10} s + 4.319 \times 10^{20}}$.

◆ Example 4: Voltage gain frequency response at high frequency

Analyze the frequency response on the transistor of example 3.

The $\dfrac{V_o}{V_i}$ is stored in workspace Av. Exercise bode(Av) to view the frequency response which is bandpass type (similar to figure 7.4(c) stretching from 10^8 to $10^{12} rad/\sec$, the graph is not shown for space reason) therefore section 7.6 quoted bandpass computing is conducted as follows:
>>w=linspace(10^8,10^12,1000); R=freqresp(Av,w); V=squeeze(R); ↵
>>A=abs(V); [M,I]=max(A); wo=w(I) ↵

wo =
      2.1119e+010   ← center frequency is $2.1119 \times 10^{10} rad/\sec$
>>wc=c_cross(w,A,M/sqrt(2)) ↵

wc =
   1.0e+010 *
    0.6606   6.1655
       ↑

     i.e. cutoff frequencies are $0.6606 \times 10^{10} rad/\sec$ and $6.1655 \times 10^{10} rad/\sec$
>>wc(2)-wc(1) ↵

ans =
      5.5050e+010   ← bandwidth is $5.5050 \times 10^{10} rad/\sec$
>>pole(Av) ↵

ans =
   1.0e+010 *

   -4.4706
   -0.9661

The last return indicates that the s plane poles of the $\dfrac{V_o}{V_i}$ are at $s =$ -4.4706×10^{10} and $s = -0.9661 \times 10^{10}$. Should the reader need zeroes, employ the command zero.

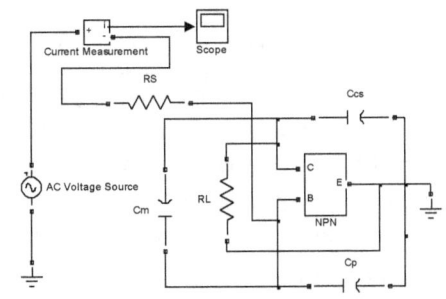

Figure 7.6(d) Model for input impedance at high frequency

✦ Example 5: Input impedance response at high frequency

Analyze the input impedance Z_i frequency response on the transistor of example 3.

For the input impedance of figure 7.6(b), we need to engage a **Current Measurement** block and figure 7.6(d) is the model of which. Start from the model of figure 7.6(c), delete the **Voltage Measurement** block and connecting line between **AC Voltage Source** and **RS**, insert a **Current Measurement** block (appendix B), save the file by some name **test1**, and follow calling like last two examples (we assume that the transistor parameters are at the workspace):

```
>>H=power_analyze('test1','ss'); ↵
>>Hm=tf(H) ↵
```

Transfer function from input "U_AC Voltage Source" to output...

U_NPN/Voltage Measurement:
$$\frac{4.052e010\ s + 3.996e020}{s^{\wedge}2 + 5.437e010\ s + 4.319e020}$$

I_Current Measurement:
$$\frac{0.004762\ s^{\wedge}2 + 6.594e007\ s + 1.537e017}{s^{\wedge}2 + 5.437e010\ s + 4.319e020}$$

Transfer function from input "I_NPN/Controlled Current Source" to output...

U_NPN/Voltage Measurement:
$$\frac{-3.436e012\ s - 2.622e008}{s^{\wedge}2 + 5.437e010\ s + 4.319e020}$$

I_Current Measurement:
$$\frac{1.636e010\ s + 9.988e005}{s^{\wedge}2 + 5.437e010\ s + 4.319e020}$$

The intended Z_i links to first column and second row in above i.e. from **AC Voltage Source** to **Current Measurement** so get it by:

```
>>Hm(2,1) ↵
```

Transfer function from input "U_AC Voltage Source" to output "I_Current Measurement":
$$\frac{0.004762\ s^{\wedge}2 + 6.594e007\ s + 1.537e017}{s^{\wedge}2 + 5.437e010\ s + 4.319e020}$$

But the above ratio is admittance i.e. **Current Measurement/AC Voltage Source** hence we obtain the intended Z_i by:

```
>>Z=1/Hm(2,1); ↵
```

Now you may analyze the frequency behavior of Z_i noticing that **Z** holds the s domain Z_i.

Above technique and approach you can apply to find output impedance too not only for BJT transistor but also for OP AMP and MOS circuits. Anyhow we intend to bring an end to the chapter with this.

Exercises

1. Figure 7.1(a) displayed R-C network has $R = 2\Omega$ and $C = 2.5\,F$. Obtain the transfer function $H(s) = \dfrac{V_0}{V_i}$ by SIMULINK circuit simulation.

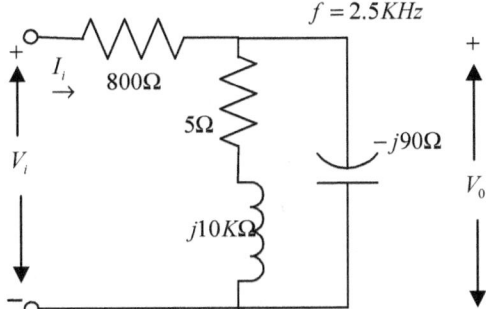

Figure E7.1(a) A passband filter circuit

2. Figure E7.1(a) displayed R-L-C network implements a passband filter circuit which has the transfer function $H(s) =$

$$\frac{V_0}{V_i} = \frac{1767\,s + 1.388 \times 10^4}{s^2 + 1775\,s + 2.235 \times 10^6}$$

from V_i to V_0. Verify the $H(s)$ by SIMULINK circuit simulation.

3. In question (2) impedance transfer function Z_i is defined as $Z_i = \dfrac{V_i}{I_i}$. Determine the $Z_i(s)$ using SIMULINK modeling.

4. An electronic system is represented by transfer function $H(s) = \dfrac{7s^3 - 13s + 12}{4s^5 - 180s - 440}$. Define the system in MATLAB.

5. In question (4) now consider $H(s) = \begin{cases} zeroes : 0, -3, -4 \\ poles : 3, 7, -8, 1 \\ gain : \text{-}2 \end{cases}$.

6. Define the voltage gain $\dfrac{V_0}{V_i}$ in s domain of Sallen and Key filter (figure 7.2(a)) in MATLAB based on $R_1 = 0.75\,K\Omega$, $R_2 = 2\,K\Omega$, $R_3 = 3\,K\Omega$, $R_4 = 4.5\,K\Omega$, $C_1 = 2.7\,mF$, and $C_2 = 5\,mF$.

7. In question (6) now consider the NMOS circuit in figure 7.2(b) with $R_1 = 1\,K\Omega$, $R_2 = 3\,K\Omega$, $R_3 = 5\,K\Omega$, $R_4 = 6\,K\Omega$, $C_1 = 3\,mF$, $C_2 = 6\,mF$, and $g_m = 0.006\Omega^{-1}$.

8. Define the lowpass counterpart of the circuit in question (7) in MATLAB.

9. (a) Determine the $H(j2)$ value for the electronic system $H(s) = \dfrac{9}{2s^2 + 0.01s + 2.1}$ by using the freqresp,

   (b) Determine the $H(-j0.1)$, $H(j0.1)$, $H(j0.5)$, and $H(j1)$ values for the electronic system in part (a) by using the freqresp,

   (c) in part (a) in magnitude and phase angle form,

   (d) in part (b) in magnitude and phase angle form, and

(e) Determine the $H(j\omega)$ values of the electronic system in part (a) as a column matrix over $-25 \le \omega \le 25$ rad/sec with a ω step 0.1 rad/sec.

10. Verify that the following electronic systems have the frequency responses over the given angular frequency range as indicated beside:

   (a) $H(s) = \dfrac{1}{2s+1}$ over $0.1 rad/sec \le \omega \le 1 Krad/sec$ as in figure A7.1(b),

   (b) $H(s) = \dfrac{13s^3 + 3}{2s^3 - 6s^2 + 6s - 1}$ over $1Hz \le f \le 400Hz$ as in figure A7.1(c),

   (c) $H(s) = \dfrac{0.2s + 9}{0.03s^2 - 0.6s + 1000}$ over $1Hz \le f \le 1KHz$ as in figure A7.1(d),

   and

   (d) $H(s) = \dfrac{0.2s^2 + 100s + 8000}{0.2s^2 + 65000s + 8000}$ over $0.01Hz \le f \le 1MHz$ as in figure A7.1(e).

11. Identify each following electronic system with regards to filtering behavior and determine the cutoff frequency if it is lowpass or highpass and the center frequency, two cutoff frequencies, and bandwidth if it is bandpass or bandstop: (a) $H(s) = \dfrac{21}{3s + 41}$, (b) $H(s) = \dfrac{92s}{9s + 1213}$, (c) $H(s) = \dfrac{23}{3.9s^2 + 0.5s + 342}$, and (d) $H(s) = \dfrac{0.5s^2 + 140s + 900000}{0.5s^2 + 1200s + 900000}$.

12. Figure E.71(b) presents two RC sections connected in series. Exercise MATLAB symbolic tools to obtain the s domain impedance like expression for $Z_{in}(s)$.

13. Figure E7.1(c) shows an integrator circuit using finite gain OP AMP. Write the necessary voltage or current equations on the circuit and use the symbolic tools of MATLAB to determine the $\dfrac{V_0}{V_i}$ in s domain.

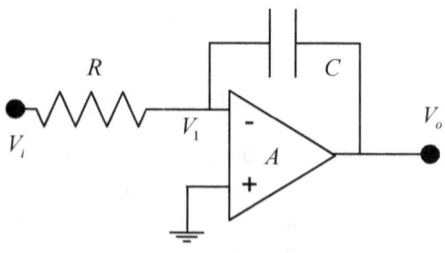

Figure E7.1(b) Series-parallel RC network

Figure E7.1(c) Integrator circuit with a finite gain OP AMP (right side)

14. Figure E7.1(d) shows an OP AMP using two RC sections as its input and feedback circuits. Write the necessary s domain impedance like expressions of the RC sections and use the symbolic tools of MATLAB to determine the voltage gain $\dfrac{V_0}{V_i}$ without and with the OP AMP finite gain A.

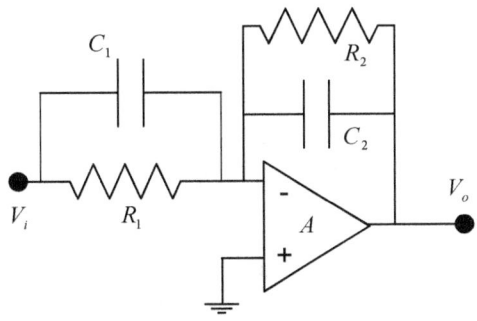

Figure E7.1(d) An OP AMP using two RC sections

15. Input impedance of the Sallen and Key filter in figure E7.2(a) is defined as $Z_i = \dfrac{V_i}{I_i}$. Write the necessary s domain equations or expressions of the filter and use the symbolic tools of MATLAB to determine the Z_i.

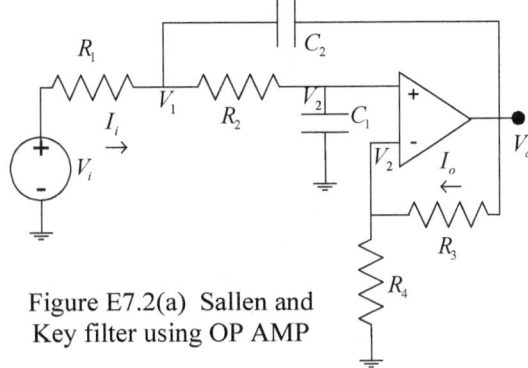

Figure E7.2(a) Sallen and Key filter using OP AMP

16. The current gain A_I of the Sallen and Key filter in figure E7.2(a) is defined as $A_I = \dfrac{I_o}{I_i}$. Write the necessary s domain equations or expressions of the filter and use the symbolic tools of MATLAB to determine the A_I.

17. In question (15) consider $R_1 = 1.5\ K\Omega$, $R_2 = 0.9\ K\Omega$, $R_3 = 2.7\ K\Omega$, $R_4 = 4\ K\Omega$, $C_1 = 4\ mF$, and $C_2 = 5\ mF$ and obtain the transfer function Z_i starting from the symbolic expression of the circuit. Graph the frequency response of Z_i.

18. In question (16) consider $R_1 = 1.5\ K\Omega$, $R_2 = 0.9\ K\Omega$, $R_3 = 2.7\ K\Omega$, $R_4 = 4\ K\Omega$, $C_1 = 4\ mF$, and $C_2 = 5\ mF$ and obtain the transfer function A_I starting from the symbolic expression of the circuit. Graph the frequency response of A_I.

19. Model the OP AMP circuit of figure E7.1(d) in SIMULINK and obtain the $A_V = \dfrac{V_0}{V_i}$ subject to $R_1 = 1.5\ K\Omega$, $R_2 = 0.9\ K\Omega$, $C_1 = 4\ mF$, and $C_2 = 5\ mF$ from the model. Analyze the frequency response of A_V.

20. Model the figure 7.6(b) shown NPN transistor in SIMULINK which is operated at high frequency. Consider the following parameters in the

modeling: $R_S = 330\Omega$, $R_L = 2.15 K\Omega$, $\beta=90$, $C_\pi=120\ fF$, $C_\mu=25\ fF$, $C_{CS}=$ 33 fF, $V_T = 26mV$, and $I_C = 2\ mA$. Obtain the $\dfrac{V_o}{V_i}$ and Z_i transfer functions for the transistor circuit from the model.

21. Determine the quality factor of the circuit in figure 7.1(c) considering that the input impedance is proportional to $\dfrac{V_o}{V_i}$.

22. In question (20) graph the frequency response of Z_i as a single plot for $\beta=75$, $\beta=100$, and $\beta=125$.

23. In question (19) graph the frequency response of A_V as a single plot for C_2 $=5\ mF$, $C_2=10\ mF$, and $C_2=15\ mF$.

24. Resistive biasing circuit of the figure E7.2(b) has the following parameters: $V_{CC}=2.5\ V$, $R_1=$ 14 $K\Omega$, $R_2=7\ K\Omega$, R_3 $=2\ K\Omega$, $R_4=200\ \Omega$,

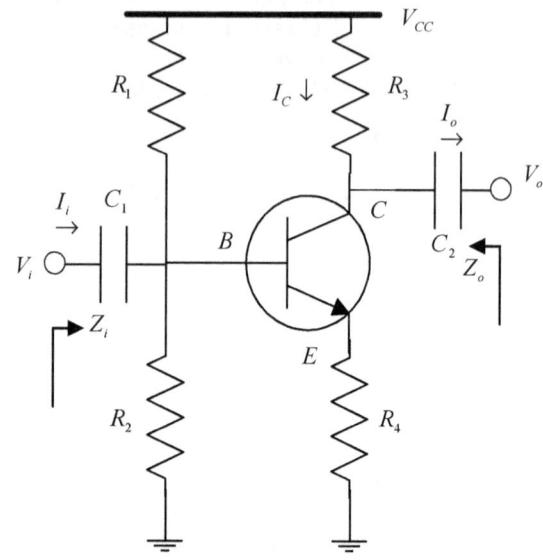

Figure E7.2(b) Resistive divider biasing circuit with input-output capacitors

$I_S = 5\times10^{-17}\ A$, $V_T=26\ mV$, $\beta=125$, $V_\gamma=0.8\ V$ (for diode model of the transistor), $C_1=5\ mF$, and $C_2=10\ mF$. Model the circuit in SIMULINK and obtain I_C without the capacitors. Remodel the circuit with capacitors and transconductance and obtain the transfer functions $A_V =\dfrac{V_o}{V_i}$, $A_I =\dfrac{I_o}{I_i}$, $Z_i =\dfrac{V_i}{I_i}$, and $Z_o =-\dfrac{V_o}{I_o}$ where the symbols have their usual meanings.

25. Based on the transfer functions obtained in question (24), analyze the frequency responses of A_V, A_I, Z_i, and Z_o.

Answers:

(1) $H(s) = \dfrac{V_0}{V_i} = \dfrac{s}{s+0.2}$ (ignore the negligible quantity)   Hint: section 7.2

(2) Hint: section 7.2

(3) $Z_i(s) = \dfrac{V_i}{I_i} = \dfrac{800s^2 + 1.42 \times 10^6 s + 1.788 \times 10^9}{s^2 + 7.854s + 2.221 \times 10^6}$

Hint: section 7.2. The model is in figure A7.1(a). Note that the model returns admittance or $\dfrac{I_i}{V_i}$

and exercise tf(1/H) for $Z_i(s)$.

Figure A7.1(a) Model for problem 3

(4) tf([7 0 -13 12],[4 0 0 0 -180 -440])
Hint: section 7.3

(5) zpk([0 -3 -4],[3 7 -8 1],-2)
Hint: section 7.3

(6) $A_V = \dfrac{1.667}{20.25s^2 + 4.925s + 1}$
Hint: section 7.3

(7) $A_V = \dfrac{-2430s^2 - 67.5s}{81s^2 + 119.3s + 37}$
Hint: section 7.3

(8) $A_V = \dfrac{s}{s+0.4444}$
Hint: section 7.3

(9) (a) $H(j2) = -1.5254 - j\,0.0052$,

(b) $H(-j0.1) = 4.3269 + j\,0.0021$, $H(j0.1) = 4.3269 - j\,0.0021$,
   $H(j0.5) = 5.6249 - j\,0.0176$, and $H(j1) = 89.1089 - j\,8.9109$,

(c) $H(j2) = 1.5254 \angle -179.81°$,

(d) $H(-j0.1) = 4.3269 \angle 0.0275°$, $H(j0.1) = 4.3269 \angle -0.0275°$,
   $H(j0.5) = 5.625 \angle -0.179°$, and $H(j1) = 89.5533 \angle -5.7106°$, and

(e) w=-25:0.1:25; R=freqresp(H,w); V=squeeze(R);.
Hint: section 7.4

(10) Hint: section 7.5

(11) (a) lowpass, cutoff frequency: 13.6342 $rad/\sec$ (b) highpass, cutoff frequency: 136.1216 $rad/\sec$ (c) bandpass, center frequency: 9.3243 $rad/\sec$, cutoff frequencies: 9.2748 $rad/\sec$ and 9.4730 $rad/\sec$, and bandwidth: 0.1982 $rad/\sec$ (d) bandstop, center frequency: 1.3112×10^3 $rad/\sec$, cutoff frequencies: 0.5605×10^3 $rad/\sec$ and 2.9627×10^3 $rad/\sec$, and bandwidth: 2.4022×10^3 $rad/\sec$.
Hint: section 7.6

(12) $Z_{in}(s) = \dfrac{R_2 R_1 C_2 s + R_2 R_1 C_1 s + R_2 + R_1}{(R_1 s C_1 + 1)(R_2 s C_2 + 1)}$ Hint: section 7.7

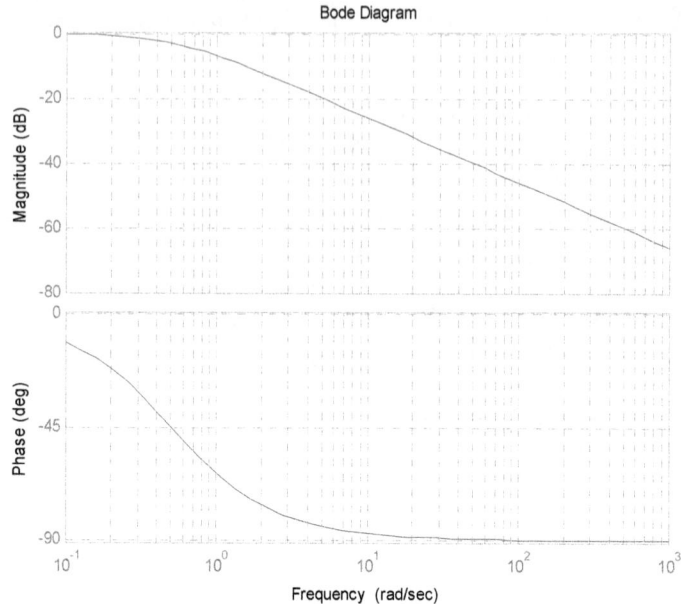

Figure A7.1(b) Low pass frequency response of the transfer function in
part a of problem 10

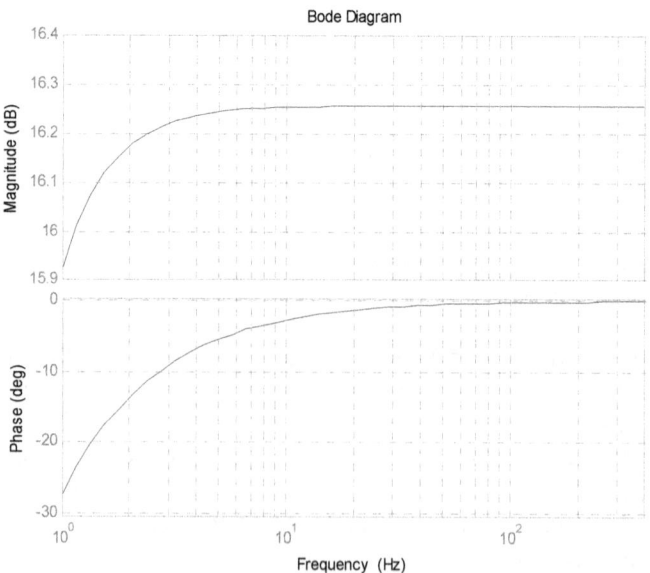

Figure A7.1(c) High pass frequency response of the transfer function
in part b of problem 10

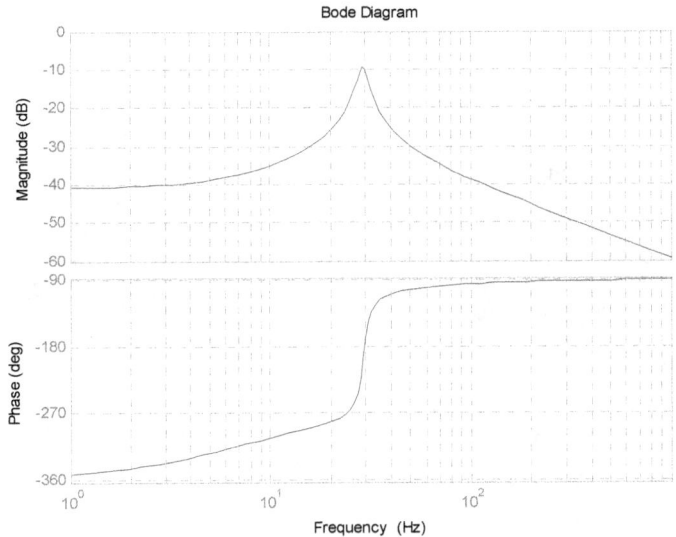

Figure A7.1(d) Pass band frequency response of the transfer
function in part c of problem 10

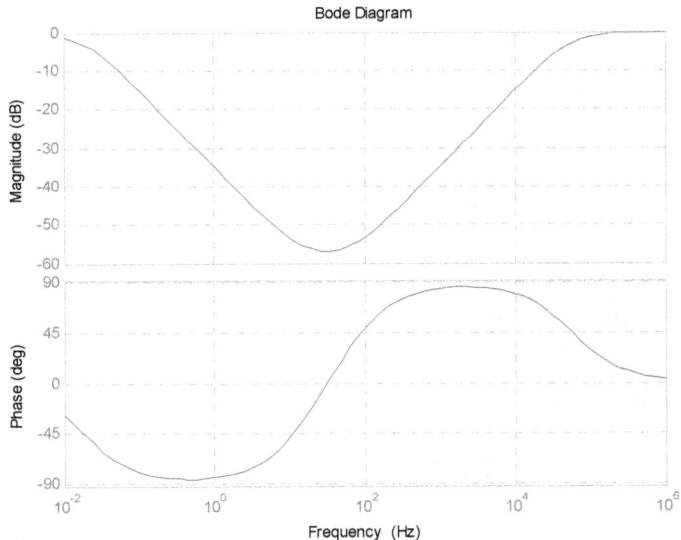

Figure A7.1(e) Stop band frequency response of the transfer
function in part d of problem 10

(13) $\dfrac{V_0}{V_i} = -\dfrac{A}{1 + sRCA + sRC}$ 　　　　Hint: section 7.7

(14) without A i.e. ideal OP AMP: $\dfrac{V_0}{V_i} = -\dfrac{R_2(1 + sC_1R_1)}{R_1(1 + sR_2C_2)}$

　　　with A : $\dfrac{V_0}{V_i} = -\dfrac{AR_2(1 + sC_1R_1)}{R_2 + R_1sR_2C_1 + R_1sR_2C_2 + R_1sR_2C_2A + R_1 + R_1A}$

Commands without A :	Commands with A :
syms R1 C1 R2 C2 s	syms R1 C1 R2 C2 s V1 Vi Vo
Z1=Zp(R1,1/s/C1);	Z1=Zp(R1,1/s/C1);
Z2=Zp(R2,1/s/C2);	Z2=Zp(R2,1/s/C2);
Av=-Z2/Z1;	e1=(Vi-V1)/Z1-(V1-Vo)/Z2;
pretty(simple(Av))	e2='V1=-Vo/A';
	O1=maple('algsubs',e2,e1);
	O2=solve(O1,Vo); Vi=1;
	Av=subs(O2); pretty(Av)

Hint: section 7.7, symbols have their usual meanings e.g. R1 for R_1.

(15) $Z_i(s) = \dfrac{R_4R_1C_1s + R_4 + R_4sC_1R_2 - sC_2R_1R_3 + s^2R_1R_2C_1C_2R_4}{s(R_4C_1 - C_2R_3 + sC_2R_4C_1R_2)}$

Governing equations:

$$\dfrac{V_1 - V_2}{R_2} - sC_1V_2 = 0, \quad V_2 = \dfrac{R_4}{R_3 + R_4}V_o, \quad \dfrac{V_i - V_1}{R_1} + (V_o - V_1)sC_2 + \dfrac{V_2 - V_1}{R_2} = 0,$$

and $I_i - \dfrac{V_i - V_1}{R_1} = 0$.

Commands you need:
```
syms s R1 C1 R2 C2 R3 R4 V1 V2 Vo Vi Ii
e1=(V1-V2)/R2-V2*s*C1;
e2='V2=Vo*R4/(R3+R4)';
e3='(Vi-V1)/R1+(Vo-V1)*s*C2+(V2-V1)/R2';
e4='Ii-(Vi-V1)/R1';
O1=maple('algsubs',e2,e1);
O2=maple('algsubs',e2,e3);
V1=solve(O1,V1);
O3=subs(O2); O4=subs(e4);
Vo=solve(O3,Vo); O5=subs(O4);
Z=solve(O5,Vi); Ii=1; Zi=subs(Z);
pretty(Zi)
```
Hint: section 7.7, symbols have their usual meanings e.g. Ii for I_i.

(16) $A_i = \dfrac{I_0}{I_i} = \dfrac{1}{s(sC_1C_2R_2R_4 + C_1R_4 - C_2R_3)}$

Governing equations:

$$I_i + (V_o - V_1)sC_2 + \dfrac{V_2 - V_1}{R_2} = 0, \quad V_2 = \dfrac{R_4}{R_3 + R_4}V_o, \quad I_o - \dfrac{V_o - V_2}{R_3} = 0, \quad \text{and}$$

$$\dfrac{V_1 - V_2}{R_2} - sC_1V_2 = 0.$$

Commands you need:
```
syms s R1 C1 R2 C2 R3 R4 V1 V2 Vo Vi Ii Io
```

```
e1=li+(Vo-V1)*s*C2+(V2-V1)/R2;
e2='V2=Vo*R4/(R3+R4)';
e3=lo-(Vo-V2)/R3;
e4=(V1-V2)/R2-s*C1*V2;
O1=maple('algsubs',e2,e1);
O2=maple('algsubs',e2,e3);
O3=maple('algsubs',e2,e4);
V1=solve(O1,V1);
O4=subs(O2); O5=subs(O3);
Vo=solve(O4,Vo); O6=subs(O5);
Al=solve(O6,lo); li=1;
Al=subs(Al); pretty(Al)
```

Hint: section 7.7, symbols have their usual meanings e.g. Vo for V_o .

(17) After executing question (15) quoted commands, execute the following:

```
R1=1.5e3; R2=0.9e3; R3=2.7e3; R4=4e3; C1=4e-3; C2=5e-3;
[N,D]=numden(subs(Zi));
Tz=tf(sym2poly(N),sym2poly(D));
bode(Tz)
```

$$Z_i = \frac{216000s^2 + 36300s + 8000}{144s^2 + 5s} \text{ held in Tz}$$

Hint: sections 7.7 and 7.8, symbols have their usual meanings e.g. C1 for C_1, the frequency response of Z_i is lowpass type – similar to figure 7.4(a), and the graph is not shown for space reason.

(18) After executing question (16) quoted commands, execute the following:

```
R1=1.5e3; R2=0.9e3; R3=2.7e3; R4=4e3; C1=4e-3; C2=5e-3;
[N,D]=numden(subs(Al));
Al=tf(sym2poly(N),sym2poly(D));
bode(Al)
```

$$A_l = \frac{2}{144s^2 + 5s} \text{ held in Al}$$

Hint: sections 7.7 and 7.8, symbols have their usual meanings e.g. Al for A_l, the frequency response of A_l is lowpass type – similar to figure 7.4(a), and the graph is not shown for space reason.

Figure A7.2(a)
Model for voltage
gain on circuit of
figure E7.1(d) –
right side figure

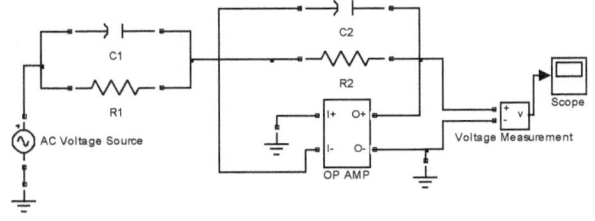

(19) Figure A7.2(a) for the model. Suppose the model file name is test then commands you need:

```
H=power_analyze('test','ss');
Hm=tf(H)
Av=Hm(1,1)
```

Mohammad Nuruzzaman

$$A_V = \frac{V_0}{V_i} = \frac{s^2 + 0.3894s + 0.03714}{s^2 + 4.5\times10^{12}s + 8.91\times10^{11}}.$$

Highpass type, graph is not shown for space reason, similar to figure 7.4(b), $\omega_c = 4.45\times10^{12}$ rad/sec, poles of A_V: 0 and -4.5×10^{12}, and zeroes of A_V: -0.2224 and -0.167.
Hint: sections 7.6, 7.8, and 7.9.

(20) $\dfrac{V_o}{V_i} = \dfrac{9.731\times10^9 s}{s^2 + 3.761\times10^{10}s + 2.321\times10^{20}}$, $Z_i = \dfrac{s^2 + 3.761\times10^{10}s + 2.321\times10^{20}}{0.00303s^2 + 4.555\times10^7 s + 1.547\times10^{17}}$

Hint: section 7.9.
(21) $Q = 1.0509$
Commands: [N,D]=tfdata(Hm,'v'); Q=sqrt(D(3)*D(1))/D(2) where Hm holds A_V from the model.
Hint: sections 7.2, 7.8, and 7.9.

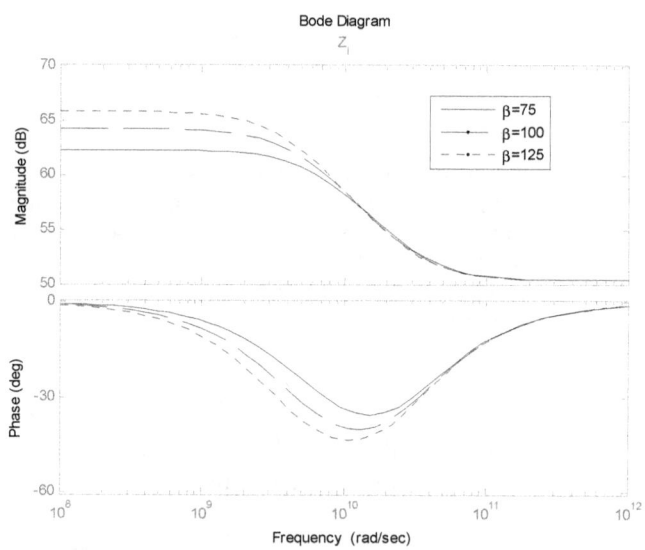

Figure A7.2(b) Frequency response of input impedance for different β

(22) Figure A7.2(b). For the first β keep the result to some variable Z1 for Z_i similarly to Z2 and Z3 for the other two then call bode as bode(Z1,Z2,Z3).   Hint: sections 7.8 and 7.9.
(23) Figure A7.2(c). For the first C_2 keep the result to some variable Av1 for A_V similarly to Av2 and Av3 for the other two then call bode as bode(Av1,Av2,Av3).   Hint: sections 7.8 and 7.9.

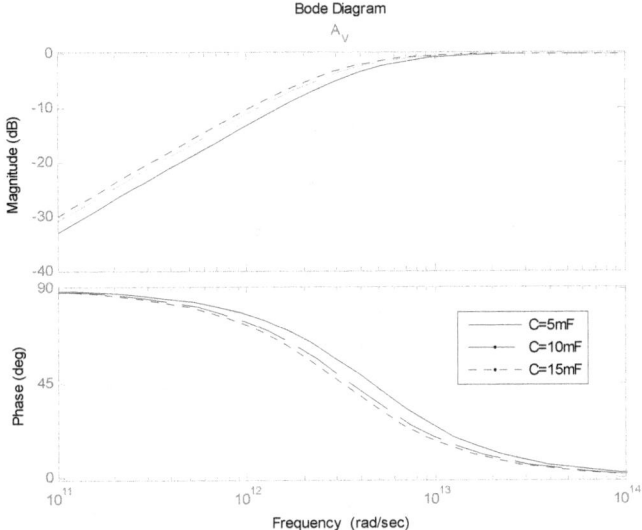

Figure A7.2(c) Voltage gain response for different C_2

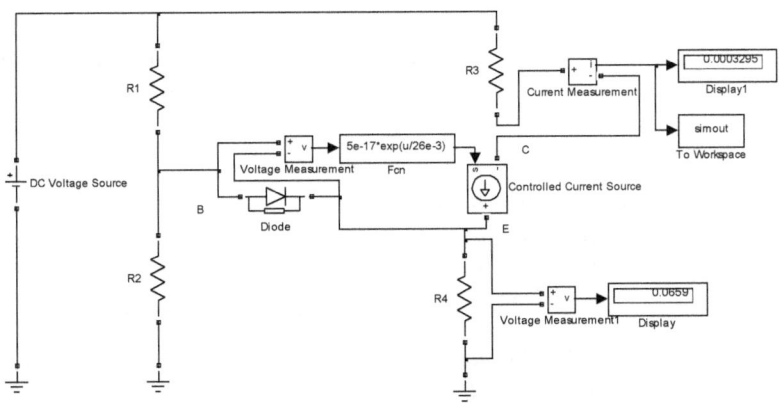

Figure A7.2(d) Model for finding I_C

(24) Run the model of figure A7.2(d) with the given parameters in order to find I_C. Then execute the following commands for transistor transconductance:

```
Ic=simout(end);
VT=26e-3; gm=Ic/VT; b=125; ri=b*VT/Ic;
```

Transconductance based model is seen in figure A7.2(e) which is implemented both for A_V and Z_i, say the file name is **test**. Then carry out the following:

```
H=power_analyze('test','ss');
```

```
Hm=tf(H)
Av=Hm(1,1)
Zi=1/Hm(3,1)
AI=Hm(4,1)/Hm(3,1)
```

and expect $A_V = \dfrac{s + 6.939 \times 10^{-18}}{s + 0.04845}$, $Z_i = \dfrac{s + 0.04845}{0.0002422s + 1.681 \times 10^{-21}}$, and $A_I = 0$.

Notice that in figure A7.2(e), Hm(1,1) is from AC Voltage Source to Voltage Measurement1 or V_o to V_o, Hm(3,1) is from AC Voltage Source to Current Measurement or V_i to I_i, Hm(4,1) is from AC Voltage Source to Current Measurement1 or V_i to I_o, and Hm(3,1) is from AC Voltage Source to Current Measurement or V_i to I_i. The current gain A_I is obtained from the division of two transfer functions and the Z_i is from the reciprocal of admittance.

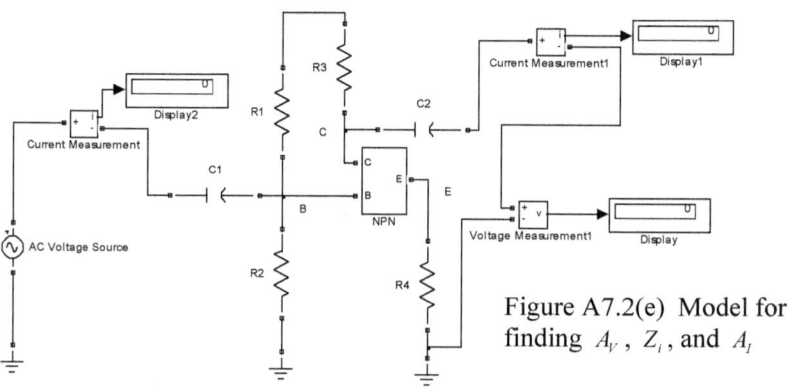

Figure A7.2(e) Model for finding A_V, Z_i, and A_I

For the Z_o another model is required. As the circuit theory says the input source must be removed or short circuited instead the source should be connected at the output where impedance is required. Figure A7.2(f) depicts the complete model for Z_o which is to be determined from AC Voltage Source to

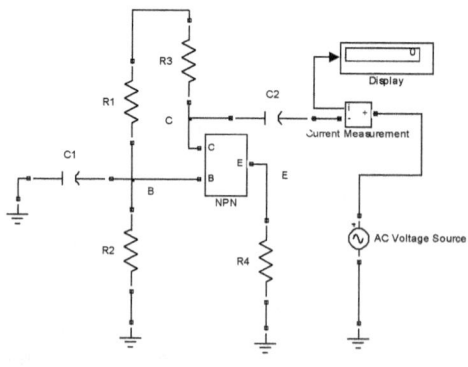

Figure A7.2(f) Model for finding Z_o

Current Measurement. Importantly the Current Measurement is flipped and follows the AC Voltage Source polarity because Z_o does not have polarity but I_o does. Unnecessary blocks in figure A7.2(e) are also deleted. Say the file name of model in figure A7.2(f) is test1 then you again need to execute the following:

```
H1=power_analyze('test1','ss');
Hm1=tf(H1)
Zo=1/Hm1(2,1)
```

where above Hm1 or H1 is just another variable and expect the output impedance to be

$$Z_o = \frac{s^2 + 0.0672s + 0.0003028}{6.25 \times 10^{-5} s^2 + 3.028 \times 10^{-6} s + 2.626 \times 10^{-24}}.$$

It is noteworthy to point out that reader should check the input-output quantities (voltage or current) linking to a transfer function from MATLAB command window because we found Z_o as 1/Hm1(2,1) or second row in Hm1, you may find it as another row e.g. 1/Hm1(1,1). It depends how the machine interprets the modeling quantities.

Hint: sections 7.6, 7.8, and 7.9.

(25) Frequency response graph is not shown for space reason in the following.

For A_V : Highpass type, similar to figure 7.4(b), $\omega_c = 0.0485 \ rad / \sec$, pole of A_V : -0.0484, and zero of A_V : 6.9389×10^{-18} (0 practically).

Commands:

```
bode(Av)
w=linspace(10^-3,10^0,1000); R=freqresp(Av,w); V=squeeze(R);
A=abs(V); L=A(end);
c_cross(w,A,L/sqrt(2))
pole(Av)
zero(Av)
```

For Z_i : Lowpass type, similar to figure 7.4(a), $\omega_c = 6.7768 \times 10^{-18} \ rad / \sec$ (0 practically), pole of Z_i : -6.9389×10^{-18} (0 practically), and zero of Z_i : -0.0484.

Commands:

```
bode(Zi)
bandwidth(Zi)
pole(Zi)
zero(Zi)
```

For A_I : Since $A_I = 0$, the frequency analysis is meaningless.

For Z_o : Lowpass type, similar to figure 7.4(a), $\omega_c = 8.471 \times 10^{-19} \ rad / \sec$ ($\approx$ 0), poles of A_V : 0 and -0.0484, and zeroes of A_V : -0.0049 and -0.0623.

Commands:

```
bode(Zo)
bandwidth(Zo)
pole(Zo)
zero(Zo)
```

Since the Z_o is a second order system, quality factor finding is possible which is $Q = 4.2313 \times 10^{-9}$ or 0 practically.

Commands:

```
[N,D]=tfdata(Zo,'v'); Q=sqrt(D(3)*D(1))/D(2)
```

Hint: sections 7.6, 7.8, and 7.9.

Chapter 8

Electronic System Projects

Basic electronics design problems trained in undergraduate course are randomly accumulated in this chapter. We treat every design problem as a mini project because each one needs MATLAB computing or SIMULINK modeling. This sort of design project is important because pedagogical approach of any particular chapter can not solve completely all design problems, some other tools or algorithms may be essential for involvement of practical parameters or factors. The project may require transistor characteristic, voltage or current gain, input or output impedance, intrinsic parameters selection, or other relevant quantity finding. In every mini project finished solution or answer along with structured explanation would provide the reader some level of confidence on electronics analysis through the wing of MATLAB or SIMULINK. These projects pave the way for insight of much more complicated electronics design despite seemingly simple. The project collection is in the sequel.

✦ **Project 1: Effect of changing parameters on a transistor:** R

Consider the figure 4.1(d) shown NPN transistor of section 4.1. We wish to find the resistor value range to keep the transistor in active region with $V_{BE} = 752\,mV$, $V_T = 26\,mV$, $I_S = 5.1 \times 10^{-16}\,A$, and $V_{CC} = 3.1\,V$.

Prerequisite: sections 4.1 and 4.14.

Solution:

We solved the problem in section 4.14 but the value of R was chosen visually from the screen with coarse resolution. If we wish to conduct that by programming tactic, the model we need is in figure 8.1(a). The **Subtract** performs $V_{CE} - V_{BE}$ operation and **Compare To Zero** checks $V_{CE} - V_{BE} > 0$. The return of **Compare To Zero** is 1 (for true) or 0 (for false)

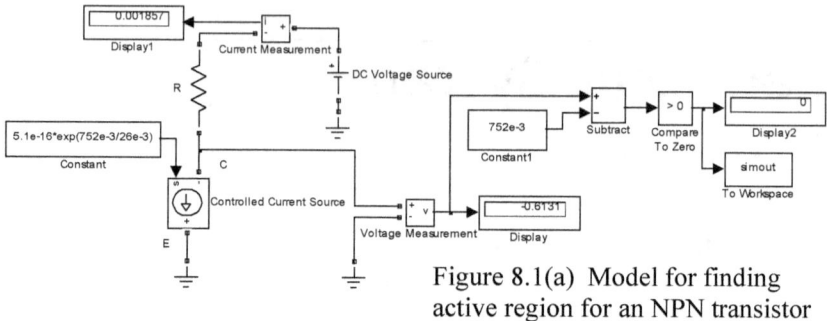

Figure 8.1(a) Model for finding
active region for an NPN transistor

which is sent to workspace through **simout**. In section 4.14 R was chosen from 1 $K\Omega$, not necessarily. If we start from $R=0$, there is some modeling problem due to internally generated indeterminate form. That problem is overcome by small numerical quantity epsilon which has MATLAB synonym **eps**. The resolution was 0.1 $K\Omega$, let us reduce it to 0.01 $K\Omega$ and up to 2 $K\Omega$ (already answer is known). Rerun the modified model in figure 8.1(a) assuming the same model name **test**:

>>R1=[eps:0.01:2]*1e3; D=[ ]; ↵
>>for R=R1,sim('test.mdl'), D=[D simout(end)]; end ↵

Be patient because model running may take a while. Appendix D.6 mentioned **find** determines integer position indexes as follows:

>>rr=find(D==1); ↵

The **rr** is a user-chosen variable and holds the integer indexes for the R where $V_{CE} - V_{BE} > 0$ is satisfied. For the cutoff we need the last value stored in rr and obtain that by:

>>R1(rr(end)) ↵

ans =
     1260

In section 4.14 it was 1200 subject to coarse resolution now it is more appropriate hence the transistor will remain active over $0\Omega < R \leq 1260\Omega$.

✦ Project 2: Effect of changing parameters on a transistor: I_S

In project 1 consider $R=1$ $K\Omega$ and if the reverse saturation current changes over $0.1\times10^{-15} A \leq I_S \leq 1\times10^{-15} A$, does the transistor remain active? We wish to investigate that by keeping the other parameters same.

Solution:

Now the I_S is variable which is inside the **Constant** in figure 8.1(a), **Value** in the parameter window of which should be **Is*exp(752e-3/26e-3)** where **Is** stands for I_S. A resolution of I_S has to be decided, say $\Delta I_S = 0.01 \times 10^{-15} A$. Set the **R** block **Resistance** as **1e3** because it was changed in project 1. Save the model by some name **P2** and run like the project 1 as follows:

```
>>Is1=[0.1:0.01:1]*1e-15; D=[ ]; ↵
>>for Is=Is1,sim('P2.mdl'), D=[D simout(end)];end ↵
```

In project 1 we had some idea about the result. Appendix E mentioned **scatter** helps us graph the D result as follows:

```
>>scatter(Is1,D) ↵
```

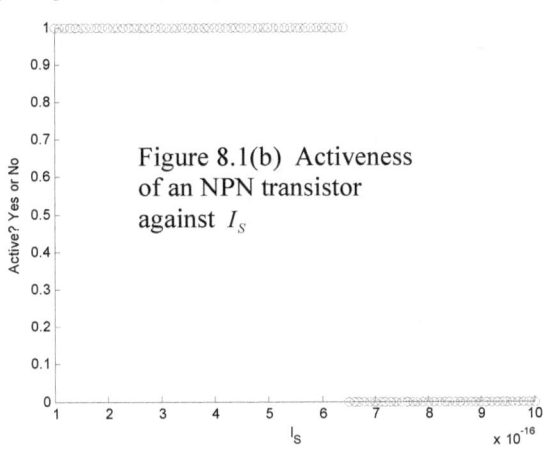

Figure 8.1(b) Activeness of an NPN transistor against I_S

Figure 8.1(b) indicates the activeness of the NPN transistor against the I_S (besides some x or y label is added). For some I_S, the transistor is active which is evident from the graph (segment related to value 1). In order to find the transition (i.e. 1 to 0) of figure 8.1(b), exercise the following:

```
>>rr=find(D==1); Is1(rr(end)) ↵
```

```
ans =
     6.4000e-016
```

The last return indicates $6.4 \times 10^{-16} A$ or $0.64 \times 10^{-15} A$ hence the transistor will remain active over $0.1 \times 10^{-15} A \le I_S \le 0.64 \times 10^{-15} A$.

♦ **Project 3: Collector current for changing I_S**

In project 2 we need to determine the collector current I_C range where the transistor is active.

Solution:

In project 2 you need another **To Workspace** block for I_C transporting. Copy-paste **To Workspace** in the model of figure 8.1(a) so that you get **To Workspace1** and connect **To Workspace1** with the **Display1** which shows I_C. For the I_C manipulation in MATLAB you need another variable **Ic** (like **D**) where **Ic** is user-chosen. The **simout** returns transistor activeness whereas **simout1** does I_C and run the commands of project 2 as follows:

```
>>Is1=[0.1:0.01:1]*1e-15; D=[ ]; Ic=[ ]; ↵
>>for Is=Is1,sim('P2.mdl'), D=[D simout(end)]; Ic=[Ic simout1(end)]; end ↵
>>rr=find(D==1); ↵
```

In order to get the lower bound of I_C the first integer index in **rr** is invoked:

```
>>Ic(rr(1)) ↵
```

ans =

    3.6403e-004   ← i.e. $I_C = 3.6403 \times 10^{-4} A$

Again for the last bound of I_C the last integer index in **rr** is invoked:

```
>>Ic(rr(end)) ↵
```

ans =

    0.0023   ← i.e. $I_C = 0.0023 A$

From the above executions what do we infer? As the I_S varies over $0.1 \times 10^{-15} A \le I_S \le 0.64 \times 10^{-15} A$, the I_C changes over $3.6403 \times 10^{-4} A \le I_C \le 0.0023 A$.

✦ Project 4: Effect of changing parameters on a transistor: V_{BE}

In project 1 we need to determine the V_{BE} range such that the transistor is in active region if V_{BE} changes over $0.7V \le V_{BE} \le 0.8V$ subject to $\Delta V_{BE} = 0.005 V$, also collector current I_C is required where the transistor is active.

Solution:

Last three project mentioned tactics and symbologies are applied, code of **Constant** is 5.1e-16*exp(Vbe/26e-3), check other parameters due to previous change, and other commands are as follows:

```
>>Vbe1=0.7:0.005:0.8; D=[ ]; Ic=[ ]; ↵
>>for Vbe=Vbe1,sim('P4.mdl'),D=[D simout(end)]; Ic=[Ic simout1(end)];end ↵
>>scatter(Vbe1,D) ↵          ← Displays graph similar to figure 8.1(b)
>>rr=find(D==1); ↵
```

Calling of **Vbe1(rr(1))**, **Vbe1(rr(end))**, **Ic(rr(1))**, and **Ic(rr(end))** at the command prompt provides 0.7, 0.755, 2.5125e-004, and 0.0021 respectively i.e. the transistor is active over $0.7V \le V_{BE} \le 0.755V$ where I_C changes over $2.5125 \times 10^{-4} A \le I_C \le 0.0021 A$.

✦ Project 5: Activeness for a resistive divider circuit

Concerning the resistive divider biasing NPN transistor of section 4.4, we intend to determine all four resistor ranges so that the transistor remains in active region.

Solution:

This is basically project 1 but testing needs four times – one for each resistor. We change one resistor and keep the other three as they are. For the NPN transistor the $V_{CE} - V_{BE}$ is also V_{CB}. Consider the R_3 and start from the model of figure 4.4(d). Figure 8.1(c) shows the required model in which **Voltage Measurement1** refers to V_{CB}. Enter the **Resistance** of R3 as R3 (just a user-chosen variable). Initially we are not sure about the range of R_3 so let us choose some coarse resolution $\Delta R_3 = 0.1\ K\Omega$ over $0\,K\Omega < R \le 1K\,\Omega$ just to avoid the run time of the model and carry out the following like project 1:

```
>>R31=[eps:0.1:1]*1e3; D=[ ]; ↵
>>for R3=R31,sim('P5.mdl'), D=[D simout(end)]; end ↵
```

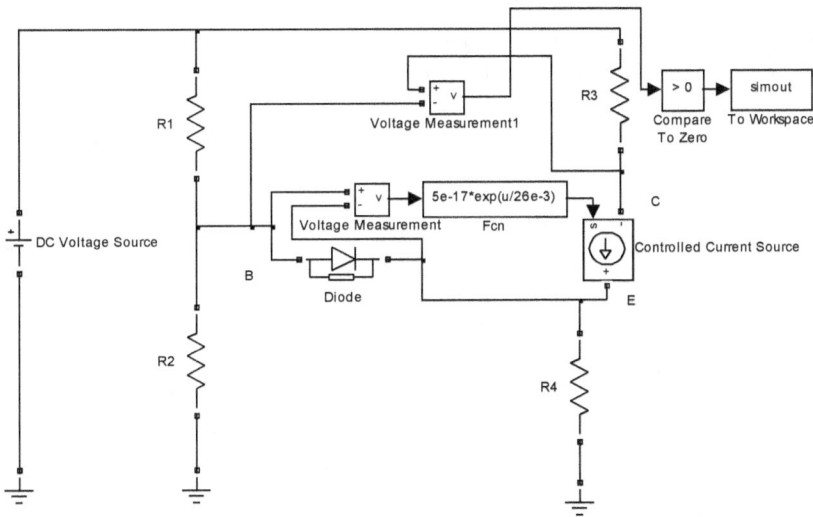

Figure 8.1(c) Activeness testing on resistively divider biased NPN transistor

Both the file name **P5.mdl** and resistor vector **R31** are user-chosen. The **scatter(R31,D)** shows graph without any transition from 1 to 0 manifesting nonactive region over $0\,K\Omega < R \le 1K\,\Omega$. Change the last bound from $1K\,\Omega$ to $2K\,\Omega$ and now **scatter(R31,D)** shows 1 to 0 transition between 1400 Ω and 1600 Ω. Take a fine resolution say $0.01K\,\Omega$ and reexecute all:

```
>>R31=[1.4:0.01:1.6]*1e3; D=[ ]; ↵
>>for R3=R31,sim('P5.mdl'), D=[D simout(end)]; end ↵
>>rr=find(D==1); R31(rr(end)) ↵
```

```
ans =
        1550
```

What can we say from this simulation? The circuit will remain active over $0\Omega < R \le 1550\Omega$ while the other parameters are unchanged. We did similar test for the other three resistors and the results are the following:

$R_1 \ge 259\Omega$ when $R_2 = 9\ K\Omega$, $R_3 = 1\ K\Omega$, and $R_4 = 100\ \Omega$,

$R_2 > 0$ when $R_1 = 16\ K\Omega$, $R_3 = 1\ K\Omega$, and $R_4 = 100\ \Omega$, and

$R_4 > 0$ when $R_1 = 16\ K\Omega$, $R_2 = 9\ K\Omega$, and $R_3 = 1\ K\Omega$.

Note that for R_1 we need rr=find(D==1); R11(rr(1)) because of its relative position in the vector R11. For the R_2 and R_4 despite increasing the last bound, no transition was found.

✦ Project 6: Self bias circuit characteristics

Figure 8.1(d) shows a self bias NPN transistor. Our objective is to determine the following for the circuit:

(a) I_C (in mA) versus V_R (in V),

(b) power dissipated in R_C (in mW) versus I_C (in mA), and

(c) power dissipated in R_C (in mW) versus R_C (in Ω)

subject to $V_{CC} = 2.5\ V$, $V_{BE} = 0.8V$, $\beta = 100$, $R_B = 10K\Omega$, and $R_E = 10K\Omega$ over $1K\Omega \le R_C \le 2K\Omega$.

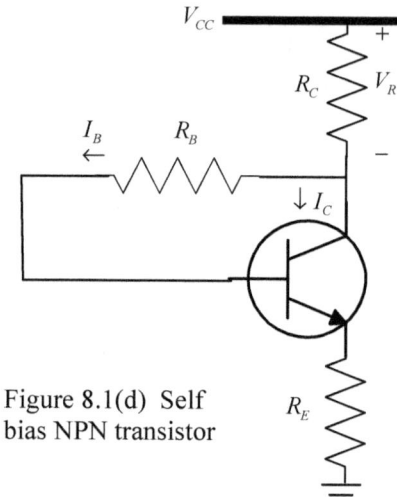

Figure 8.1(d) Self bias NPN transistor

Prerequisite: sections 4.1, 4.5, and 4.14 and project 1.

Solution:

The circuit modeling is shown in figure 8.1(e). Blocks you need: one DC Voltage Source, one Current Measurement, one Voltage Measurement, one Fcn, one To Workspace, one Series RLC Branch, one Diode, one Gain, and one

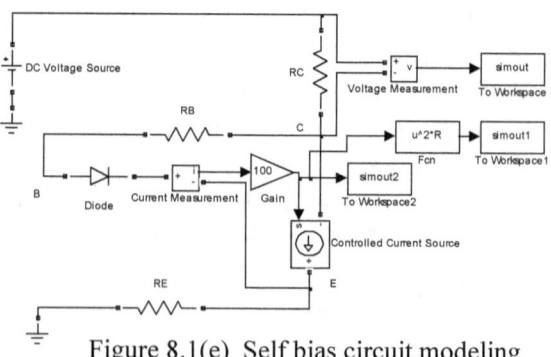

Figure 8.1(e) Self bias circuit modeling

-214-

Controlled Current Source (get the blocks in a new model file and enter the parameters as given). The **simout**, **simout1**, and **simout2** refer to V_R, power dissipated in R_C, and I_C all in standard unit respectively. Change the parameter of **Series RLC Branch** (section 2.1) to turn it to resistor with rename **RB**. Copy-paste and/or rotate like figure 8.1(e) to get the other two resistors which are renamed as **RE** and **RC**. Set the data format of **To Workspace** as **Array** and copy-paste it to get the other two. The code of **Fcn** which is **u^2*R** indicates electrical power I^2R. Make sure the **Resistance** of **RC** is **R** in the parameter window because that is changing in the problem, say $\Delta R_C = 0.1K\Omega$. Run the model from prompt assuming file name is **P6.mdl**:

```
>>R1=[1:0.1:2]*1e3; Vr=[ ]; Ic=[ ]; P=[ ]; ↵
>>for R=R1,sim('P6.mdl'), Vr=[Vr simout(end)]; ↵
>>P=[P simout1(end)]; Ic=[Ic simout2(end)]; end ↵
```

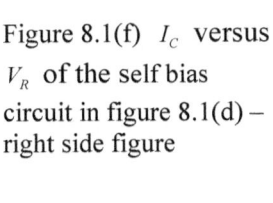

Figure 8.1(f) I_C versus V_R of the self bias circuit in figure 8.1(d) – right side figure

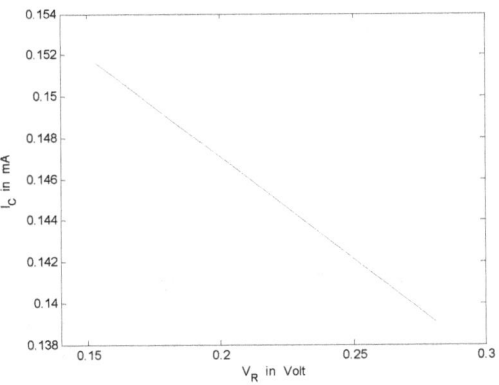

Figure 8.1(g) P versus I_C of the self bias circuit in figure 8.1(d) – right side figure

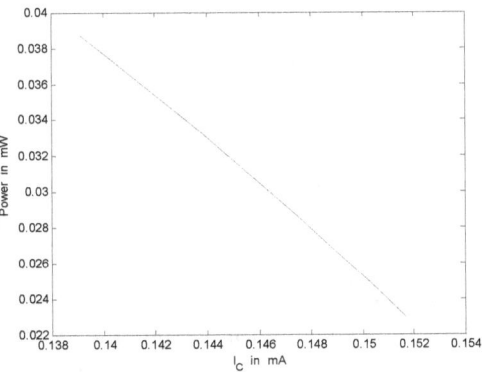

In above the variables **R1, Vr, Ic,** and **P** correspond to R_C, V_R, I_C, and power $I_C^2 R_C$ samples respectively. The **plot** (appendix E) is exercised to obtain each self bias trace. The **P** contents are in W which are converted to mW by

multiplying 1000 or using **1000*P** inside the **plot**, applied in other quantities too in the sequel:

>>plot(Vr,1000*Ic) ↵   ← I_C (in mA) versus V_R (in V), figure 8.1(f)

>>plot(1000*Ic,1000*P) ↵ ← Power dissipated in R_C (in mW) versus I_C (in mA), figure 8.1(g)

>>plot(R1/1000,1000*P) ↵ ← Power dissipated in R_C (in mW) versus R_C (in $K\Omega$), figure 8.1(h)

In addition some x or y labels are attached in above figures e.g. in the last figure xlabel('R_C in K\Omega') is exercised for the horizontal label.

Figure 8.1(h)   *P* versus R_C of the self bias circuit in figure 8.1(d) – right side figure

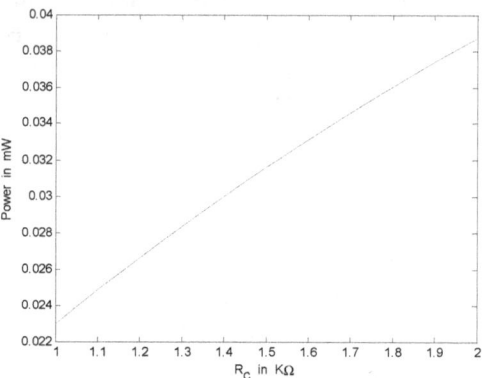

✦ Project 7: CE amplifier circuit design

Figure 4.2(a) shows a CE amplifier core circuit. Design the circuit with V_{CC} =2.5 V , V_{BE} =0.8 V , V_T =26 mV , and maximum power consumption of 5 mW while maximum voltage gain is achieved.

```
1 -    R=linspace(eps,2e3,1000);
2 -    Ic=5e-3/2.5;
3 -    Vcb=2.5-Ic*R-0.8;
4 -    Av=Ic*R/26e-3;
5 -    subplot(121),plot(R,Vcb),xlabel('R_L in \Omega'),ylabel('V_C_B in V'),grid
6 -    subplot(122),plot(R,Av),xlabel('R_L in \Omega'),ylabel('A_V'),grid
7 -    In=find(Vcb>=0);
8 -    Rm=R(In(end))
9 -    Avm=Av(In(end))
10
```

Figure 8.2(a) MATLAB script file for the project 7

Solution:

The circuit design means finding the range of R_L fulfilling given criteria. Although SIMULINK modeling is presented in section 4.1, computing approach is introduced here given the nature of the problem. In order to keep the transistor in active region, the condition $V_{CB}=$ $V_{CC}-I_C R_L -V_{BE}$ >0 must be satisfied. No range of R_L is given, first choose one range, then check the conditionality, and replace by another range if conditionality is not satisfied say $0 < R_L \leq 2K\Omega$. The $R_L = 0$ is avoided for modeling complication, can be done starting from small numerical quantity epsilon (**eps** in MATLAB). The voltage gain is given by $|A_V| = \dfrac{I_C R_L}{V_T}$ from CE theory.

Figure 8.2(a) shows the commands we need for the project, explanation of which is the following: line 1 – R_L vector generation and assigned to user-chosen **R** with 1000 samples from 0 to 2 $K\Omega$ (appendix D.9 for **linspace**), line 2 – computing maximum I_C from P_{max}/V_{CC}, line 3 – computing $V_{CB} = V_{CC} - I_C R_L - V_{BE}$, line 4 –

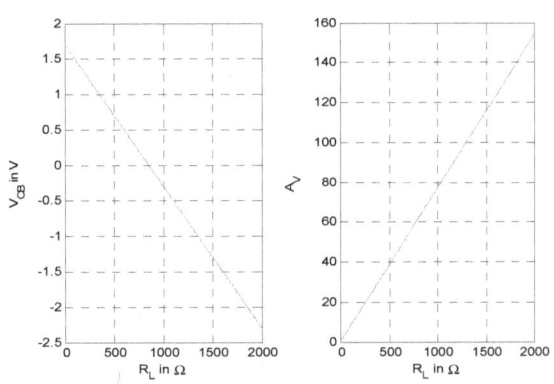

Figure 8.2(b) V_{CB} versus R_L and $|A_V|$ versus R_L of CE amplifier

computing $|A_V|$, and lines 5 and 6 – graphing V_{CB} versus R_L and $|A_V|$ versus R_L using **subplot** (appendix E) along with x-y labelings, outcome is the figure 8.2(b). In V_{CB} versus R_L plot we find $V_{CB} > 0$ or positive portion which is essential for the amplifier. The integer indexes of such samples in **R** are determined and assigned to user-chosen **In** by **find** (appendix D.6) in line 7.

The last integer in **In** corresponds to the last bound of R_L (i.e. the bound is **R(In(end))**) as well as the achievable maximum $|A_V|$ in figure 8.2(b), both of which are determined and assigned to user-chosen **Rm** and **Avm** in lines 8 and 9 respectively.

Rm =
    848.8488

Avm =
    65.2961

After running the script file you find above i.e. $0 < R_L \leq 848.8488\Omega$ and $|A_V|_{max} = 65.2961$.

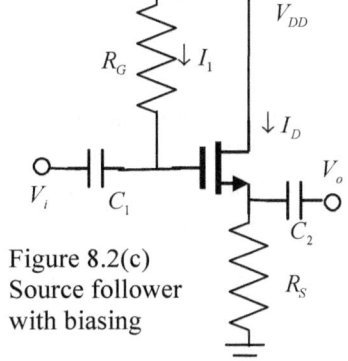

Figure 8.2(c) Source follower with biasing

◆ **Project 8: Source follower MOS circuit design**

Figure 8.2(c) shows a source follower NMOS transistor with $\mu_n C_{ox} = 110\,\mu A/V^2$, $V_{TH} = 0.5\,V$, and $V_{DD} = 2\,V$. Design the circuit parameters to obtain a voltage gain 0.8 and power rating $10\,mW$. The voltage drop across R_G should be 0 but assume a small value e.g. $0.01\,V$. The current I_1 should also be 0 but choose one thousandth of I_D.

Prerequisite: appendix G and sections 6.1 and 6.2.

Solution:

We just need to solve the equations from given specifications. Let us write the related equations of the source follower assuming the two coupling capacitors as short:

power rating is $(I_1+I_D)V_{DD}=10\,mW$,

$I_1=I_D/1000$,

voltage drop across R_G is $I_1 R_G=0.01\,V$,

from MOS theory voltage gain $A_V=\dfrac{R_S}{\dfrac{1}{g_m}+R_S}=0.8$,

drain current in saturation region is $I_D=\dfrac{1}{2}\mu_n C_{ox}\dfrac{W}{L}(V_{GS}-V_{TH})^2$,

V_{DD} to ground voltage drop is $(I_1+I_D)R_S+V_{GS}+0.01=V_{DD}$, and

from MOS circuit theory transconductance is $g_m=\dfrac{2I_D}{V_{GS}-V_{TH}}$.

Figure 8.2(d)
MATLAB script
file for the
project 8 – right
side figure

```
1 -   e1='(I1+Id)*2=10e-3';
2 -   e2='I1=Id/1000';
3 -   e3='I1*Rg=0.01';
4 -   e4='Rs/(Rs+1/g)=0.8';
5 -   e5='Id=0.5*110e-6*WL*(Vgs-0.5)^2';
6 -   e6='(Id+I1)*Rs+Vgs+0.01=2';
7 -   e7='2*Id/(Vgs-0.5)=g';
8 -   S=solve(e1,e2,e3,e4,e5,e6,e7);|
-
```

All we need is write equation codes of the electronic circuit that are presented sequentially in figure 8.2(d), line 1 of which is the code for $(I_1+I_D)V_{DD}=10\,mW$, line 2 of which is the code for $I_1=I_D/1000$, and so forth. The symbology we exercised is l1 for I_1, Rs for R_S, Vgs for V_{GS}, WL for $\dfrac{W}{L}$, Rg for R_G, Id for I_D, and g for g_m. The e1, e2, etc are user-chosen variables and hold the equations consecutively. Keep in mind that the number of unknowns which happen to be (R_S, V_{GS}, I_D, I_1, g_m, R_G, and $\dfrac{W}{L}$) must be equal to the number of equations. The solution is found and is stored in user-chosen S. In order to view the solution for R_G, call the following at the prompt:

>>S.Rg ↵

ans =

2002.

i.e. $R_G = 2002\Omega$. Similar calling

provides $R_S = 198.73\Omega$, $\dfrac{W}{L} =$

368.66, and $V_{GS} = 0.9963\,V$ and the other three quantities we do not need as far as design is concerned but can be found in a similar fashion.

✦ **Project 9: Common source MOS circuit design**

Figure 8.3(a) shows a common source NMOS transistor with $\mu_n C_{ox} = 100\,\mu A/V^2$, $V_{TH} = 0.4\,V$, and $V_{DD} = 1.5\,V$. Design the circuit parameters to obtain a voltage gain 5, input impedance $50\,K\Omega$, power rating $5\,mW$, and voltage drop $0.4\,V$ across R_S.

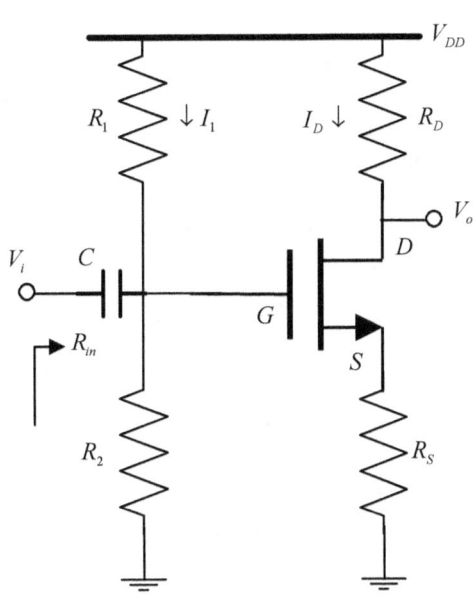

Figure 8.3(a) Resistive divider biasing circuit of an NMOS transistor

Prerequisite: project 8.

Solution:

Considering short for the coupling capacitor, write the relevant equations of the common source MOS circuit as follows:

power rating is $(I_1 + I_D)V_{DD} = 10\,mW$,

from MOS theory voltage gain $|A_V| = g_m R_D = 5$,

from MOS theory input impedance $R_{in} = R_1 \parallel R_2 = \dfrac{R_1 R_2}{R_1 + R_2} = 50\,K\Omega$,

voltage drop across R_S is $I_D R_S = 0.4\,V$,

from MOS circuit theory transconductance is $g_m = \dfrac{2I_D}{V_{GS} - V_{TH}}$,

drain current in saturation region is $I_D = \dfrac{1}{2}\mu_n C_{ox}\dfrac{W}{L}(V_{GS} - V_{TH})^2$,

gate to ground voltage drop is $\dfrac{R_2}{R_1 + R_2}V_{DD} = V_{GS} + 0.4$, and current through R_1 is $I_1 = \dfrac{V_{DD} - V_{GS} - 0.4}{R_1}$.

In this problem we have 9 unknowns but 8 equations so we have to choose some reasonable V_{GS}, say $V_{GS} = 1\,V$. Figure 8.3(b) shows the codes necessary for the solution with the symbology of project 8. Lines 1 through 9 are just the straightforward codes of the equations and the solution is put to S in line 10. Since the return is in symbolic form as well as structure array, double is used to obtain decimal value of concerned variables in lines 11 and 12 because later calculation needs those. Note that computing takes place in decimal not in symbolic form. Like name variable is chosen for easy identification e.g. Rd=double(S.Rd) in which the left Rd could have been any other user-chosen variable.

The conditionality (section 6.10) of saturation region is $V_{GS} > V_{TH}$ and $V_{DS} - V_{GS} + V_{TH} > 0$ also $V_{DS} = V_{DD} - 0.4 - I_D R_D$ for the circuit. Line 13 computes V_{DS} and line 14 evaluates conditionality through operators (appendix D.1) and assigns it to user-chosen A. After running the file we found A=0 i.e. the MOS is not in saturation with $V_{GS} = 1\,V$, say now $V_{GS} = 0.7\,V$ hence modify the line 1 of figure 8.3(b) as e1='Vgs=0.7';, run the script file, and find A=1 manifesting fulfillment of the conditionality.

Figure 8.3(b)
MATLAB script
file for the
project 9 – right
side figure

```
1 -    e1='Vgs=1';
2 -    e2='(I1+Id)*1.5=5e-3';
3 -    e3='Rd*g=5';
4 -    e4='R1*R2/(R1+R2)=50e3';
5 -    e5='Id*Rs=0.4';
6 -    e6='g=2*Id/(Vgs-0.4)';
7 -    e7='Id=0.5*100e-6*WL*(Vgs-0.4)^2';
8 -    e8='R2/(R1+R2)*1.5=Vgs+0.4';
9 -    e9='I1=(1.5-Vgs-0.4)/R1';
10 -   S=solve(e1,e2,e3,e4,e5,e6,e7,e8,e9);
11 -   Rs=double(S.Rs);  Id=double(S.Id);
12 -   Vgs=double(S.Vgs);  Rd=double(S.Rd);
13 -   Vds=1.5-0.4-Id*Rd;
14 -   A=(Vgs>0.4)&(Vds-Vgs+0.4)>0
15
```

Call the Rs to see its content:
>>Rs ↵

Rs =
        120.2116
i.e. $R_S = 120.2116\,\Omega$ similarly we get $R_1 = 68181.82\,\Omega$, $R_2 = 187500\,\Omega$, $R_D = 225.3967\,\Omega$, and $\dfrac{W}{L} = 739.4370$. Once again for R_1 we call S.R1, do so for the R_2, R_D, and $\dfrac{W}{L}$.

✦ Project 10: Symbolic expression derivation: voltage gain of single stage electronic circuit

Figure 8.3(c) shows a single stage NPN transistor. We wish to obtain symbolic voltage gain expression for the stage incorporating the common emitter forward current gain β and transconductance g_m.

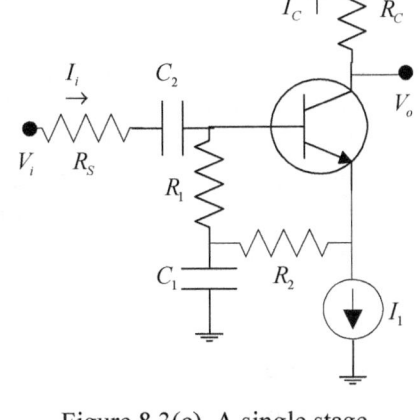

Prerequisite: section 7.7 and appendices D.11, D.12, and G.

Solution:

The approach we are contemplating is write the KVL or KCL equations after drawing the transconductance based model and eliminate variables until we obtain the voltage gain.

Figure 8.3(c) A single stage NPN transistor

Figure 8.3(d) shows transconductance based model of the single stage circuit in figure 8.3(c). Let us write the voltage and/or current related equations as follows (I_i and I_c are not used):

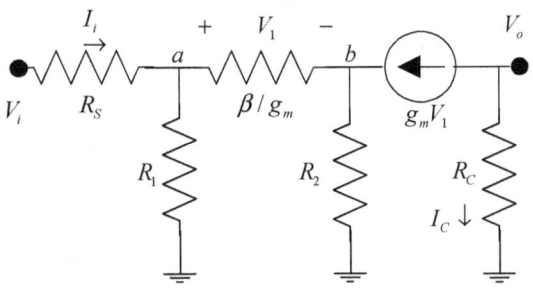

Figure 8.3(d) Transconductance based model of the single stage

equation 1; voltage across R_C:

$$V_o = -g_m V_1 R_C,$$

equation 2; KCL at node b:

$$\frac{g_m V_1}{\beta} + g_m V_1 = \frac{V_b}{R_2},$$

equation 3; KCL at node a:

$$\frac{g_m V_1}{\beta} + \frac{V_a - V_i}{R_S} + \frac{V_a}{R_1} = 0, \text{ and}$$

equation 4; voltage difference between nodes a and b:

$$V_1 = V_a - V_b.$$

Script file for the solution is provided in figure 8.3(e) where the like name symbology exercised is the following: $g \Leftrightarrow g_m$, $V1 \Leftrightarrow V_1$, $Rs \Leftrightarrow R_S$, etc. In the figure 8.3(e) the first line is the declaration of concerned variables. You may declare other variables but that is not necessary. Lines 2 through 5 are just

-221-

the equations assigning to user-chosen variables **e1**, **e2**, etc respectively. The V_1 is substituted in the other three equations and the results are put to user-chosen **O1**, **O2**, and **O3** respectively. The V_a is expressed in terms of other variables (line 9) from **O3** (could have been **O1** or **O2**) and the result is put to **Va** again which is eliminated from **O1** and **O2** in lines 10 and 11 respectively. In a similar fashion V_b is expressed and eliminated in lines 12 and 13 respectively. The **O6** holds V_o and V_i related expression which is solved for V_o and the outcome is in **O7** to which $V_i = 1$ is substituted and finally **Av** holds the A_V. It is worth pointing out semicolon is not used at the end in most lines (e.g. line 6) just to view the contents in MATLAB command window however at last you expect the return as

```
1 -   syms Va Vb Vo Vi V1
2 -   e1='Vo=-g*V1*Rc';
3 -   e2='g*V1/b+g*V1=Vb/R2';
4 -   e3='g*V1/b+(Va-Vi)/Rs+Va/R1=0';
5 -   e4='V1=Va-Vb';
6 -   O1=maple('algsubs',e4,e1)
7 -   O2=maple('algsubs',e4,e2)
8 -   O3=maple('algsubs',e4,e3)
9 -   Va=solve(O3,Va)
10 -  O4=subs(O1)
11 -  O5=subs(O2)
12 -  Vb=solve(O4,Vb)
13 -  O6=subs(O5)
14 -  O7=solve(O6,Vo)
15 -  Vi=1;
16 -  Av=subs(O7);
17 -  pretty(Av)
```

Figure 8.3(e) Script file for the single stage voltage gain

$$
\begin{array}{c}
R1\ b\ g\ Rc \\
\hline
- \quad g\ R2\ R1\ b\ +\ g\ R2\ R1\ +\ g\ R2\ b\ Rs\ +\ g\ R2\ Rs\ +\ g\ R1\ Rs\ +\ R1\ b\ +\ b\ Rs
\end{array}
$$

i.e. $A_V = \dfrac{V_o}{V_i} = -\dfrac{R_1\,\beta\,g_m R_C}{g_m R_2 R_1 \beta + g_m R_2 R_1 + g_m R_2\,\beta R_S + g_m R_2 R_S + g_m R_1 R_S + R_1\,\beta + \beta R_S}$ that is what we are after. We did not include intermediate display of command window for the space reason.

♦ **Project 11: Symbolic expression derivation: voltage gain of two transistor electronic circuit**

We wish to obtain the symbolic voltage gain expression for the two transistor circuit in figure 8.4(a) incorporating identical forward current gain β and transconductances g_{m1} and g_{m2} (left and right transistors respectively).

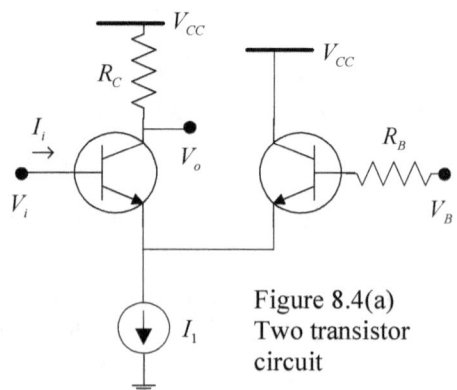

Figure 8.4(a)
Two transistor
circuit

Prerequisite: project 10.

Solution:

Figure 8.4(b) is the transconductance and forward current gain concerning model. Relevant KVL or KCL equations are as follows:

equation 1; KCL at node a :

$$V_1 g_{m1} + \frac{g_{m1}V_1}{\beta} + V_2 g_{m2} + \frac{V_2 g_{m2}}{\beta} = 0 ,$$

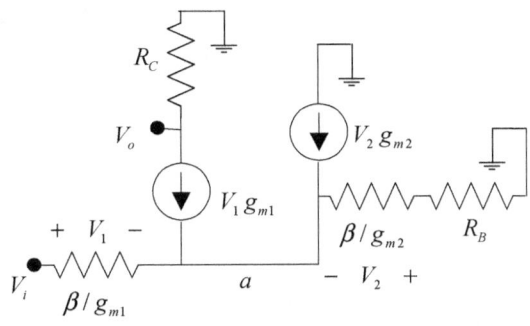

Figure 8.4(b) Transconductance based model of the circuit in figure 8.4(a)

equation 2; voltage across R_C :

$$V_o = -R_C V_1 g_{m1} ,$$

equation 3; node a voltage from left side transistor: $V_a = V_i - V_1$, and

equation 4; node a voltage from right side transistor:

$$V_a = -\frac{V_2 g_{m2}}{\beta}(R_B + \beta / g_{m2}) .$$

Figure 8.4(c) is the script file exercising akin symbology of project 10. If we equate equations 3 and 4, we remove one equation that is what is done in line 4 of figure 8.4(c). Since the variable to be substituted must be on the left side of an equation, we expressed V_1 in terms of others in line 3 of the script file. After running the file, you expect the Av content ignoring the other variable popup as follows:

```
1 -    syms V1 V2 Vo Vi
2 -    e1='g1*V1+g1*V1/b+V2*g2+V2*g2/b=0';
3 -    e2='V1=-Vo/g1/Rc';
4 -    e3='Vi-V1=-V2*g2/b*(Rb+b/g2)';
5 -    O1=maple('algsubs',e2,e1)
6 -    O2=maple('algsubs',e2,e3)
7 -    V2=solve(O1,V2)
8 -    O3=subs(O2)
9 -    O4=solve(O3,Vo)
10 -   Vi=1;
11 -   Av=subs(O4);
12 -   pretty(Av)
```

Figure 8.4(c) Script file for the two transistor circuit in figure 8.4(b)

```
        g1 Rc g2 b
  - ---------------------
      g2 b + g1 Rb g2 + g1 b
```

i.e. $A_V = \dfrac{V_o}{V_i} = -\dfrac{g_{m1}R_C g_{m2}\beta}{g_{m2}\beta + g_{m1}R_B g_{m2} + g_{m1}\beta}$.

✦ Project 12: Symbolic expression derivation: output impedance of emitter follower

In the emitter follower of circuit in figure 8.4(d) the output impedance is defined as $Z_o = \dfrac{V_o}{I_o}$. Obtain the symbolic Z_o using transconductance model.

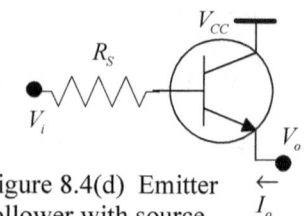

Figure 8.4(d) Emitter follower with source impedance

Prerequisite: project 10.

Solution:

In projects 10-11 we have exercised substitution approach which needs concentration regarding variables. Another approach for symbolic solution is write the equations as they are and just solve them and from the solution, pick the required one. In this sort of problem the number of equations must be one less than the number of unknowns which is also true for the last two projects. Anyhow equation writing needs the transconductance based model which you find in figure 8.4(e). The KVL and/or KCL equations are the following:

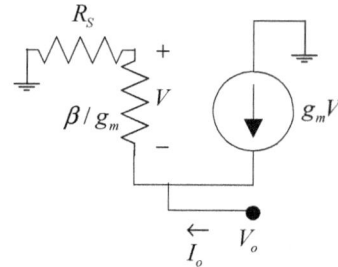

Figure 8.4(e) Transconductance based model of circuit in figure 8.4(d)

equation 1; KCL at the output node:

$$I_o + g_m V + \frac{g_m V}{\beta} = 0 \text{ and}$$

equation 2; KVL from output node to ground through resistors:

$$V_o = -\frac{g_m V}{\beta} R_S - V .$$

In the two equations concerned variables are V_o, I_o, and V, out of which V is the unwanted. We consider V_o and V as unknowns so that V_o or V is expressible in terms of I_o and the others. The V_o-I_o expression will lead the Z_o for which you need the code of figure 8.4(f) maintaining the symbology of project 10. Now there is no substitution except $I_o = 1$. MATLAB returns the following:

```
1 -    syms V Vo Io
2 -    e1='Io+g*V+g*V/b=0';
3 -    e2='Vo=-Rs*V*g/b-V';
4 -    S=solve(e1,e2,Vo,V)
5 -    Io=1;
6 -    Zo=subs(S.Vo);
7 -    pretty(Zo)
```

Figure 8.4(f) Script file for circuit in figure 8.4(d)

```
S =
    V: [1x1 sym]
    Vo: [1x1 sym]
```

That is V_o and V are available in S (each one is in terms of I_o) but V is nonessential so just ignore it. Finally we obtain:

```
Rs g + b
---------
g (b + 1)
```

i.e. $Z_o = \dfrac{V_o}{I_o} = \dfrac{R_S g_m + \beta}{g_m(\beta+1)}$. Note that in most text books you find $Z_o = \dfrac{R_S}{\beta+1} + \dfrac{1}{g_m}$

because $\dfrac{\beta}{\beta+1} \approx 1$.

❖ Project 13: Symbolic expression derivation: output impedance of common base stage

Figure 8.5(a) shown electronic circuit output impedance is defined as $Z_o = \dfrac{V_o}{I_o}$. Obtain the output impedance of the common base stage using transconductance based model with identical β and different g_m.

Prerequisite: projects 10-12.

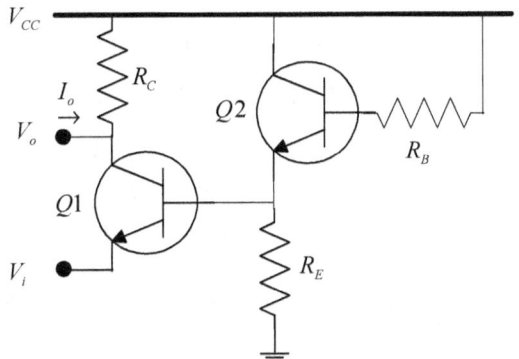

Figure 8.5(a) Circuit with common base stage

Solution:

Figure 8.5(b) shows the transconductance based model. Essential KVL or KCL equations are the following:

equation 1; KCL at node a:

$$g_{m2}V_2 + \frac{g_{m2}V_2}{\beta} = I_1 + \frac{g_{m1}V_1}{\beta},$$

equation 2; equality of voltages between first transconduct-

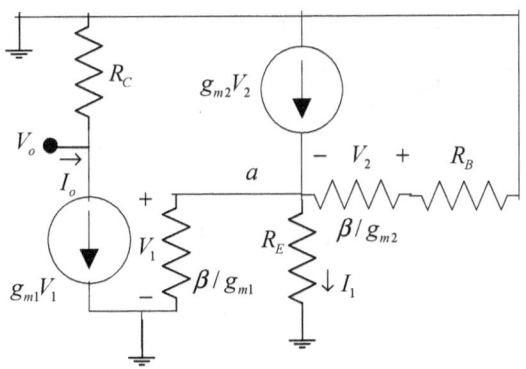

Figure 8.5(b) Transconductance based model of the circuit in figure 8.5(a)

ance and emitter resistor: $V_1 = R_E I_1$,

equation 3; KCL at the output node: $I_o = \dfrac{V_o}{R_C} + g_{m1}V_1$, and

equation 4; KVL around second transconductance-emitter resistor-base resistor:

$$V_1 + V_2 = -\dfrac{V_2 g_{m2}}{\beta} R_B.$$

```
1 -    syms Vo V1 V2 I1 Io
2 -    e1='g2*V2+g2*V2/b=I1+g1*V1/b';
3 -    e2='V1=Re*I1';
4 -    e3='Io=Vo/Rc+g1*V1';
5 -    e4='V1+V2=-V2*g2/b*Rb';
6 -    S=solve(e1,e2,e3,e4,Vo,V1,V2,I1)
7 -    Io=1;
8 -    Zo=subs(S.Vo);
9 -    pretty(Zo)
```

Figure 8.5(c) Script file for circuit in figure 8.5(b)

Figure 8.5(c) is the script file you need. Exercising the symbology of project 10, you find $Z_o = R_C$ after running the file. Besides in the command window you also find:

S =

   I1: [1x1 sym]
   V1: [1x1 sym]
   V2: [1x1 sym]
   Vo: [1x1 sym]

i.e. V_1, V_2, V_o, and I_1 are in terms of I_o and other variables, out of which just Vo is required as far as Z_o is concerned.

✦ Project 14: Symbolic expression derivation: current gain of common emitter circuit

In figure 8.3(c) the current gain is defined as $A_I = \dfrac{I_C}{I_i}$. Obtain the gain using transconductance based model.

Prerequisite: projects 10-13.

Solution:

Figure 8.3(d) depicted circuit is applicable here too but the modified KVL and/or KCL equations are the following:

equation 1; current through R_C: $I_C = -g_m V_1$,

equation 2; KCL at node b: $\dfrac{g_m V_1}{\beta} + g_m V_1 = \dfrac{V_b}{R_2}$,

equation 3; KCL at node a: $\dfrac{g_m V_1}{\beta} + \dfrac{V_a - V_i}{R_S} + \dfrac{V_a}{R_1} = 0$,

equation 4; current from input to node a: $I_i = \dfrac{V_i - V_a}{R_S}$, and

equation 5; voltage difference between nodes a and b: $V_1 = V_a - V_b$.

Figure 8.5(d) shows the codes you need for the current gain maintaining symbology of project 10. The concerned variables are V_1, V_a, V_b, V_i, I_i, and I_c and the script file running returns the outputs so:

S =

    Ic: [1x1 sym]
    V1: [1x1 sym]
    Va: [1x1 sym]
    Vb: [1x1 sym]
    Vi: [1x1 sym]

Only I_i and I_c related expression is required as far as current gain is concerned, others we just ignore. The AI or $A_I = \dfrac{I_C}{I_i}$ outcome is as follows:

```
 1 -   syms Ic Ii Va Vb Vi V1
 2 -   e1='Ic=-g*V1';
 3 -   e2='g*V1/b+g*V1=Vb/R2';
 4 -   e3='g*V1/b+(Va-Vi)/Rs+Va/R1=0';
 5 -   e4='Ii=(Vi-Va)/Rs';
 6 -   e5='V1=Va-Vb';
 7 -   S=solve(e1,e2,e3,e4,e5,Ic,V1,Vi,Va,Vb)
 8 -   Ii=1;
 9 -   AI=subs(S.Ic);
10 -   pretty(AI)
```

Figure 8.5(d) Script file for the current gain

```
           g R1 b
  - ---------------------------
    g R2 + g b R2 + g R1 + b
```

i.e. $A_I = \dfrac{I_C}{I_i} = -\dfrac{g_m R_1 \beta}{g_m R_2 + g_m \beta R_2 + g_m R_1 + \beta}$ – our objective is accomplished.

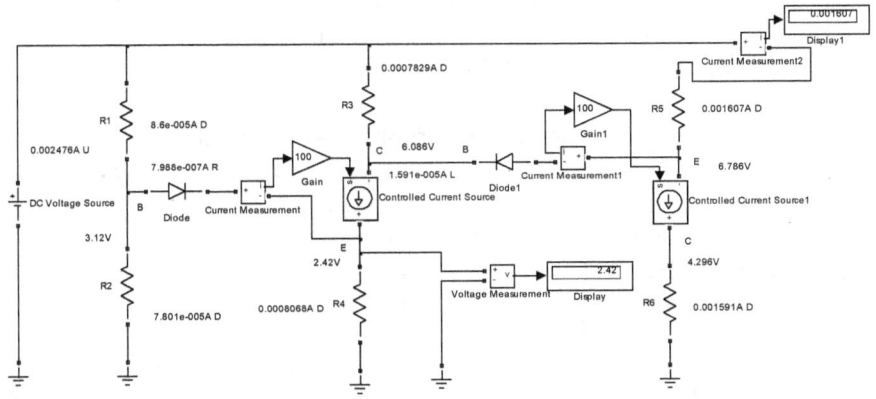

Figure 8.6(a) Voltages and currents in a cascaded circuit

✦ Project 15: Probable voltages and currents in a cascaded circuit

In figure 4.10(a) a two stage cascade of NPN and PNP transistors is depicted. We intend to determine probable node voltages and currents of the cascade using SIMULINK modeling where $V_{CC} = 10\,V$, $R_1 = 80\,K\Omega$, $R_2 = 40\,K\Omega$, $R_3 = 5\,K\Omega$, $R_4 = 3\,K\Omega$, $R_5 = 2\,K\Omega$, $R_6 = 2.7\,K\Omega$, $V_\gamma = 0.7\,V$ (for diode model of the transistor), and $\beta = 100$ for each transistor.

Prerequisite: section 4.11.

Solution:

Figure 4.10(b) depicted model we need for the simulation. Start from the model, doubleclick every resistor or other element, and enter its parameter for example 80e3 as **Resistance** for R_1. Get one set **Voltage Measurement-Display** for any node voltage and one set **Current Measurement-Display** for any branch current. Every single voltage or current will be determined one at a time from the **Display** reading then an annotation (chapter 1) is included beside the element. Figure 8.6(a) presents the findings. For instance **Voltage Measurement** is connected with R_4 and the **Display** is showing $2.42\,V$ that is why we added **2.42V** above R4 in the model. For other voltages only delete the line connected to the + port of the **Voltage Measurement** and then connect the port to another node, run the model, and annotate the voltage thus finish labeling all node voltages. Note that all voltages are with respect to the circuit ground.

The same we also conduct for the current for example **Current Measurement2-Display1** is showing **0.001607** which is the current through R_5 and annotated as **0.001607A D** meaning the current in Amperes pointing downwards (R for rightwards, L for leftwards, and U for upwards). Delete both connection lines of the **Current Measurement2**, reconnect the circuit to the form in figure 4.10(b), disconnect another branch, insert **Current Measurement2** in a similar fashion, run the model, annotate the current, and finish labeling all branch currents.

⬥ Project 16: Frequency response of a two stage circuit

Figure 8.6(b) shows a two stage circuit with coupling capacitors where $V_{CC} = 2.5\,V$, $R_1 = 100\,K\Omega$, $R_2 = 50\,K\Omega$, $R_3 = 1\,K\Omega$, $R = 1\,K\Omega$, $C_1 = 200\,nF$, $C_2 = 200\,nF$, $V_\gamma = 0.8\,V$ (for diode model of each transistor), and $\beta = 100$ for each transistor. Determine the two collector currents by SIMULINK modeling

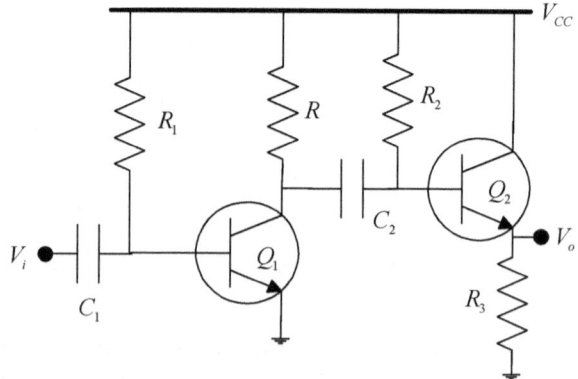

Figure 8.6(b) A two stage circuit with coupling capacitors

and then analyze the frequency response of the circuit using transconductance based model and considering the voltage gain $A_V = \dfrac{V_o}{V_i}$.

Prerequisite: sections 2.12, 4.5-4.7, 4.10, 7.2, 7.8, and 7.9.

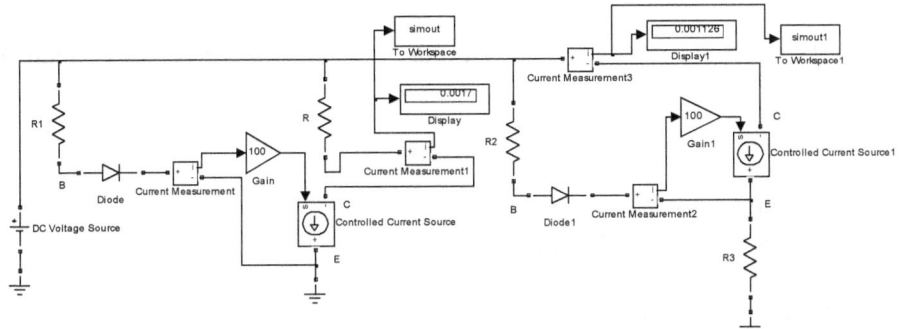

Figure 8.6(c) SIMULINK model of the circuit in figure 8.6(b)

Solution:

Figure 8.6(c) depicted model we need for the two collector currents to determine the transconductances of coarse without the coupling capacitors. Enter Diode, Resistor, Gain, and DC Voltage Source parameters as given. The collector currents of Q_1 and Q_2 are

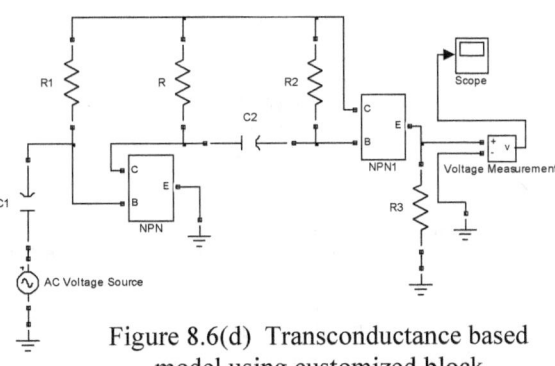

Figure 8.6(d) Transconductance based model using customized block

available to workspace through variables simout and simout1 after running the model and get the two currents to user-chosen variables Ic1 and Ic2 respectively by:

>>Ic1=simout(end); Ic2=simout1(end); ↵

Transconductances and input resistances of Q_1 and Q_2 are computed and assigned to user-chosen variables (e.g. g_{m1} of Q_1 to gm1) by:

>>gm1=Ic1/26e-3; gm2=Ic2/26e-3; ri1=100/gm1; ri2=100/gm2; ↵

Now we need to reconstruct the transconductance based model like figure 8.6(d) using customized NPN block and incorporating capacitors. Enter R and Gain inside NPN as ri1 and gm1 respectively, do so for the NPN1, and enter each Capacitance as 200e-9 by doubleclicking. Suppose the later model file name is test, then carry out the following:

```
>>H=power_analyze('test','ss'); ↵
>>Hm=tf(H); Av=Hm(1,1) ↵
```
Transfer function from input "U_AC Voltage Source" to output "U_Voltage Measurement":

```
0.009589 s^2 + 0.9401 s + 4.275e-013
---------------------------------------------
     s^2 + 3461 s + 4.759e005
```

In above return user-chosen Av holds $A_V = \dfrac{V_o}{V_i}$ transfer function upon which we conduct the following:
```
>>bode(Av) ↵
```

The outcome is the depiction of figure 8.6(e) which appears to be highpass type. We know pole magnitude of the transfer function refers to the cutoff frequency, let us find that by:
```
>>abs(pole(Av)) ↵
```

```
ans =
        1.0e+003 *
        3.3179
        0.1434
```

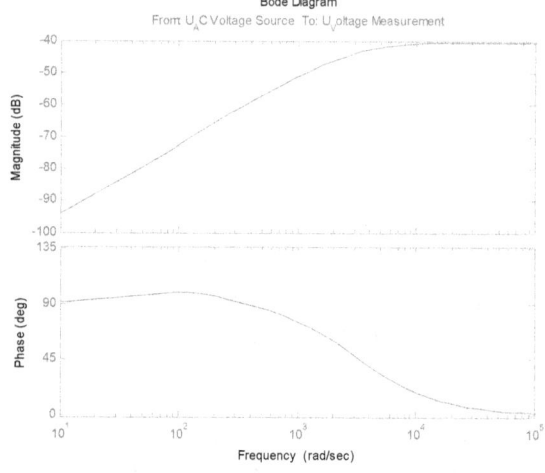

Figure 8.6(e) Frequency response of the two stage circuit in figure 8.6(b)

Clearly above two frequencies are in radians/sec but there should be one frequency for the highpass response hence the figure 8.6(e) shown response is quasi highpass because a lowpass cutoff is hidden. However the two frequencies in Hertz we get by:
```
>>abs(pole(Av))/2/pi ↵
```

```
ans =
        528.0554
        22.8268
```

Hence figure 8.6(e) shown quasi highpass response has the cutoff at $f_c = 528.0554\ Hz$. The response should not come as a surprise because first stage introduces one filtering while the second another.

Above few demonstrations indicate how effectively MATLAB and/or SIMULINK work out the electronics problems. We hope the reader will be enthusiastic in study, analysis, and design of electronic circuit issues incorporating facile illustrations of the text.

Appendices

Mohammad Nuruzzaman

Appendix A

Coding in MATLAB/SIMULINK

MATLAB executes the code of an expression in terms of string which is the set of keyboard characters placed consecutively. One distinguishing feature of MATLAB is that the workspace variable itself is a matrix. The strings adopted for computation are divided into two classes – scalar and vector. The scalar computation results the order of the output matrix same as that of the variable matrix. On the contrary, the order for the vector computation is determined in accordance with the matrix algebra rules. Some symbolic functions and their MATLAB counterparts are presented in table A.1. The operators for arithmetic computations are as follows:

addition	+
subtraction	−
multiplication	*
division	/
power	^

The operation sequence of different operators in a scalar or vector string observes the following order:

enclosing braces	( )	first,
power operator	^	then,
division operator	/	next,
multiplication operator	*	after that,
addition operator	+	then, and
subtraction operator	−	finally.

The syntax of the scalar computation urges us to use .*, ./, and .^ in lieu of *, /, and ^ respectively. The operators *, /, and ^ are never preceded by . for the vector computation. The vector string is the MATLAB code of any symbolic expression or function often found in mathematics. In the sequel we present some examples on writing an expression in MATLAB.

◆ **Write MATLAB codes both in scalar and vector forms on following functions**

A. $\sin^3 x \cos^5 x$ B. $2 + \ln x$ C. $x^4 + 3x - 5$ D. $\dfrac{x^3 - 5}{x^2 - 7x - 7}$

E. $\sqrt{|x^3| + \sec^{-1} x}$ F. $(1 + e^{\sin x})^{x^2 + 3}$ G. $\dfrac{\cosh x + 3}{\sqrt{\dfrac{x+4}{\log_{10}(x^3 - 6)}}}$

H. $\dfrac{1}{(x-3)(x+4)(x-2)}$ I. $\dfrac{1}{1 + \dfrac{1}{1 + \dfrac{1}{x}}}$ J. $\dfrac{a}{x+a} + \dfrac{b}{y+b} + \dfrac{c}{z+c}$

K. $\dfrac{u^2 v^3 w^9}{x^4 y^7 z^6}$

In tabular form, they are coded as follows:

Example	String for scalar computation	String for vector computation
A	sin(x).^3.*cos(x).^5	sin(x)^3*cos(x)^5
B	2+log(x)	2+log(x)
C	x.^4+3*x-5	x^4+3*x-5
D	(x.^3-5)./(x.^2-7*x-7)	(x^3-5)/(x^2-7*x-7)
E	sqrt(abs(x.^3)+asec(x))	sqrt(abs(x^3)+asec(x))
F	(1+exp(sin(x))).^(x.^2+3)	(1+exp(sin(x)))^(x^2+3)
G	(cosh(x)+3)./sqrt((x+4)./log10(x.^3-6))	(cosh(x)+3)/sqrt((x+4)/log10(x^3-6))
H	1./(x-3)./(x+4)./(x-2)	1/(x-3)/(x+4)/(x-2)
I	1./(1+1./(1+1./x))	1/(1+1/(1+1/x))
J	a./(x+a)+b./(y+b)+c./(z+c)	a/(x+a)+b/(y+b)+c/(z+c)
K	u.^2.*v.^3.*w.^9./x.^4./y.^7./z.^6	u^2*v^3*w^9/x^4/y^7/z^6

Electronic circuit programming circumstance dictates the type of code – whether scalar or vector should be employed.

Table A.1   Some mathematical functions and their MATLAB counterparts

Mathematical notation	MATLAB notation	Mathematical notation	MATLAB notation	Mathematical notation	MATLAB notation
$\sin x$	sin(x)	$\sin^{-1} x$	asin(x)	π	pi
$\cos x$	cos(x)	$\cos^{-1} x$	acos(x)	$A+B$	A+B
$\tan x$	tan(x)	$\tan^{-1} x$	atan(x)	$A-B$	A–B
$\cot x$	cot(x)	$\cot^{-1} x$	acot(x)	$A \times B$	A*B
$\cosec x$	csc(x)	$\sec^{-1} x$	asec(x)	e^x	exp(x)
$\sec x$	sec(x)	$\cosec^{-1} x$	acsc(x)	A^B	A^B
$\sinh x$	sinh(x)	$\sinh^{-1} x$	asinh(x)	$\ln x$	log(x)
$\cosh x$	cosh(x)	$\cosh^{-1} x$	acosh(x)	$\log_{10} x$	log10(x)
$\sec hx$	sech(x)	$\sec h^{-1} x$	asech(x)	$\log_2 x$	log2(x)
$\cosech x$	csch(x)	$\cosech^{-1} x$	acsch(x)	Σ	sum
$\tanh x$	tanh(x)	$\tanh^{-1} x$	atanh(x)	Π	prod
$\coth x$	coth(x)	$\coth^{-1} x$	acoth(x)	$\lvert x \rvert$	abs(x)
10^A	1e A e.g. 1e3	10^{-A}	1e- A e.g. 1e-3	$\sqrt{x}$	sqrt(x)

* In the six trigonometric functions for example sin(x), the x is in radian. If the x is in degree, we use sind(x). The other five functions also have the syntax cosd(x), tand(x), cotd(x), cscd(x), and secd(x) when the x is in degree. The default return from asin(x) is in radian, if you need the return to be in degree, use the command asind(x). Similar degree return is also possible from acosd(x), atand(x), acotd(x), asecd(x), and acscd(x).

Numerical examples to point out the difference between scalar and vector computations are in the following.

We have the matrices $A = \begin{bmatrix} 3 & 5 \\ 7 & 8 \end{bmatrix}$, $B = \begin{bmatrix} 5 & 2 & 1 \\ 0 & 1 & 7 \end{bmatrix}$, and $C = \begin{bmatrix} 3 & 2 & 9 \\ 4 & 0 & 2 \end{bmatrix}$. The scalar computation is not possible between the matrices A and B because of their unequal order, nor is between the matrices A and C for the same reason. On the contrary the scalar multiplication can be conducted between the B and C for having the same order, which is $B.*C = \begin{bmatrix} 15 & 4 & 9 \\ 0 & 0 & 14 \end{bmatrix}$ (element by element multiplication).

Matrix algebra rule says that any matrix A of order $M \times N$ can only be multiplied with another matrix B of order $N \times P$ so that the resulting matrix has the order $M \times P$. For the last paragraph cited A and B, we have $M=2$, $N=2$, and $P=3$ and obtain the vector-multiplied matrix as $A \times B =$ $\begin{bmatrix} 3 \times 5 + 5 \times 0 & 3 \times 2 + 5 \times 1 & 3 \times 1 + 5 \times 7 \\ 7 \times 5 + 8 \times 0 & 7 \times 2 + 1 \times 8 & 7 \times 1 + 8 \times 7 \end{bmatrix} = \begin{bmatrix} 15 & 11 & 38 \\ 35 & 22 & 63 \end{bmatrix}$, which has the MATLAB code A*B not A.*B. Similar interpretation follows for the operators * and /.

Whenever writing the scalar codes A.*B, A./B, and A.^B, we make it certain that both the A and B are identical in matrix size. The 3*A means all elements of matrix A are multiplied by 3 and we do not use 3.*A. Also do we not use A./3 but do A/3. The signs + and - are never preceded by the operator . in scalar codes. The command 4./A means 4 is divided by all elements in A. The A.^4 means power on all elements of A is raised by 4 and so on.

♦ Scale factors or units

In science and engineering physical quantity measurement always requires the understanding of units. Measured unit of a physical quantity can be far apart from the standard unit very often in the power of 10 that is why scale factors are important. Table A.2 presents the engineering scale factor units and their MATLAB equivalences.

Table A.2 Engineering unit scale factors and their MATLAB counterparts

Scale factor	Symbol	As power of 10	MATLAB code
giga	G	10^9	e9
mega	M	10^6	e6
kilo	K	10^3	e3
milli	m	10^{-3}	e-3
micro	μ	10^{-6}	e-6
nano	n	10^{-9}	e-9
pico	p	10^{-12}	e-12

For example the time $10.7\,m\sec$ is coded as 10.7e-3. Again a distance of $4.7\,km$ is entered by writing 4.7e3 in standard unit.

Appendix B

SIMULINK block links for modeling electronic circuits

When you open a new or work in a previously saved SIMULINK model file, the very next step is to know the exact location or link of a block which will be employed in modeling electronic problems. Table B accumulates icon appearance, brief function, and location or link of SIMULINK blocks exercised in the text.

Table B. SIMULINK blocks and their links for modeling electronic circuits

Block name	Icon appearance	Function of block or operation	Link or location
AC Current Source	AC Current Source	It simulates the current generating action of an ideal AC current source of different amplitude, phase, and frequency	SimPowerSystems → Electrical Sources → AC Current Source
AC Voltage Source	AC Voltage Source	It simulates the voltage generating action of an ideal AC voltage source of different amplitude, phase, and frequency	SimPowerSystems → Electrical Sources → AC Voltage Source
Add	Add	It sums two or more signal values connected to its input ports	SIMULINK → Math Operations → Add
Compare To Constant	<= 3 Compare To Constant	It compares every input sample value to a user-chosen constant and returns 1/0	SIMULINK → Logic and Bit Operations → Compare To Constant
Compare To Zero	<= 0 Compare To Zero	It compares every input sample value to zero and returns 1/0	SIMULINK → Logic and Bit Operations → Compare To Zero
Complex to Magnitude-Angle	Complex to Magnitude-Angle	It converts complex function or data to magnitude and phase angle form	SIMULINK → Math Operations → Complex to Magnitude-Angle
Connection Port	1 Connection Port	It simulates electric circuit node for subsystem forming	SimPowerSystems → Elements → Connection Port

Continuation of the last table:

Block name	Icon appearance	Function of block or operation	Link or location
Constant	1 Constant	It generates user-defined constant values	SIMULINK → Sources → Constant
Controlled Current Source	Controlled Current Source	It transforms any user-defined time domain function to electrical current source of identical function	SimPowerSystems → Electrical Sources → Controlled Current Source
Controlled Voltage Source	Controlled Voltage Source	It transforms any user-defined time domain function to electrical voltage source of identical function	SimPowerSystems → Electrical Sources → Controlled Voltage Source
Current Measurement	+ - i Current Measurement	It simulates the action of an ammeter both for alternating and direct currents	SimPowerSystems → Measurements → Current Measurement
DC Voltage Source	DC Voltage Source	It simulates an ideal direct current voltage source of constant value	SimPowerSystems → Electrical Sources → DC Voltage Source
Derivative	du/dt Derivative	It performs the numerical differentiation of the signal to its input port in continuous sense	SIMULINK → Continuous → Derivative
Diode	Diode	It models a diode from user-supplied forward biased voltage both in ideal and practical senses	SimPowerSystems → Power Electronics → Diode
Display	0 Display	It shows instantaneous value of the signal at the end of simulation which it is connected to	SIMULINK → Sinks → Display

Continuation of the last table:

Block name	Icon appearance	Function of block or operation	Link or location
Divide	× ÷ Divide	It performs user-defined multiplication or division of input signals	SIMULINK → Math Operations → Divide
Fcn	f(u) Fcn	It performs user-defined mathematical operations on the signal to its input port assuming that the input signal name is u	SIMULINK → User-Defined Functions → Fcn
Gain	1 Gain	It multiplies the signal to its input port according to user-supplied gain	SIMULINK → Math Operations → Gain
Ground	⏚	It simulates a reference node in electrical circuits	SimPowerSystems → Elements → Ground
Logical Operator	AND Logical Operator	It performs logical AND comparison of two input signals assuming input in terms of 1/0 and returns 1/0	SIMULINK → Logic and Bit Operations → Logical Operator
Mean Value	In Mean Mean Value	It finds average value of a periodic or nonperiodic signal	SimPowerSystems → Extra Library → Measurements → Mean Value
Mux		It multiplexes two or more signals from single vector to group vector form	SIMULINK → Commonly Used Blocks → Mux
Neutral	node 10	It simulates a node point in electrical circuits	SimPowerSystems → Elements → Neutral
NMOS	D G S NMOS	It simulates NMOS transistor using transconductance	Author-devised, can be obtained through email link of page ii

Continuation of the last table:

Block name	Icon appearance	Function of block or operation	Link or location
NPN	NPN	It simulates NPN transistor using transconductance	Author-devised, can be obtained through email link of page ii
OP AMP	OP AMP	It simulates operational amplifier	Author-devised, can be obtained through email link of page ii
Out1	Out1	It simulates SIMULINK output port for subsystem forming	SIMULINK → Ports & Subsystems → Out1
PMOS	PMOS	It simulates PMOS transistor using transconductance	Author-devised, can be obtained through email link of page ii
PNP	PNP	It simulates PNP transistor using transconductance	Author-devised, can be obtained through email link of page ii
Product	Product	It multiplies two or more signal values connected to its input ports	SIMULINK → Math Operations → Product
Ramp	Ramp	It generates straight line functions of various characteristics	SIMULINK → Sources → Ramp
Real-Imag to Complex	Real-Imag to Complex	It converts real and imaginary function or data to complex one	SIMULINK → Math Operations → Real-Imag to Complex
Req	Req	It determines equivalent resistance of any passive resistive circuit	Author-devised, can be obtained through email link of page ii

Continuation of the last table:

Block name	Icon appearance	Function of block or operation	Link or location
RMS	>signal rms > RMS	It finds root mean square value of a periodic signal	SimPowerSystems → Extra Library → Measurements → RMS
Scope	Scope	It shows functional variation (s) of some signal (s) which it is connected to	SIMULINK → Sinks → Scope
Series RLC Branch	Series RLC Branch	It places user-defined series resistance, inductance, and/or capacitance in an electrical circuit	SimPowerSystems → Elements → Series RLC Branch
Subtract	+ - Subtract	It subtracts two or more signal values connected to its input ports	SIMULINK → Math Operations → Subtract
Sum	+ +	It adds two or more signal values connected to its input ports	SIMULINK → Math Operations → Sum
To Workspace	simout To Workspace	It exports signal data from SIMULINK to MATLAB workspace which it is connected to	SIMULINK → Sinks → To Workspace
Universal Bridge	g A B C Universal Bridge	It models diode / thyristor based bridge circuit of single or three phase	SimPowerSystems → Power Electronics → Universal Bridge
Voltage Measurement	+ - v Voltage Measurement	It simulates the action of a voltmeter both for AC and DC voltages	SimPowerSystems → Measurements → Voltage Measurement
XY Graph	XY Graph	It graphs y versus x type data present in a model	SIMULINK → Sinks → XY Graph

Appendix C
MATLAB functions exercised in the text

Function name	Purpose	Page
abs	extracts magnitude values from $H(j\omega)$ samples	171
algsubs	substitutes one algebraic expression into another	182
angle	extracts phase angles in radians from $H(j\omega)$ samples	171
axis	forces axis setting of last graph to user-defined axis setting	96
bandwidth	computes the cutoff frequency of a lowpass electronic system	176
bode	computes and graphs the bode plot of $H(s)$	173
c_cross	author written script file which finds functional value crossing from x and y samples	254
double	converts rational or symbolic numbers to double precision one or decimal	220, 272
end	terminates the execution of a for-loop or if-else checking	246
eps	lowest numerical quantity in MATLAB treated as epsilon	92,172
ezplot	draws y versus x type graph from y expression and x interval	263
factor	factorizes an integer	7
find	determines the element position in a matrix subject to condition	252
for-end	beginning and ending statements of a for-loop	249
freqresp	computes and graphs the frequency response of $H(s)$	169
grid	attaches horizontal and vertical grid lines in a drawn trace	93
help	finds assistance or description of an embedded function	25
hold	retains the last graph for subsequent plotting	95
imag	extracts imaginary values from $H(j\omega)$ samples	171
legend	includes distinctive words for multiple traces on a drawn graphics	94,189

Continuation of the last table:

Function name	Purpose	Page
linspace	generates row or column matrix of linearly spaced elements	255
lookfor	searches all possible embedded functions from user-supplied partial or full function name	25
maple	embedded function for executing MATLAB commands in maple package	259
max	finds the maximum from numerical data supplied as a row or column matrix	251
maximize	determines maximum from an expression	143
min	finds the minimum from numerical data supplied as a row or column matrix	251
numden	separates numerator and denominator from an algebraic expression	185
plot	graphs y versus x data in continuous sense	261
pole	extracts poles from $H(s)$	185
power_analyze	embedded function for obtaining the state space model from an electrical circuit	163
powerdemo	demonstrates on library of power system	25
powersys	embedded library of power system	25
pretty	displays a mathematical readable form from coded expression	180
rad2deg	converts radians to degrees	171
real	extracts real values from $H(j\omega)$ samples	171
scatter	graphs y versus x data using user-defined discrete symbols	267
sim	simulates a SIMULINK model from MATLAB	92
simple	simplifies an algebraic expression in different forms	202
simplify	simplifies an algebraic expression	180
solve	solves multiple algebraic expressions or equations	272
squeeze	compresses three dimensional array if empty dimension exists	170
subplot	divides a graphics window into user-defined subwindows	264
subs	substitutes value or expression into another algebraic expression	182
sym2poly	finds numerical polynomial coefficients from a polynomial expression	186
syms	embedded word for declaring variables	180

Continuation of the last table:

Function name	Purpose	Page
tf	defines transfer function from numerator-denominator polynomial coefficients, used too for conversion	163
tfdata	extracts transfer function data – numerator and denominator polynomial coefficients	165
xlabel	attaches user-defined horizontal axis label in a drawn trace	93
ylabel	attaches user-defined vertical axis label in a drawn trace	93
zero	extracts zeroes from $H(s)$	185
Zp	author written function file which computes parallel equivalent of two impedances	257
zpk	defines transfer function from pole-zero-gain	166

Appendix D

MATLAB functions/statements for electronic circuit study

While working on electronic circuits in MATLAB, we come across lots of built-in MATLAB functions or programming statements. In order to employ these elements for solving electronic circuit problems, we need to understand their input and output argument types and purpose of the elements. Functions or program elements exercised in the text with brief descriptions are in the following.

D.1 Comparative and logical operators

Comparative operators are used for comparison on two scalar elements, one scalar and one matrix elements, or two identical size matrix elements. There are six comparative operators as presented in table D.1.

Table D.1   Equivalence of comparative operators

Comparative operation	Mathematical notation	MATLAB notation
equal to	$=$	==
not equal to	$\neq$	~=
greater than	$>$	>
greater than or equal to	$\geq$	>=
less than	$<$	<
less than or equal to	$\leq$	<=

The output of expression pertaining to the comparative operators is logical – either true (indicated by 1) or false (indicated by 0). For example when A=3 and B=4, the comparisons A=B, $A \neq B$, A>B, $A \geq B$, A<B, and $A \leq B$ should be false(0), true(1), false(0), false(0), true(1), and true(1) respectively. We implement these comparative operations as presented in table D.2.

Table D.2   Scalar comparative operation

>>A=3; B=4; ↵ >>A==B ↵ ans = 0 >>A~=B ↵ ans = 1	>>A>B ↵ ans = 0 >>A>=B ↵ ans = 0	>>A<B ↵ ans = 1 >>A<=B ↵ ans = 1

There are two operands A and B in table D.2, each of which is a single scalar. Each of the operands can be a matrix in general. In that case the logical decision takes place element by element on all elements in the matrix. For instance if $A = \begin{bmatrix} 5 & 8 \\ 5 & 7 \end{bmatrix}$ and $B = \begin{bmatrix} 2 & 1 \\ -2 & 9 \end{bmatrix}$, A>B should be $\begin{bmatrix} 5>2 & 8>1 \\ 5>-2 & 7>9 \end{bmatrix} = \begin{bmatrix} 1 & 1 \\ 1 & 0 \end{bmatrix}$. Again if the A happens to be a scalar (say A=4), the single scalar is compared to all elements in the B therefore $A \leq B$ should be $\begin{bmatrix} 4 \leq 2 & 4 \leq 1 \\ 4 \leq -2 & 4 \leq 9 \end{bmatrix} =$

$\begin{bmatrix} 0 & 0 \\ 0 & 1 \end{bmatrix}$. In a similar fashion the B also operates on A however the scalar and matrix related comparative implementation is presented in the table D.3.

Some basic logical operations are NOT, OR, and AND. The characters ~, |, and & of the keyboard are adopted for the logical NOT, OR, and AND respectively. In all logical outputs the 1 and 0 stand for true and false respectively. All logical operators apply to the matrices in general. For the matrix $A=\begin{bmatrix} 0 & 0 \\ 0 & 1 \end{bmatrix}$, NOT(A) operation should provide $\begin{bmatrix} 1 & 1 \\ 1 & 0 \end{bmatrix}$ (see table D.4). The logical OR and AND operations on the like positional elements of the two matrices $A=\begin{bmatrix} 1 & 1 \\ 0 & 1 \end{bmatrix}$ and $B=\begin{bmatrix} 0 & 1 \\ 1 & 1 \end{bmatrix}$ must return $\begin{bmatrix} 1 & 1 \\ 1 & 1 \end{bmatrix}$ and $\begin{bmatrix} 0 & 1 \\ 0 & 1 \end{bmatrix}$ respectively. Table D.4 shows both implementations.

Table D.3   Scalar and matrix comparative operations

when A and B are matrices,	when A is scalar and B is matrix,

>>A=[5 8;5 7]; ↵
>>B=[2 1;-2 9]; ↵
>>A>B ↵
 	
ans =
    1   1
    1   0
	>>A=4; ↵
	>>B=[2 1;-2 9]; ↵
	>>A<=B ↵
	
	ans =
	    0 0
	    0 1

Table D.4   Basic logical operations on matrix elements

for NOT(A) operation,	for A OR B,	for A AND B,	for A XOR B,

>>A=[0 0;0 1]; ↵
>>~A ↵
 	
ans =
    1   1
    1   0
	>>A=[1 1;0 1]; ↵
	>>B=[0 1;1 1]; ↵
	>>A|B ↵
	
	ans =
	    1 1
	    1 1
		>>A&B ↵
		
		ans =
		    0 1
		    0 1
			>>xor(A,B) ↵
			
			ans =
			    1 0
			    1 0

If the A or the B is a single 1 or 0, it operates on all elements of the other.

Sometimes we need to check the interval of the independent variable of mathematical functions for instance $-6 \le x \le 8$. The interval is split in two parts $-6 \le x$ and $x \le 8$. In terms of the logical statement one expresses the $-6 \le x \le 8$ as (-6<=x)&(x<=8).

There is no operator for the XOR logical operation instead the MATLAB function xor syntaxed by xor(A,B) implements the operation as presented in the table D.4.

D.2 Simple if/if-else/nested if syntax

Conditional commands are exercised by the **if-else** statements (reserve words). Also comparisons and checkings may need **if-else** statements. We can have different **if-else** structures namely simple-if, if-else, or nested-if depending on programming circumstances, some of which we discuss in the following.

⊟ Simple if

The program syntax of simple-if is as follows:

> if *logical expression*
> > *Executable MATLAB command(s)*
> end

Logical expression usually requires the use of comparative operators which are explained in appendix D.1. If the logical expression beside the **if** is true, the command between the **if** and **end** is executed otherwise not. In tabular form a simple-if implementation is as follows:

Example: If $x \geq 1$, we compute $y = \sin x$. When $x = 2$, we should see $y = \sin 2 = 0.9093$.	Executable M-file: x=2; if x>=1 y=sin(x); end	Steps: Save the statements in a new M-file (section 1.1) by the name test and execute the following: >>test ↵	Check from the command window after running the M-file: >>y ↵ y = 0.9093

⊟ If-else

General program syntax for the **if-else** structure is as follows:

> if *logical expression*
> > *Executable MATLAB command(s)*
> else
> > *Executable MATLAB command(s)*
> end

If the logical expression beside the **if** is true, the command between **if** and **else** is executed else the command between **else** and **end** is executed. In tabular form, an **if-else-end** implementation is the following:

Example: When $x = 1$, we compute $y = \sin\dfrac{x\pi}{2} = 1$ otherwise $y = \cos\dfrac{x\pi}{2} = 0$.	Executable M-file: x=1; if x==1 y=sin(x*pi/2); else y=cos(x*pi/2); end	Steps: Save the statements in a new M-file by the name test and execute the following: >>test ↵	Check from the command window after running the M-file: >>y ↵ y = 1

If we had x=2; in the first line of M-file in last exercise, we would see y= $\cos\pi$ =−1.

⊡ **Nested-if**

The third type of if structure is the nested-if whose general program syntax is attached in the right side text box. Clearly the syntax takes care of multiple logical expressions which we demonstrate by one example as shown in the following table.

> if *logical expression*
>   *Executable MATLAB command(s)*
> elseif *logical expression*
>   *Executable MATLAB command(s)*
>   ⋮
> elseif *logical expression*
>   *Executable MATLAB command(s)*
> else
>   *Executable MATLAB command(s)*
> end

| Example: The best example can be taking the decision of grades out of 100 based on the achieved number of a student. The grading policy is stated as if the achieved number of a student is greater than or equal 90, greater than or equal to 80 but less than 90, greater than or equal to 70 but less than 80, greater than or equal to 60 but less than 70, greater than or equal to 50 but less than 60, and less than 50, then the grade is decided as A, B, C, D, E, and F respectively. | Executable M-file:

N=77;
if N>=90
  g='A';
elseif (N<90)&(N>=80)
  g='B';
elseif (N<80)&(N>=70)
  g='C';
elseif (N<70)&(N>=60)
  g='D';
elseif (N<60)&(N>=50)
  g='E';
else
  g='F';
end | In the executable M-file, the N and g refer to the number achieved and the grade respectively. If the number N is 77, the grade g should be C. Any character is argumented under the single inverted comma.

Steps: Save the left statements in a new M-file by the name **test** and execute the following:
>>test ↵ | Check from the command window after running the M-file:
>>g ↵

g =

C |

D.3 Data accumulation

Sometimes it is necessary that we perform appending operation on an existing matrix at MATLAB workspace.

⬥ **Appending rows**

Assume that the $A = \begin{bmatrix} 1 & 3 & 5 \\ 2 & 6 & 8 \\ 9 & 5 & 0 \\ 4 & 7 & 8 \end{bmatrix}$ is formed by appending two row matrices [9 5 0] and [4 7 8] with the matrix $B = \begin{bmatrix} 1 & 3 & 5 \\ 2 & 6 & 8 \end{bmatrix}$.

We first enter the matrix B (section 1.1) into MATLAB and append one row after another by using the command as presented below:

for entering B,	for appending the first row,	for appending the second row,
>>B=[1 3 5;2 6 8] ⏎	>>B=[B;[9 5 0]] ⏎	>>A=[B;[4 7 8]] ⏎
B =		A =
1 3 5	B =	1 3 5
2 6 8	1 3 5	2 6 8
	2 6 8	9 5 0
	9 5 0	4 7 8

The command B=[B;[9 5 0]] in above execution says that the row [9 5 0] is to be appended with the existing B (inside the third bracket) and that the result is again assigned to B. You can append as many rows as you want. The important point is the number of elements in each row that is to be appended must be equal to the number of columns in the matrix B.

✦ Appending columns

Suppose $C = \begin{bmatrix} 1 & 3 & 5 & 9 & 3 \\ 2 & 6 & 8 & 0 & 1 \\ 9 & 5 & 0 & 1 & 9 \end{bmatrix}$ is formed by appending two column matrices $\begin{bmatrix} 9 \\ 0 \\ 1 \end{bmatrix}$ and $\begin{bmatrix} 3 \\ 1 \\ 9 \end{bmatrix}$ with matrix $D = \begin{bmatrix} 1 & 3 & 5 \\ 2 & 6 & 8 \\ 9 & 5 & 0 \end{bmatrix}$. We get the matrix D into MATLAB and append one column after another as follows:

for entering D,	for appending the first column,	for appending the second column,
>>D=[1 3 5;2 6 8;9 5 0] ⏎	>>D=[D [9 0 1]'] ⏎	>>C=[D [3 1 9]'] ⏎
D =		C =
1 3 5	D =	1 3 5 9 3
2 6 8	1 3 5 9	2 6 8 0 1
9 5 0	2 6 8 0	9 5 0 1 9
	9 5 0 1	

The column matrix [9 0 1]' and D in above execution have one space gap within the third bracket. In the second of above implementation, the resultant matrix is again assigned to D. Append as many columns as you want just remember that the number of elements in each column that is to be appended must be equal to the number of rows in the matrix D.

✦ Data accumulation by using the two appending techniques

Suppose initially there is nothing in the f matrix, which in MATLAB we write by the statement f=[ ]; (an empty matrix is assigned to f). An empty matrix does not have any size and completely empty, it follows the null symbol $\emptyset$ of matrix algebra. Let us say k=2 and perform the assignment as follows:

```
>>f=[ ]; k=2; ⏎
```

Now if we execute f=[f k] time and again first f=[f k] returns 2, second f=[f k] returns [2 2], third f=[f k] returns [2 2 2], and so on. This is called row directed data accumulation. Column directed data accumulation occurs by executing f=[f;k] each time.

The demonstrated k is just a scalar but it can be a return from some function, scalar, row matrix, column matrix, or rectangular matrix.

D.4 For-loop syntax

A for-loop performs similar operations for a specific number of times and must be started with the **for** and terminated by an **end** statements. Following the **for** there must be a counter. The counter of the for-loop can be any variable that counts integer or fractional values depending on the increment or decrement. If the MATLAB command statements between the **for** and **end** of a for-loop are few words lengthy, one can even write the whole for-loop in one line. The programming syntax and some examples on the for-loop are as follows:

✦ **Program syntax**

> for *counter*=starting value:increment or decrement of the
>   counter value:final value
> *Executable MATLAB command(s)*
> end

✦ **Example 1**

Our problem statement is to compute $y = \cos x$ for $x = 10^0$ to 70^0 with the increment 10^0. Let us assign the computed values to some variable y where y should be [cos10^0 cos20^0 cos30^0 cos40^0 cos50^0 cos60^0 cos70^0]=[0.9848   0.9397   0.866   0.766 0.6428   0.5   0.342].

In the programming context, y(1) means the first element in the row matrix y, y(2) means the second element in the row matrix y, and so on. MATLAB code for the $\cos x$ is **cosd(x)** where x is in degree. The for-loop counter expression should be k=1:1:7 or k=1:7 to have the control on the position index in the row matrix y (because there are 7 elements or indexes in y). Since the computation needs 10 to 70, one generates that by writing k*10. Following is the implementation:

Executable M-file:	*Or, as a one line:*

`for k=1:1:7`
`        y(k)=cosd(k*10);`
`end`
	`for k=1:1:7 y(k)=cosd(k*10); end`

Steps we need:
Open a new M-file (section 1.1), type the executable M-file statements in the M-file editor, save the editor contents by the name **test** in your working path, and call the **test** as shown below.

```
>>test ↵
>>y ↵
```

y =

    0.9848 0.9397 0.8660 0.7660   0.6428 0.5000   0.3420

✦ Example 2

A for-loop helps us accumulate data (appendix D.3) controlled by the consecutive loop index. In this example we accumulate some data row directionally according to the for-loop counter index.

For $k=1$, 2, and 3, we intend to accumulate the k^2 side by side. At the end we should be having [1   4   9] assigned to some variable f – this is our problem statement.

for the right shifting,	for the left shifting,
>>f=[ ]; for k=1:3 f=[f k^2]; end ↵	>>f=[ ]; for k=1:3 f=[k^2 f]; end ↵
>>f ↵	>>f ↵
f =	f =
1 4 9	9 4 1

The for-loop for the accumulation is presented above. The accumulation may occur as right or left shifting. Corresponding to the right shifting, the vector code (appendix A) for k^2 is k^2. The statement f=[ ]; means that an empty matrix is assigned to f outside the for-loop but at the beginning. The k variation in our problem is put as the for-loop counter. How the for-loop accumulates is shown below:

```
When k=1,  f=[f k^2]; returns  f=[[ ] 1^2];  ⇒ f=1;
When k=2,  f=[f k^2]; returns  f=[1 2^2];   ⇒ f=[1  4];
When k=3,  f=[f k^2]; returns  f=[1 4 3^2]; ⇒ f=[1  4  9];
```

The accumulation is happening from the left to the right. A single change provides the shifting from the right to the left which is f=[k^2 f];. The complete code and its execution result are also shown above by the heading 'for the left shifting'.

✦ Example 3

Another accumulation can be column directed that is we wish to see the output like $\begin{bmatrix} 1 \\ 4 \\ 9 \end{bmatrix}$ in example 2.

We just insert the row separator of a rectangular matrix (done by the operator ;) in the command f=[f k^2];. Again the shifting can happen either from the up to down or from the down to up. Both implementations are shown below:

for the down shifting,	for the up shifting,
>>f=[ ]; for k=1:3 f=[f;k^2]; end ↵	>>f=[ ]; for k=1:3 f=[k^2;f]; end ↵
>>f ↵	>>f ↵
f =	f =
1	9
4	4
9	1

✦ Example 4

Many electronic problems need writing multiple for-loops. Usually one loop is for one dimensional function, two loops are for two dimensional function, and so on. One dimensional function data takes the form of a row or column matrix.

Suppose we have the one dimensional data as $y=[9\ 6\ 7\ 4\ 6]$. We wish to access to every data in y. A single for-loop helps us conduct that as shown below:

>>y=[9 6 7 4 6]; for k=1:length(y) v=y(k); end ↵

First we assign the data to workspace y as a row matrix. The command length finds the number of elements in the row matrix y. The y(k) means the k-th element in the y which we assign to workspace v (any user-chosen variable). Every single data of the y is available sequentially in the v. The contents of y can be a column matrix too.

D.5 Finding the maximum/minimum numerically

Given a matrix, one finds the maximum element from the matrix by using the command max (min for the minimum). Let us say we have three matrices $R=[1\ -2\ 3\ 9]$, $C=\begin{bmatrix} 23 \\ -20 \\ 30 \\ 8 \end{bmatrix}$, and $A=\begin{bmatrix} 2 & 4 & 7 \\ -2 & 7 & 9 \\ 3 & 8 & -8 \end{bmatrix}$ whose maxima are

9, 30, and 9 (from all elements in the matrix) and minima are −2, −20, and −8 respectively. We find the maxima first entering (section 1.1) the respective matrices as follows:

for the row matrix,	for the column matrix,	for the rectangular matrix,
>>R=[1 -2 3 9]; ↵	>>C=[23;-20;30;8]; ↵	>>A=[2 4 7;-2 7 9;3 8 -8]; ↵
>>max(R) ↵	>>max(C) ↵	>>max(max(A)) ↵
ans =	ans =	ans =
9	30	9
>>min(R) ↵	>>min(C) ↵	>>min(min(A)) ↵
ans =	ans =	ans =
-2	-20	-8

Font equivalence is maintained by using the same letter for example A⟺ A in last implementation. If the matrix is a row or column one, we apply one max or min. For a rectangular matrix, the max or min separately operates on each column that is why two max or min functions are required. The functions are equally applicable on decimal number elements.

In the row matrix R, the maximum 9 is occurring as the fourth element in the matrix. Suppose we intend to find the position index (that is 4) of the maximum element in the R. Now we need two output arguments – one for the maximum and the other for its index. Its implementation is shown in the right side attached text box of this paragraph in which the two output arguments M and I correspond to the maximum and its integer index respectively.

The function min keeps this type of integer index returning option in a similar fashion.

```
for index finding in R,
>>[M,I]=max(R) ⏎

M =
        9
I =
        4
```

D.6 Position indexes of matrix elements with conditions

MATLAB command find looks for the position indexes of matrix elements subject to some logical condition whose general format is [R C]= find(condition) where the indexes returned to the R and C are meant to be for the row and column directions respectively. The R and C are user-chosen workspace variables. Let us consider $A = \begin{bmatrix} 11 & 10 & 11 & 10 \\ 12 & 10 & -2 & 0 \\ -7 & 17 & 1 & -1 \end{bmatrix}$ which we enter by the following:

>>A=[11 10 11 10;12 10 -2 0;-7 17 1 -1]; ⏎ ← A is assigned to A

We would like to know what the position indexes of A where the elements are greater than 10 are. In matrix A the left-upper most element has the position index (1,1). The elements of A being greater than 10 have the position indexes (1,1), (2,1), (3,2), and (1,3). MATLAB finds the required index in accordance with columns. Placing the row and column indexes vertically, we have $\begin{bmatrix} 1 \\ 2 \\ 3 \\ 1 \end{bmatrix}$ and $\begin{bmatrix} 1 \\ 1 \\ 2 \\ 3 \end{bmatrix}$ respectively. The output arguments R and C of the find receive these two column matrices respectively. The input argument of the find must be a logical statement, any element in A greater than 10 is written as A>10 (appendix D.1). The position indexes are found as shown in the right side attached text box.

```
where elements of A are
greater than 10,
>>[R C]=find(A>10) ⏎

R =
        1
        2
        3
        1
C =
        1
        1
        2
        3
```

where elements of
$A = 10$:
>>[R C]=find(A==10) ↵

R =
1
2
1
C =
2
2
4

where elements of
$A \leq 0$:
>>[R C]=find(A<=0) ↵

R =
3
2
2
3
C =
1
3
4
4

for the row matrix D:
>>D=[-10 34 1 2 8 4]; ↵
>>R=find(D>=8) ↵

R =
2   5
for the column matrix E:
>>E=[-2 8 -2 7]'; ↵
>>C=find(E~=-2) ↵

C =
2
4

To exercise more conditions, what are the position indexes in the matrix A where the elements are equal to 10? The answer is (1,2), (2,2), and (1,4). Again the position indexes where the elements are less than or equal to zero are (3,1), (2,3), (2,4), and (3,4).

The comparative operators $>$, $<$, $\geq$, $\leq$, and $\neq$ have the MATLAB counterparts >, <, >=, <=, and ~= respectively.

So far we considered a rectangular matrix for demonstration on position index finding. Let us see how the find works for a row or column matrix. Let us take $D = [-10 \quad 34 \quad 1 \quad 2 \quad 8 \quad 4]$ from which we find the position indexes of the elements where they are greater than or equal to 8. Obviously they are the 2nd and 5th elements. Here we do not need to place two output arguments to the find.

Again let us find the position indexes of the elements in column

matrix $E = \begin{bmatrix} -2 \\ 8 \\ -2 \\ 7 \end{bmatrix}$ where the elements are not equal to –2. The 2nd and 4th

elements are not equal to –2.

The output of find is a row one for the row matrix input and a column one for the column matrix input.

Presented above are the executions on all these conditional findings.

D.7 Matrix of ones, zeroes, and constants

MATLAB built-in commands ones and zeros implement user-defined matrix of ones and zeroes respectively. Each function conceives two input arguments, the first and second of which are the required numbers of rows and columns respectively. Let us say we intend to form the matrices A

$= \begin{bmatrix} 1 & 1 & 1 \\ 1 & 1 & 1 \\ 1 & 1 & 1 \\ 1 & 1 & 1 \end{bmatrix}$, $B = \begin{bmatrix} 1 & 1 & 1 \\ 1 & 1 & 1 \\ 1 & 1 & 1 \end{bmatrix}$, and $C = \begin{bmatrix} 1 & 1 & 1 & 1 \\ 1 & 1 & 1 & 1 \end{bmatrix}$. Their orders are 4×3, 3×3, and 2×4

respectively and the implementations are as follows:

for A,
   >>A=ones(4,3) ↵

A =

1	1	1
1	1	1
1	1	1
1	1	1

for B,
   >>B=ones(3) ↵

B =

1	1	1
1	1	1
1	1	1

for C,
   >>C=ones(2,4) ↵

C =

1	1	1	1
1	1	1	1

Either the number of rows or columns will do if the matrix is a square. For the row and column matrices of ones for example of length 6, the commands would be **ones(1,6)** and **ones(6,1)** respectively.

    Formation of the matrix of zeroes is quite similar to that of the matrix of ones. Replacing the function **ones** by **zeros** does the formation. Matrix of zeroes like $A = \begin{bmatrix} 0 & 0 & 0 \\ 0 & 0 & 0 \\ 0 & 0 & 0 \\ 0 & 0 & 0 \end{bmatrix}$, $B = \begin{bmatrix} 0 & 0 & 0 \\ 0 & 0 & 0 \\ 0 & 0 & 0 \end{bmatrix}$, and $C = \begin{bmatrix} 0 & 0 & 0 & 0 \\ 0 & 0 & 0 & 0 \end{bmatrix}$ (whose

orders are 4×3, 3×3, and 2×4) we form by the commands **A=zeros(4,3)**, **B=zeros(3)**, and **C=zeros(2,4)** respectively. A row and a column matrices of 6 zeroes are formed by the commands **zeros(1,6)** and **zeros(6,1)** respectively.

    A matrix of constants is obtained by first creating a matrix of ones of the required size and then multiplying by the constant number. For example the matrix $\begin{bmatrix} 0.2 & 0.2 & 0.2 \\ 0.2 & 0.2 & 0.2 \\ 0.2 & 0.2 & 0.2 \\ 0.2 & 0.2 & 0.2 \end{bmatrix}$ is generated by the command **0.2*ones(4,3)**.

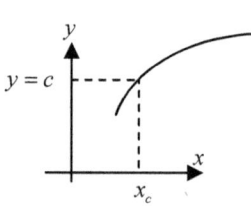

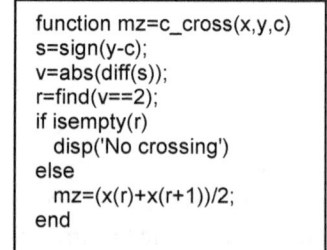

```
function mz=c_cross(x,y,c)
s=sign(y-c);
v=abs(diff(s));
r=find(v==2);
if isempty(r)
    disp('No crossing')
else
    mz=(x(r)+x(r+1))/2;
end
```

Figure D.1 Crossing of y
through a particular value c

Figure D.2 Author-written function file for finding the crossing point of a function

D.8 Functional value crossing

    Suppose we have a graph y versus x. The y may cross a particular value c. Figure D.1 shows the value crossing of the graph when $x = x_c$. We wish to determine x_c.

    What data should be available? – undoubtedly the samples of y and x along with c.

Author-written function file c_cross conducts this sort of finding. Figure D.2 presents the complete code of the file. Type the codes in a script file (section 1.1.2) and save the file by the name c_cross in your working path. The c_cross has three input arguments; x samples as a row matrix, y samples as a row matrix, and c which are indicated by (x,y,c).

Consider the function $y = xe^{-x}$ over $0 \le x \le 3$. Its graph you may view by exercising the command ezplot('x*exp(-x)',[0 3]) (appendix E). The graph is not shown for space reason. We wish to determine the value of x when $y = 0.1$. In order to make the samples available we have to choose some step size say $\Delta x = 0.01$, smaller is better. Generate the x samples by:

```
>>x=0:0.01:3; ↵
```

Generate the y samples by scalar code (appendix A):

```
>>y=x.*exp(-x); ↵
```

Just call the c_cross as:

```
>>c_cross(x,y,0.1) ↵

ans =
        0.1150
```

Above return says that $x_c = 0.115$. There is no y value as $c = 0.4$ and the response is:

```
>>c_cross(x,y,0.4) ↵
No crossing
```

There are two crossings with $c = 0.2$ over the given bounds which we find by:

```
>>c_cross(x,y,0.2) ↵

ans =
        0.2550    2.5450
```

i.e. the first x_c is 0.255 and the second x_c is 2.545. You may obtain the function file through email link of page ii.

D.9 Linear or logarithmic samples from specific number

From user-defined bounds and sample number one can generate linearly or logarithmically spaced vectors. Both are addressed below.

Linear vector:

Linearly spaced vector elements form an arithmetic progression. If the first element in the vector is a and common difference of the progression is d, the vector becomes $[a \quad a+d \quad a+2d \quad .. \quad a+(N-1)d]$ where N is the number of elements in the vector. Clearly d is equal to $\dfrac{Last\ element - First\ element}{N-1}$. The function linspace (abbreviation for the linear space) forms a linearly spaced vector for which the syntax is linspace(first element, last element, number of points from the first to last) and whose output is a row matrix.

Let us form a row vector R from 3 to 13 with 6 points therefore $d=2$ and R should be [3   5   7   9   11   13] and do so by:

>>R=linspace(3,13,6) ↵   ← R holds the required R and is user-chosen

R =
   3 5 7 9 11 13

Again a column vector C is to be formed from −7 to 3 with 5 points so that

$$d=\frac{5}{2} \text{ and } C=\begin{bmatrix} -7 \\ -\frac{9}{2} \\ -2 \\ \frac{1}{2} \\ 3 \end{bmatrix} \text{ which one obtains by:}$$

>>C=sym(linspace(-7,3,5))' ↵   ← C holds the required C and is user-chosen

C =
   -7
  -9/2
   -2
  1/2
   3

The command **sym** turns a number from decimal fractional to rational e.g. 4.5 to 9/2. A row matrix is changed to column by using the transpose operator ' (chapter 1). We have to conduct it for **C** because the **linspace** return is a row one.

Logarithmic vector:

Logarithmically (base of the logarithm is 10) spaced vector elements form a geometric progression. If first element in the vector is a and common ratio of the progression is r with length N, the vector is given by [a ar ar^2.. ar^{N-1}]. The function **logspace** (abbreviation for <u>log</u>arithmically <u>space</u>d) generates a logarithmically spaced vector with the syntax **logspace**(power of the first element, power of the last element, number of points from the first to last) and whose output is a row vector.

We wish to form a logarithmically spaced vector where power of the elements will be from 3 to 4 and the number of elements will be 5 therefore $a=10^3$, $N=5$, and $r=10^{\frac{1}{4}}$ which results the vector to be $L=$ [10^3 $10^3 10^{\frac{1}{4}}$ $10^3 10^{\frac{2}{4}}$ $10^3 10^{\frac{3}{4}}$ $10^3 10^{\frac{4}{4}}$]=[1000   1778   3162   5623   10000] (neglecting the fractional parts). Following is the execution:

>>L=logspace(3,4,5)' ↵

L =
   1.0e+004 *

0.1000
0.1778
0.3162
0.5623
1.0000

If the power of 10 is higher, the return from the **logspace** will be of higher digits that is why the return is in exponential form. Concerning the execution, **1.0e+004** * means $1.0 \times 10^4 \times$ and each of the return elements is multiplied by 10^4. We assigned the column vector to **L** (any user-chosen variable).

D.10 Series-parallel circuit resistance computation

Simple MATLAB function file eases the equivalent resistance finding of series-parallel circuit. Referring to appendix F, we write a function file which has some input and some output data.

Figure D.3 shows two resistors connected in parallel. Their equivalent resistance is simply $Z_{eq} = \dfrac{Z_1 Z_2}{Z_1 + Z_2}$ which needs

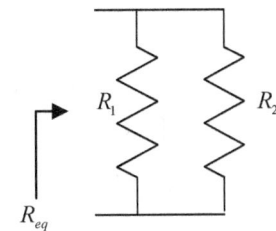

Figure D.3 Two resistors in parallel

the code writing of figure D.4. As a first step open a new M-file (subsection 1.1.2) editor, type the two line codes of figure D.4 for the parallel resistor equivalent in the newly opened M-file, and save the file by the name **Zp** (can be any name of your choice) in your working path or folder. In the M-file of figure D.4 the variable relevancy is as follows:
$Z1 \Leftrightarrow R_1$, $Z2 \Leftrightarrow R_2$, and $Z \Leftrightarrow R_{eq}$.

```
function Z=Zp(Z1,Z2)
Z=Z1*Z2/(Z1+Z2);
```

Figure D.4 Function file for the parallel equivalent of two resistors

When $R_1 = 4\Omega$ and $R_2 = 6\Omega$, we should get the computed equivalent resistance as $R_{eq} = \dfrac{12}{5}$ =2.4Ω. Let us verify that from just written function file **Zp** as follows:

>>Zp(4,6) ↵

ans =
        2.4

Figure D.5 Three parallel resistors in series with another resistor

If we wrote the command R=Zp(3,6);, the equivalent resistance would have been assigned to R where R is any user-chosen variable name outside the function file. The use of the command **sym** on the given resistance returns the symbolic or rational output. For just mentioned two resistors, the rational form equivalent resistance is $\frac{12}{5}\,\Omega$ and we see the rational output as follows:

```
>>Zp(sym(4),sym(6)) ↵

ans =

12/5
```

As another example the series-parallel circuit of figure D.5 has equivalent resistance $R_{eq} = \frac{638}{107} = 5.9626\,\Omega$ with $R_1 = 4\,\Omega$, $R_2 = 6\,\Omega$, $R_3 = 5\,\Omega$, and $R_4 = 7\,\Omega$ which we wish to find from MATLAB.

The R_2, R_3, and R_4 of figure D.5 are in parallel and whole parallel equivalent of the three resistors needs to be added with R_1 to compute the equivalent resistance R_{eq}. Let us execute the following to get R_{eq}:

```
>>R1=4; R2=6; R3=5; R4=7; ↵
>>Req=R1+Zp(Zp(R2,R3),R4) ↵

Req =
        5.9626
```

In above implementation the R1 is equivalent to R_1, so are the R2, R3, and R4 and also Req$\Leftrightarrow R_{eq}$. The names R1, R2, etc are all user-given. The first line in above implementation is to assign the resistance values to like names. In the second line of above implementation Zp(R2,R3) is equivalent to calculating $R_2 \| R_3$. The Zp(Zp(R2,R3),R4) means the calculation of $(R_2 \| R_3) \| R_4$. In order to find the rational form output, we perform the following at the command prompt:

```
>>R1=sym(4); R2=sym(6); R3=sym(5); R4=sym(7); ↵
>>Req=R1+Zp(Zp(R2,R3),R4) ↵

Req =

638/107
```

Each resistor value now needs to be declared symbolically by using the sym before the equivalent resistance computing e.g. R1=sym(4); for R_1.

D.11 Mathematics readable form by pretty

There is a function called pretty which returns any functional expression as close as mathematical form provided that its input argument is in vector string or code (appendix A) form. For example in vector string form, the x^3 is coded as x^3. When x^3 is the input argument of the pretty, the return of the pretty is x^3 provided that the independent variable x is defined before by using the command syms.

The **pretty** is applied only for the display reason, no computation is conducted on the expression. If a function has the code **2*x**, you will see the function as **2 x**. It does not work on numeric values but usage of the command **sym** on the numeric values shows the rational form. For example **pretty(sym(3.3))** displays $\frac{33}{10}$. Just to show by another example, **1/(x+1)** is $\frac{1}{x+1}$ which can be executed as:

>>**syms x** ⏎   ← Declaring the related x of $\frac{1}{x+1}$ by the **syms**

>>**y=1/(x+1);** ⏎   ← Assigning the code of $\frac{1}{x+1}$ to **y** where **y** is a user-chosen
                                   variable

>>**pretty(y)** ⏎   ← Applying the **pretty** on the codes stored in **y**
       1
    ------
    x + 1

D.12 Difference between MATLAB and maple functions

MATLAB is full of libraries and library functions, one specific library is called **maple** which is very convenient for symbolic computation. Any **maple** statement is executed in MATLAB with the syntax **maple**(statement under quote). In **maple** there is a function by the name **line** in a family called **geometry** so hierarchically we organize the **line** function in **maple** as shown in the figure D.6.

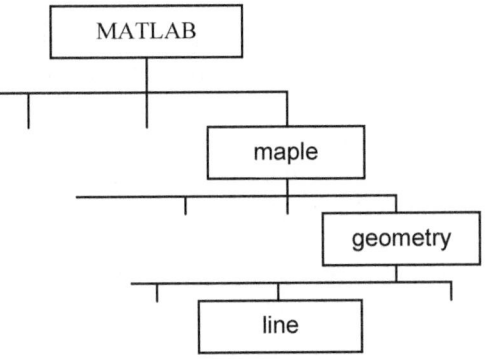

Figure D.6 Hierarchy of the **line** function in MATLAB

Recall the function **sin** of section 1.3 which exists both in **maple** and MATLAB. Their calling differences on some **x** are shown below:
         >>**sin(x)** ⏎   ← when called in MATLAB
         >>**maple('sin(x)')** ⏎   ← when called in **maple**

D.13 Numerator-denominator separation from an expression

Given an algebraic expression, one separates its numerator and denominator components using the function **numden** (abbreviation for <u>num</u>erator and <u>den</u>ominator) which is conducted completely using the symbolic concept. We need to declare the independent variable using **syms** beforehand. The function has two output arguments – one for the numerator and the other for the denominator. The input argument of the function as well as its output argument exercises vector string form (appendix A). For example, the expression $\dfrac{x^2}{y} + yx + \dfrac{5y}{7x}$ has the numerator and denominator $7x^3 + 7x^2y^2 + 5y^2$ and $7xy$ respectively. There are two independent variables in the expression – x and y. Again the trigonometric function $\dfrac{\cos A}{\cos A - 5} - \dfrac{\sin A}{\sin A + 3}$ has the numerator and denominator as $3\cos A + 5\sin A$ and $(\cos A - 5)(\sin A + 3)$ respectively in which the single independent variable is A. Both implementations are presented below:

Numerator and denominator separation for the first expression:	Numerator and denominator separation for the second expression:
>>syms x y ↵	>>syms A ↵
>>[n d]=numden(x^2/y+y*x+5/7*y/x); ↵	>>f1=cos(A)/(cos(A)-5);f2=sin(A)/(sin(A)+3); ↵
>>pretty(n) ↵	>>[n d]=numden(f1-f2); ↵
3 2 2 2	>>pretty(n) ↵
7 x + 7 y x + 5 y	3 cos(A) + 5 sin(A)
>>pretty(d) ↵	>>pretty(d) ↵
7 y x	(cos(A) - 5) (sin(A) + 3)

Font equivalence is maintained in all executions for example $x \Leftrightarrow x$. The vector string of the output numerator and denominator are returned to the workplace n and d (can be any variable of your choice) respectively. Command pretty (appendix D.11) just displays the strings in readable form. It is permissible that we enter long expression part by part. For example the first and second parts of the second expression are assigned to the workspace f1 and f2 (can be any name of your choice) respectively. The whole expression is composed of f1-f2 on which the numden is exercised.

Appendix E

Some graphing functions of MATLAB

One of MATLAB's nicest features is you can have your graphics drawn while programming electronic system related problems. There are so many easy accessible built-in graphics functions that one finds it very interesting when the input-output argumentation style of these functions is understood. Some graphing functions which we applied frequently in previous chapters are addressed for syntax details in the sequel.

♦ y versus x data

The command plot graphs y versus x data. Let us say we have the attached (on the right side in this paragraph) tabular data. We intend to graph these data as y versus x graph.

Tabular data of y versus x type:							Command to graph the y vs x data:
x	-6	-4	0	4	5	7	>>x=[-6 -4 0 4 5 7]; ↵
y	9	3	-3	-5	2	0	>>y=[9 3 -3 -5 2 0]; ↵ >>plot(x,y) ↵

Commands to graph the data are also presented beside the tabular data on the right side in the last paragraph. We first assign the x and y data to workspace x and y (some user-chosen variables) respectively and then call the command plot to see the figure E.1(a). The plot has two input arguments, the first and second of which are the x and y data both as a row or column matrix of identical size respectively.

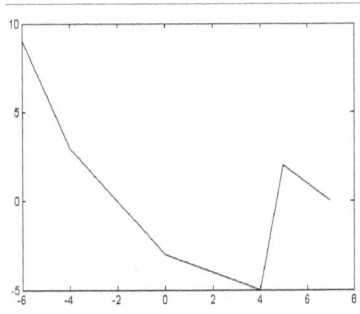

Figure E.1(a) y vs x plot of the tabular data

In order to graph the mathematical expression by using the plot, one first needs to calculate the functional values by using the scalar code (appendix A) and then applies the command. During the calculation, computing step selection is mandatory which is completely user-defined.

For instance we wish to graph the function $f(x) = x^2 - x + 2$ over $-2 \le x \le 3$.

Let us choose some x step 0.1. The x vector as a row matrix is generated by x=-2:0.1:3; (section 1.1). At every element in x vector, the functional value is computed and assigned to workspace f by f=x.^2-x+2;. The f is any user-chosen variable. Now we call the grapher as plot(x,f) to see the trajectory (not shown for space reason).

The command **plot** just draws the graph, no graphical features such as x axis label or title are added to the graph. It is the user who is supposed to add these graphical features.

✦ Multiple y data versus common x data

The **plot** keeps many options, one of which is just discussed. We graph several y data versus common x data with the help of **plot** but with different number of input arguments. Let us choose the right side attached table for graphing.

Tabular data for multiple y versus common x :						
x	-6	-4	0	4	5	7
y_1	9	3	-3	-5	2	0
y_2	0	-2	1	0	5	7.7
y_3	-1	2	8	1	0	-3

We intend to plot the y_1, y_2, and y_3 on common x data. To do so,

>>x=[-6 -4 0 4 5 7]; ↵   ← Assigning the x data as a row matrix to x
>>y1=[9 3 -3 -5 2 0]; ↵   ← Assigning the y_1 data as a row matrix to y1
>>y2=[0 -2 1 0 5 7.7]; ↵   ← Assigning the y_2 data as a row matrix to y2
>>y3=[-1 2 8 1 0 -3]; ↵   ← Assigning the y_3 data as a row matrix to y3
>>plot(x,y1,x,y2,x,y3) ↵   ← Applying the command **plot**

The **plot** now has six input arguments − two for each graph, the first and second of which are the common x and y data to be plotted respectively. If there were four y data, the command would be plot(x,y1,x,y2,x,y3,x,y4). Once the data is plotted for several y, identifying the y traces is obvious which is carried out by the command **legend**. The command legend('y1', 'y2','y3') puts identifying

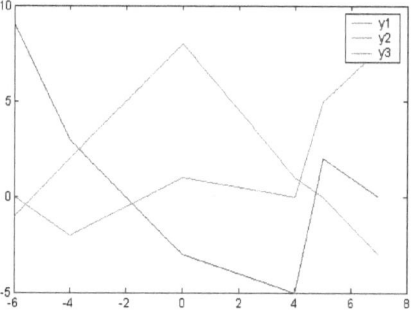

Figure E.1(b) Multiple y vs x for the tabular data

marks/colors among various graphs. The input argument of the **legend** is any user-given word but under quote and separated by a comma. The number of y traces must be equal to the number of input arguments of **legend**. We gave the names y1, y2, and y3 for the three y traces respectively. In doing so, we end up with the figure E.1(b). You can even move the legend on the plot area by using mouse.

You see all graphics throughout the text as black and white because we did not include color graphics in the text (for expense reason). But MATLAB displays figures in color plots, which you can easily identify.

Another situation can be that we have several functions and intend to plot those on common x variation. For instance we wish to graph $y_1 = x^3 - x^2 + 4$ and $y_2 = x^2 - 7x - 5$ over the common $-1 \le x \le 3$.

Under these circumstances, the step selection of x data is compulsory. Without calculating the functional values of given y curves, we can not graph the functions for which we exercise the scalar code. Let us choose the x step as 0.1. We first generate the common x vector as a row matrix by writing x=-1:0.1:3; and then calculate the y_1 and y_2 (y1$\Leftrightarrow y_1$ and y2$\Leftrightarrow y_2$) data by writing y1=x.^3-x.^2+4; y2=x.^2-7*x-5; and eventually the graph appears by executing plot(x,y1,x,y2), graph is not shown for space reason. Thus you can graph three or more functions.

◆ Functions of the form $y = f(x)$

If any function is of the form $y = f(x)$ and the $f(x)$ versus x is to be graphed, the built-in **ezplot** is the best option which uses a syntax **ezplot**(functional vector code under quote according to appendix A, interval bounds as a two element row matrix) where the first and second elements in the row matrix are beginning and ending bounds of the interval respectively. The **ezplot** graphs $y = f(x)$ in the default interval

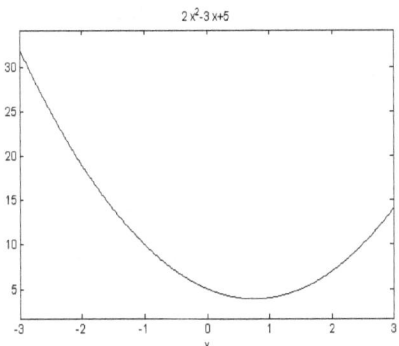

Figure E.1(c) Plot of $y = 2x^2 - 3x + 5$ versus x over $-3 \le x \le 3$

$-2\pi \le x \le 2\pi$ when no interval description is argumented.

We intend to graph the function $y = 2x^2 - 3x + 5$ over the interval $-3 \le x \le 3$. We first give $2x^2 - 3x + 5$ MATLAB vector code and then assign that to y as follows:
```
>>y='2*x^2-3*x+5'; ↵
```
In above implementation the y is any user-chosen variable. The interval $-3 \le x \le 3$ is entered by [-3 3]. To obtain the plot of y in the given interval, we execute the following at the command prompt:
```
>>ezplot(y,[-3,3]) ↵
```
Above command results the figure E.1(c).

◆ Multiple graphs in the same window

The function **subplot** splits a figure window in subwindows based on the user definition. It accepts three positive integer numbers as the input arguments, the first and second of which indicate the number of subwindows in the horizontal and the number of subwindows in the vertical directions respectively. For example 22 means two subwindows horizontally and two subwindows vertically, 32 means three subwindows horizontally and two subwindows vertically, ... and so on. The third integer in the input argument numbered consecutively offers control on the subwindows so generated. If the first two digits are 32, there should be 6 subwindows

Commands for the figure E.1(d):
>>subplot(121) ↵   ← It handles the first graph
>>ezplot('x') ↵   ← Plotting $y = x$
>>subplot(122) ↵   ← It handles the second graph
>>ezplot('exp(-x)') ↵   ← Plotting $y = e^{-x}$

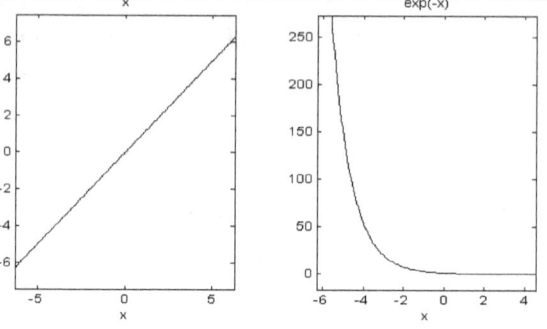

Figure E.1(d) Plots of $y = x$ and $y = e^{-x}$ side by side in the same window

and they are numbered and controlled by using 1 through 6. When you plot some graph in a subwindow, as if you are handling an independent figure window.

We wish to graph $y = x$ and $y = e^{-x}$ side by side as two different plots by using earlier mentioned **ezplot** but in the same window. If we imagine the subfigures as matrix elements, we have a figure matrix of size 1×2 (one row and two columns). That is why the first two integers of the input argument of **subplot** should be 12. Attached commands in the upper right text box of last paragraph show the figure E.1(d). The third integers 1 and 2 in the **subplot** give the control on the first and second subfigures respectively.

As another example we wish to plot $y = x$ and $y = e^{-x}$ in the upper row and only $y = (1 - e^{-x})$ in the lower row subfigures in the same window

Commands for the figure E.1(e):
>>subplot(221) ↵   ← Subfigure selection for $y = x$
>>ezplot('x') ↵   ← Plotting $y = x$
>>subplot(222) ↵   ← Subfigure selection for $y = e^{-x}$
>>ezplot('exp(-x)') ↵   ← Plotting $y = e^{-x}$
>>subplot(212) ↵   ← Subfigure selection for $y = (1 - e^{-x})$
>>ezplot('1-exp(-x)') ↵ ← Plotting $y = (1 - e^{-x})$

whose implementation needs above attached text box commands and whose final output is the figure E.1(e). We are supposed to have four figures when the integer input argument of **subplot** is 22 (two for rows and two for columns). The arguments 221, 222, 223, and 224 provide handle on the four figures consecutively. The figures could have been plotted on 223 and 224 are absent so we ignore them. The argument 21 creates two subfigures (two rows and one column) handled by 211 and 212, but 211 is absent so we ignore that too.

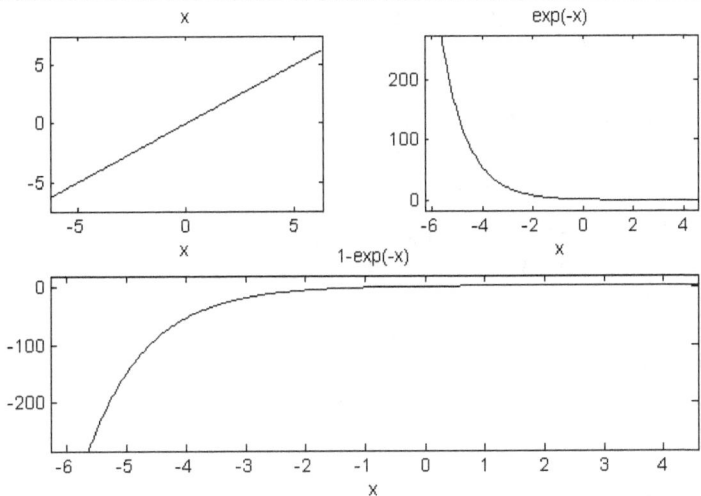

Figure E.1(e) Plots of $y = x$ and $y = e^{-x}$ in the upper row and

$y = (1 - e^{-x})$ in the lower row in the same window

Let us see the input arguments of **subplot** for different subfigures (each third brace set [ ] is one subfigure in the following tabular representation) as follows:

Subfigures needed	First two input integers of **subplot**	Third input integer of **subplot**	Commands we need
[ ] [ ]	22	[1] [2]	subplot(221) subplot(222)
[ ] [ ]		[3] [4]	subplot(223) subplot(224)
[ ] [ ]	22 for upper two (lower two remain empty)	[1] [2]	subplot(221) subplot(222)
[ ]	21 for the lower one (upper one remains empty)	[2]	subplot(212)
[ ]	21 for the upper one (lower one remains empty)	[1]	subplot(211)
[ ] [ ]	22 for the lower two (upper two remain empty)	[3] [4]	subplot(223) subplot(224)
[ ] [ ]	22 for the left two (right two remain empty)	[1] [ ] [2]	subplot(221) subplot(223)
[ ] [ ]	12 for the right one (left one remains empty)	[3] [ ]	subplot(122)
[ ] [ ]	22 for right two (left two remain empty)	[ ] [2] [1] [4]	subplot(222) subplot(224)
[ ] [ ]	12 for the left one (right one remains empty)		subplot(121)

♣ Symbol on a drawn graph

Let us say $r = e^{-2\theta} \sin 2\,\theta$ is to be plotted over $0 \le \theta \le \pi$. Earlier quoted **ezplot** graphs the function by **ezplot('exp(-2*t)* sin(2*t)',[0 pi])** – t is used for θ, figure E.1(f) shows it. We would like to drop the text **The plot of** $e^{-2\theta} \sin 2\theta$ on the graph. The command **gtext** gives the provision for dropping a mouse-driven text (written under quote) on a drawn graphics. Execute the following at the command prompt:

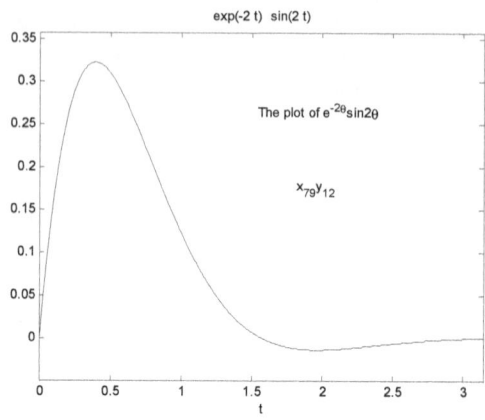

Figure E.1(f) Plot of $r = e^{-2\theta} \sin 2\theta$

>>gtext('The plot of e^-^2^\thetasin2\theta') ↵

Table E.1   MATLAB codes for various symbols (not in alphabetical order)

MATLAB code	Symbol	MATLAB code	Symbol	MATLAB code	Symbol	MATLAB code	Symbol
\omega	ω	\gamma	γ	\mu	μ	\Xi	Ξ
\Omega	Ω	\Gamma	Γ	\nu	ν	\xi	ξ
\phi	ϕ	\delta	δ	\surd	$\sqrt{}$	\oplus	$\oplus$
\Phi	Φ	\Delta	Δ	\in	$\in$	\alpha	α
\zeta	ζ	\epsilon	ε	\chi	χ	\sim	$\sim$
\pi	π	\eta	η	\leq	$\le$	\iota	ι
\Pi	Π	\Psi	Ψ	\geq	$\ge$	\infty	∞
\beta	β	\psi	ψ	\pm	$\pm$	\exists	$\exists$
\theta	θ	\kappa	κ	\int	$\int$	\cap	$\cap$
\Theta	Θ	\Sigma	Σ	\copyright	$\copyright$	\subset	$\subset$
\lambda	λ	\sigma	σ	\nabla	∇	\ni	$\ni$
\Lambda	Λ	\neq	$\neq$	\upsilon	υ	\oslash	$\varnothing$
\partial	∂	\rho	ρ	\tau	τ	\otimes	$\otimes$

Then go to the figure window and you find the mouse pointer activated and a crosshair is appearing. Choose any convenient position in the graph and click the left button of mouse to see the inside text as in figure E.1(f). The symbol θ is written by the command **\theta** in graphics. Any superscript is placed by the command ^, as explanation we can say $e^{-} \Leftrightarrow$ e^-, $e^{-2} \Leftrightarrow$ e^-^2, $e^{-2\theta} \Leftrightarrow$ e^-^2\theta, … etc. What if we have a subscript (performed by the operator _) for example let us drop the symbolic text $x_{79} y_{12}$ on the last graph for which we execute the command **gtext('x_7_9y_1_2')** at the command prompt. After that go to the figure window, choose any position in the plot to drop the subscript text, and click the left button of mouse to find the text as in figure E.1(f). Multiple subscript writing follows the syntax similar to that of the superscript. The reader may need to know the MATLAB codes for frequently encountered Greek symbols, which are presented in Table E.1.

♦ Scatter data plot using small circles

Instead of having a graph as continuous line, it is possible to have the graph in terms of bold dots or round circles like the figure E.1(g). The function **scatter** returns this sort of graph for which the common syntax is **scatter**(x data as a row matrix, y data as a row matrix, size of the circle, color of the circle). The function also accepts the first two input arguments. The size of the circle is any user-given integer number. The larger is the number, the bigger is the size for example 75, 100, etc.

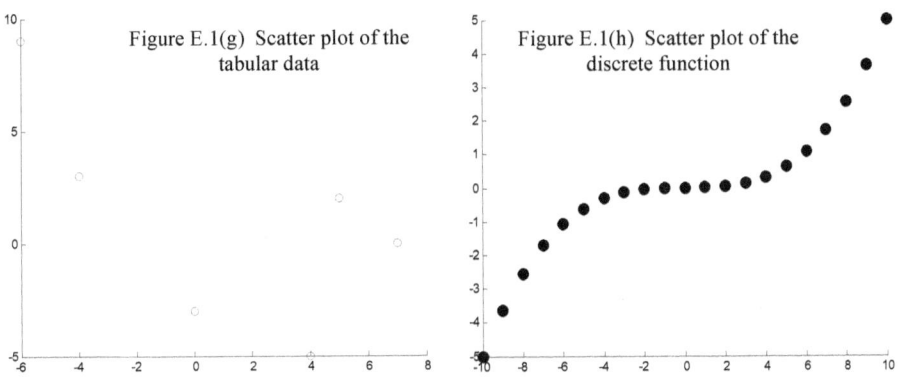

Figure E.1(g) Scatter plot of the tabular data

Figure E.1(h) Scatter plot of the discrete function

Let us graph the page 261 quoted tabular data (x and y) as the scatter plot. The command we need is the following:
>>x=[-6 -4 0 4 5 7]; y=[9 3 -3 -5 2 0]; scatter(x,y) ↵

Upon execution of above command, we see the figure E.1(g). The color of the circle is blue by default but any three element row matrix sets the user-defined color. The three element row matrix refers to red, green, and blue components respectively each one within 0 and 1. Black color means all zero, white means all 1, red means other two components zero, and so on. The circle displayed in the figure E.1(g) is all empty but one fills the circle by using the reserve word **filled** under quote and included as another input argument to the **scatter**. Let us say we intend to scatter graph with circle size 100 and the circles should be filled with black color. The necessary command is **scatter(x,y,100,[0 0 0],'filled')**, the graph is not shown for the space reason.

This type of graph is suitable for representation of the function which is discrete in nature. For instance the discrete function $y[n] = \dfrac{n^3}{200}$ over $-10 \le n \le 10$ is to be plotted with black circles of size 100 where n is integer. We form a row matrix n to generate the interval with start value -10, increment 1, and end value 10 by writing **n=-10:10;**. The scalar code of appendix A computes the $y[n]$ values and assigns those to workspace y. However the complete code to bring about the graph of figure E.1(h) is presented below:

    >>n=-10:10; y=n.^3/200; scatter(n,y,100,[0 0 0],'filled') ↲

Appendix F

Creating a function file

A function file is a special type of M-file (section 1.1) which has some user-defined input and output arguments. Both arguments can be single or multiple. The first line in a function file always starts with the reserve word **function**. A function file must be in your working path or its path must be defined in MATLAB. Depending on problem, a function file is written by the user and can be called from the MATLAB command prompt or from another M-file. For convenience, long and clumsy programs are split into smaller modules and these modules are written in a function file. The basic structure of a function file is as follows:

MATLAB Prompt   function file

$$>> g = \text{call } f \qquad \Longrightarrow \qquad g(\underbrace{y_1, y_2, \dots y_m}_{\text{output arguments}}) = f(\underbrace{x_1, x_2, x_3, \dots x_n}_{\text{input arguments}})$$

We present following examples for illustration of function files keeping in mind that the arguments' order and type of the caller and function file are identical.

⌗ **Example 1**

Let us say the computation of $f(x) = x^2 - x - 8$ is to be implemented as a function file. When $x = -3$ and $x = 5$, we should be having 4 and 12 respectively.

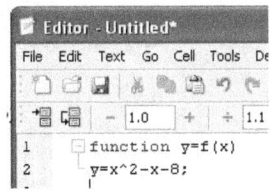

Figure F.1(a)   Single input – single output function file

The vector code (appendix A) of the function is **x^2-x-8** assuming scalar **x** and obviously the **x** is for x. We have one input (which is x) and one output (which is $f(x)$). Open a new M-file editor, type the codes of figure F.1(a) exactly as they appear in the M-file, and save the file by the name f. The assignee **y** and independent variable **x** can be any variable of your choice, which are the output and input arguments of the function respectively. Again the file and function name **f** can be any user-chosen name only the point is the chosen function or file name should not exist in MATLAB. Let us call the function **f(x)** to verify the programming as shown in the right side text box. You can write dozens of MATLAB executable statements in the file but whatever is assigned to the last **y** returns the function **f(x)** to **g**. Writing the = sign between the **y** and **f(x)** in the function file is compulsory.

Calling for example 1:
```
>>g=f(-3) ↵   ← call f(x) for x = -3

g =
    4
```
for $x = 5$,
```
>>g=f(5) ↵

g =
    12
```

⬛ Example 2

Example 1 presents one input-one output function how if we handle multiple inputs and one output? The input argument variables are separated by commas in a function file. A three variable function $f(x_1, x_2, x_3) = x_1^2 - 2x_1 x_2 + x_3^2$ is to be computed by a function file. The input arguments (assuming all scalar) are x_1, x_2, and x_3 and

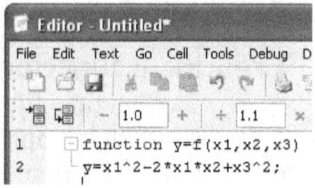

Figure F.1(b) Multiple inputs – single output function file

the output argument is the functional value of the function. The x_1 is written as **x1**, and so is the others. Follow the M-file procedure of example 1 but the code should be as shown in figure F.1(b). Let us inspect the function (with the specific $x_1 = 3$, $x_2 = 4$, and $x_3 = 5$, the output value of the three variable function must be $f(3,4,5) = 3^2 - 2 \times 3 \times 4 + 5^2 = 10$) as presented in the text box below.

```
Calling for example 2: when input arguments are all scalar:
>>g=f(3,4,5) ↵    ← calling f(x₁, x₂, x₃) for x₁=3, x₂=4, and x₃=5

g =
      10
Calling for the example 2: when input arguments are all column matrix:
>>x1=[2 3 4]'; ↵   ← x₁ values are assigned to x1 as a column matrix
>>x2=[-2 2 5]'; ↵  ← x₂ values are assigned to x2 as a column matrix
>>x3=[1 0 3]'; ↵   ← x₃ values are assigned to x3 as a column matrix
>>f(x1,x2,x3) ↵    ← calling f(x₁, x₂, x₃) using column matrix input arguments

ans =
      13
      -3
     -15
```

The **function** not only works for the scalar inputs but also does for matrices in general for example a set of input argument values are $x_1 = \begin{bmatrix} 2 \\ 3 \\ 4 \end{bmatrix}$, $x_2 = \begin{bmatrix} -2 \\ 2 \\ 5 \end{bmatrix}$, and $x_3 = \begin{bmatrix} 1 \\ 0 \\ 3 \end{bmatrix}$ for which the $f(x_1, x_2, x_3)$ values should be $\begin{bmatrix} 13 \\ -3 \\ -15 \end{bmatrix}$ respectively. The computation needs the scalar code (appendix A) of $f(x_1, x_2, x_3)$ regarding x_1, x_2, and x_3. The modified second line statement of the figure F.1(b) now should be **y=x1.^2-2*x1.*x2+x3.^2;**. On making the modification and saving the file, let us carry out the commands which are placed in above text box of this page too. If it is necessary, the output can be assigned to user-supplied workspace variable **v** by writing **v=f(x1,x2,x3)** at the command prompt. The return from the function file also follows the same input matrix order. If the input arguments of **f(x1,x2,x3)** are rectangular matrix, so is the output. The input arguments of the function file do not have to be the mathematics symbol. Suppose x_1=**ID**, x_2=**Value**, and x_3=**Data**, one could have written the first and second lines of the function file in the figure F.1(b) as **function y=f(ID,Value, Data)** and **y=ID.^2-2*ID.*Value+ Data.^2**; respectively.

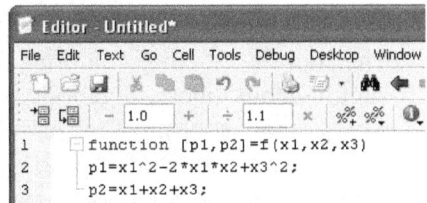

Figure F.1(c)   Function file for three input and two output arguments

⊟ Example 3

To illustrate a multi-input and multi-output function file, let us consider that p_1 and p_2 are to be found from three variables x_1, x_2, and x_3 (all are scalars) employing the expressions $p_1 = x_1^2 - 2x_1 x_2 + x_3^2$ and $p_2 = x_1 + x_2 + x_3$ whose function file (type the codes in a new M-file editor and save the file by the name **f**) is presented in figure F.1(c).

Choosing x_1=4, x_2=5, and x_3=6, one should get p_1=12 and p_2=15 for which right side text box commands are conducted at the command prompt. More

Function file calling for the example 3:
>>[p1,p2]=f(4,5,6) ⏎ ← calling the function file f for p_1 and p_2 using x_1=4, x_2=5, and x_3=6
p1 =
12
p2 =
15

than one output arguments (which are here p_1 are p_2 and represented by **p1** and **p2** respectively) are separated by commas and placed inside the third bracket following the word **function** in figure F.1(c).

When we call the function from the command prompt, the output argument writing is similar to that of the function file (that is why we write **[p1,p2]** as output arguments at the command prompt). The output argument variable names do not have to be **p1** and **p2** and can be any name of user's choice. If there were three output arguments p_1, p_2, and p_3, the output arguments in the function file would be written as **[p1,p2,p3]** and their calling would happen in a like manner.

Note: We saved different function files by the same name **f** just for simplicity and maintaining unifying approach. By this action any previously saved file by the name **f** disappears. What we suggest is save the function file by other name like **f1** and call accordingly for instance the first line of figure F.1(c) would be **function [p1,p2]=f1(x1,x2,x3)** and calling would take place as **[p1,p2]=f1(4,5,6)** for the last illustration.

Appendix G

Algebraic equation solver

By virtue of master function **solve**, we find the solution of a single or multiple algebraic equations when the equations are its input arguments. The notion of the solution is symbolic and a substantial number of simultaneous linear, algebraic, or trigonometric equations can be solved by using the **solve**. The common syntax of the implementation is **solve**(equation-1, equation-2, ... so on in vector string form – appendix A, unknowns of the equations separated by a comma but put under quote).

The return from **solve** is in general a structure array which is beyond the discussion of the text (reference 32). Very briefly a structure array is composed of several members. In order to view the solution from the **solve**, one needs to call a member of the array. If **s** is a structure array and **u** is one of its members, we call the member by the command **s.u**. One can assign the **s.u** to some other workspace variable if it is necessary. Following points must be considered while using the **solve**:

(a) We solve equations related to variables which have power 1 or more. One variable equation of power 1 for example $2x - 7 = 5x$ is written as **2*x-7=5*x** in code form, two variable equations $7x + y - 7 = 0$ and $2y + 4 = 5x$ are written as **7*x+y-7=0** and **2*y+4=5*x** respectively, and so on.

(b) Since the solution approach is completely symbolic, we enter rational value of the equation coefficient in case of decimal data for example the equation $2.4x - 7.5 = 5x$ had better be written as **24*x/10-75/10=5*x**.

(c) The **i** or **j** is unit imaginary number in MATLAB. Such use sometimes turns the **solve** non-executable. If any variable **i** is in the equations, we use a dummy variable for example **c**.

(d) The return from **solve** is usually in rational form for instance **24/10** instead of 2.4. If we need the decimal value, we employ the command **double** on the return.

(e) The equations are assigned under quote while entering as input arguments to the **solve** or assigning to some variables.

Let us go through the solution finding on following three examples.

◆ Example 1

The solution of the equation $2x + 7 = 9x$ is $x = 1$ which we wish to find. We execute the following for the solution at the command prompt:

```
>>s=solve('2*x+7=9*x','x') ↵

s =
```

The **s** in last execution is any user-chosen variable. The **s=1** return indicates the $x=1$ solution. The independent variable in given equation is x that is why the second input argument of **solve** is **'x'**.

✦ Example 2

The equation set $\begin{Bmatrix} 6x - y = -8 \\ 9x = 8y + 5 \end{Bmatrix}$ has the solution $x = -\dfrac{23}{13} = -1.7692$ and

$y = -\dfrac{34}{13} = -2.6154$ and our objective is to obtain the solution.

The given two equations have the codes **6*x-y=-8** and **9*x=8*y+5** respectively. The related variables in the two equations are x and y therefore we carry out the following at the command prompt:

>>s=solve('6*x-y=-8','9*x=8*y+5','x','y') ↵

```
s =
         x: [1x1 sym]
         y: [1x1 sym]
```

The **s** in above execution is also any user-chosen variable. The **solve** returns the solution to the **s**. As we mentioned earlier, the return from **solve** is a structure array and its members are the **x** and **y** (related variables in given equations). The return is an object (called symbolic object, indicated by the **sym**) rather than data. Should we pick the solution of x and y from the **s**, we need to exercise the commands **s.x** and **s.y** respectively. Let us see what we obtain as the solution:

For the x value:

>>s.x ↵

ans =

-23/13
>>double(s.x) ↵

ans =
-1.7692

For the y value:

>>s.y ↵

ans =

-34/13
>>double(s.y) ↵

ans =
-2.6154

The result is as expected. We could have assigned the return to some variable for example **s.x** or **double(s.x)** to a by writing **a=s.x** or **a=double(s.x)**.

✦ Example 3

For multiple equations it is not feasible that we enter all equations as one line to **solve**. Instead we first assign the given equations to some user-chosen variables and then call the **solve** with these variable names as the input arguments. The equation set $\{x - y - 3.2z + 2u = -8, 8.5y - 7z + u = 5, x - 4y + 2z = 76, -3.4x + 6z + 7u = -12\}$ has the solution $u =$

$$\frac{29598}{1387}=21.3396, \quad x=\frac{348525}{2774}=125.6399, \quad y=\frac{95869}{2774}=34.5598, \text{ and } z=\frac{245775}{5548}$$

$=44.2997$ which we find by exercising ongoing function and symbology as follows:

```
>>e1='x-y-32*z/10+2*u=-8'; ↵        ← assigning the first equation to e1
>>e2='85*y/10-7*z+u=5'; ↵           ← assigning the second equation to e2
>>e3='x-4*y+2*z=76'; ↵              ← assigning the third equation to e3
>>e4='-34*x/10+6*z+7*u=-12'; ↵      ← assigning the fourth equation to e4
>>s=solve(e1,e2,e3,e4,'x','y','z','u') ↵  ← calling the solve on e1, e2, e3, and e4
```

```
s =                          ← s holds the solution as a structure array
    u: [1x1 sym]             ← u is a member of s
    x: [1x1 sym]             ← x is a member of s
    y: [1x1 sym]             ← y is a member of s
    z: [1x1 sym]             ← z is a member of s
```

The e1, e2, e3, and e4 are all user-chosen variables in above. The next step is to see the values returned by the **solve**:

for the rational value of x :

```
>>s.x ↵

ans =

348525/2774
```

for the rational value of z :

```
>>s.z ↵

ans =

245775/5548
```

for the decimal value of x :

```
>>double(s.x) ↵

ans =
      125.6399
```

for the decimal value of z :

```
>>double(s.z) ↵

ans =
      44.2997
```

for the rational value of y :

```
>>s.y ↵

ans =

95869/2774
```

for the rational value of u :

```
>>s.u ↵

ans =

29598/1387
```

for the decimal value of y :

```
>>double(s.y) ↵

ans =
      34.5598
```

for the decimal value of u :

```
>>double(s.u) ↵

ans =
      21.3396
```

When we assign the equations, we use the quote but inside the **solve** the assignees do not have quote for example e1 not 'e1'. All four values as a four element row matrix are seen by:

```
>>[s.x s.y s.z s.u] ↵

ans =

[ 348525/2774, 95869/2774, 245775/5548, 29598/1387]
```

All four values in decimal form as a row matrix are seen as follows:

```
>>double([s.x s.y s.z s.u]) ↵

ans =
      125.6399   34.5598   44.2997   21.3396
```

Obviously the return is in the order we typed in.

References

>> >> **Electric or Electronic Circuit Fundamentals** >> >>

[1] Robert L. Boylestad, *"Introductory Circuit Analysis"*, 9th Edition, 2000, Prentice Hall, Inc., Pearson Education, Upper Saddle River, New Jersey.

[2] Leonard S. Bobrow, *"Elementary Linear Circuit Analysis"*, 1987, Second Edition, Harcourt Brace Jovanovich College Publishers, New York.

[3] Clayton Paul, *"Fundamentals of Electric Circuit Analysis"*, 2001, John Wiley and Sons. Inc., New York.

[4] Behzad Razavi, *"Fundamentals of Microelectronics"*, 2008, John Wiley & Sons, Inc., New Jersey.

[5] B. Streetman and S. Banerjee, *"Solid-State Electronic Device"*, Fifth Edition, 1999, Prentice-Hall.

[6] B. Razavi, *"Design of Analog CMOS Integrated Circuits"*, 2001, McGraw-Hill.

[7] B. Razavi, *"RF Microelectronics"*, 1998, Prentice-Hall, New Jersey.

[8] R. Schaumann and M. E. van Valkenberg, *"Design of Analog Filters"*, 2001, Oxford University Press.

[9] G. Roberts and A. S. Sedra, *"SPICE"*, 1997, Oxford University Press.

[10] Adel S. Sedra and Kenneth C. Smith, *"Microelectronic Circuits"*, 2004, Oxford University Press, New York.

[11] M. H. Rashid, *"Microelectronic Circuits: Analysis and Design"*, 1999, PWS, Boston.

[12] J. Millman and A. Grabel, *"Microelectronics"*, 1987, McGraw-Hill, New York.

>> >> **MATLAB or SIMULINK Basics** >> >>

[13] Nuruzzaman, M., *"Tutorials on Mathematics to MATLAB"*, 2003, AuthorHouse, Bloomington, Indiana.

[14] Nuruzzaman, M., *"Modeling and Simulation in SIMULINK for Engineers and Scientists"*, 2005, AuthorHouse, Bloomington, Indiana.

[15] Duffy, Dean G., *"Advanced Engineering Mathematics with MATLAB"*, Second Edition, 2003, Chapman & Hall, CRC, Boca Raton.

[16] Hanselman, Duane C. and Littlefield, Bruce R., *"Mastering MATLAB 5: A Comprehensive Tutorial"*, 1998, Prentice Hall, Upper Saddle River, New Jersey.

[17] Shampine, Lawrence F. and Reichelt, Mark W., *"The MATLAB ODE Suite"*, 1996, The Math-Works, Inc., Natick, MA.

[18] Marcus, Marvin, *"Matrices and MATLAB - A Tutorial"*, 1993, Prentice Hall, Englewood Cliffs, N. J.

[19] Ogata, Katsuhiko, *"Solving Control Engineering Problems with MATLAB"*, 1994, Englewood Cliffs, N. J. Prentice Hall.

[20] Part-Enander, Eva, *"The MATLAB Handbook"*, 1998, Harlow: Addisson Wesley.

[21] Saadat, Hadi., *"Computational Aids in Control Systems Using MATLAB"*, 1993, McGraw-Hill, New York.

[22] Gander, Walter. and Hrebicek, Jiri., *"Solving Problems in Scientific Computing Using MAPLE and MATLAB"*, 1997, Third Edition, Springer Verlag, New York.

[23] Biran, Adrian B and Breiner, Moshe, *"MATLAB for Engineers"*, 1997, Addison Wesley, Harlow, Eng.

[24] D. M. Etter, *"Engineering Problem Solving with MATLAB"*, 1993, Prentice Hall, Englewood Cliffs, N. J.

[25] Ogata, Katshuiko, *"Designing Linear Control Systems with MATLAB"*, 1994, Prentice Hall, Englewood Cliffs, N. J.

[26] Bishop, Robert H., *"Modern Control Systems Analysis and Design Using MATLAB"*, 1993, Addsison Wesley, Reading, MA.

[27] Moscinski, Jerzy and Ogonowski, Zbigniew., *"Advanced Control with MATLAB and Simulink"*, 1995, E. Horwood, Chichester, Eng.

[28] Alberto Cavallo, Roberto Setola, and Francesco Vasca, *"Using MATLAB Simulink and Control Systems Toolbox - A Practical Approach"*, 1996, Prentice Hall, London.

[29] Kuo, Benjamin C. and Hanselman, Duanec., *"MATLAB Tools for Control System Analysis and Design"*, 1994, Prentice Hall, Englewood Cliffs, N. J.

[30] Math Works Inc., *"MATLAB Reference Guide"*, Math Works Inc., 1993, Natick, Massachusetts.

[31] Theodore F. Bogart, *"Computer Simulation of Linear Circuits and Systems"*, 1983, John Wiley and Sons, Inc., New York.

[32] Nuruzzaman, M., *"Technical Computation and Visualization in MATLAB for Engineers and Scientists"*, February, 2007, AuthorHouse, Bloomington, Indiana.

[33] Nuruzzaman, M., *"Electric Circuit Fundamentals in MATLAB and SIMULINK"*, October 2007, BookSurge Publishing, Charleston, South Carolina.

[34] Nuruzzaman, M., *"Signal and System Fundamentals in MATLAB and SIMULINK"*, July 2008, BookSurge Publishing, Charleston, South Carolina.

[35] Nuruzzaman, M., *"Modern Approach to Solving Electromagnetics in MATLAB"*, January 2009, BookSurge Publishing, Charleston, South Carolina.

[36] Nuruzzaman, M., *"Finite Difference Fundamentals in MATLAB"*, July, 2013, CreateSpace, South Carolina.

[37] Nuruzzaman, M., *"Digital Audio Fundamentals in MATLAB"*, July, 2010, CreateSpace, California.

[38] Nuruzzaman, M., *"Control System Analysis & Design in MATLAB and SIMULINK"*, June, 2014, Lulu Press, Inc., North Carolina.

Subject Index

solver time 13
start time 23
started in SIMULINK 20
stop time 13,23
submenu 3
submenu APPS 3

T_____
text color 10
time interval 23
tutorial on SIMULINK 21
types of block 19

U_____
untitled model file 13

V_____
variable data 12
variable deletion 10
variables present 11
vector entering 4

W_____
workspace 4
workspace browser 2
workspace variables 11

ELECTRONICS:

A_____
active filter 167
active region 114,209
adder 127
amplifier design 216
angular frequency 170
aspect ration 142
average signal value 77

B_____
bandpass 176
bandpass bandwidth 177
bandpass cutoff 176
bandstop 176
bandstop bandwidth 177

bandstop cutoff 176
bandwidth 176
bandwidth bandpass 177
bandwidth bandstop 177
bandwidth highpass 177
bandwidth lowpass 177
base emitter voltage 94,212
basic diode circuit 62
bias voltage 61
biasing a MOS 146
biasing circuit 97
bipolar junction transistor 87
BJT 87
bode bandpass 176
bode bandstop 176
bode highpass 176
bode lowpass 176
bode multiple 175
bode plot 175
bounds of voltage gain 104

C_____
cascade modeling 153
cascaded circuit 227
cascaded transistor 109
CB 113
CC 113
CE 113
CE amplifier 216
center frequency 177
channel length modulation 147
characteristic I/V 78,93,142
characteristic of self bias 214
circuit design 216
circuit on RLC 161
circuits on clipping 65
circuits on diode 65
clipping circuit 65
collector current 94,211
collector emitter voltage 94
common base 113
common base stage 225
common collector 113
common emitter 113
common emitter current gain 99,226
component spectrum 170
controlled current source 87,144
current gain 105,151,226

Mohammad Nuruzzaman